International Conference on Quantum Science and Applications (ICQSA-2016)

Journal of Physics: Conference Series Volume 766

Eskisehir, Turkey
25 – 27 May 2016

Editors:

A. Algin
M. Arik
A. M. Gavrilik

ISBN: 978-1-5108-3183-4
ISSN: 1742-6588

Printed from e-media with permission by:

Curran Associates, Inc.
57 Morehouse Lane
Red Hook, NY 12571

Some format issues inherent in the e-media version may also appear in this print version.

Printed by Curran Associates, Inc. (2016)

For permission requests, please contact the Institute of Physics
at the address below.

Institute of Physics
Dirac House, Temple Back
Bristol BS1 6BE UK

Phone: 44 1 17 929 7481
Fax: 44 1 17 920 0979

techtracking@iop.org

Additional copies of this publication are available from:

Curran Associates, Inc.
57 Morehouse Lane
Red Hook, NY 12571 USA
Phone: 845-758-0400
Fax: 845-758-2633
Email: curran@proceedings.com
Web: www.proceedings.com

TABLE OF CONTENTS

Journal of Physics: Conference Series **766** (2016) 011001 doi:10.1088/1742-6596/766/1/011001

Preface

This special volume of *Journal of Physics: Conference Series* is dedicated to the proceedings of "International Conference on Quantum Science and Applications (ICQSA-2016)". The conference was organized by the Centre for Quantum Research and Applications at Eskisehir Osmangazi University, Eskisehir, Turkey. It was held in Eskisehir Osmangazi University Congress and Culture Centre during May 25-27, 2016 (http://icqsa2016.ogu.edu.tr). It gathered actively 143 participants from different disciplines in natural and applied sciences coming from 16 different countries from all over the word. It was the first international conference in its content on the scientific research fields of quantum science and applications in Turkey. It also consisted of 12 plenary lectures and 119 contributed oral presentations covering interdisciplinary fields of research.

The ICQSA-2016 conference focused on recent modern theoretical and experimental developments of quantum science in multi-disciplinary aspects of areas of mathematics, physics, statistics, chemistry, biology, computer science, electronics, informatics, medicine, education, etc. It served as an interaction platform for students, researchers, public, and private sector delegates for sharing new scientific and technological ideas on quantum applications in science and technology.

The topics of the conference were: Quantum theory and quantum computing, quantum information theory and its applications, quantum statistics and its applications, quantum thermodynamics, quantum cryptography, classical and quantum symmetries, quantum calculus in science, engineering, medicine, education, etc., classical and quantum integrable systems, modeling and numerical methods in quantum systems, other modern mathematical methods in science and technology.

Each of the submitted papers for this special volume of the proceedings of ICQSA-2016 has been reviewed by external referees. There are 36 accepted papers. Besides, it is worth pointing out that a great deal of interest shown by the participants of the ICQSA-2016 conference gives hope for the continuation of this meeting in the following years.

The conference organization was sponsored by Eskisehir Osmangazi University (ESOGU), Eskisehir Tepebasi Municipality, the Scientific and Technological Research Council of Turkey (TUBITAK) under the project number 113F226. Other supporters of this organization were: Eskisehir Metropolitan Municipality, Eskisehir Governorship Provincial Directorate of Culture and Tourism, Pegem Academy Publishing Company. In addition, this special volume is supported by the Scientific and Technological Research Council of Turkey (TUBITAK) under the project number 113F226.

Finally, our special thanks go to the Scientific and Organizing Committee members, to the sponsors and supporters, and to all of the participants.

We hope this special volume will provide a useful resource for students, experts, and other participants, who are interested in interdisciplinary fields of quantum science and applications.

Guest Editors: A. Algin[1], M. Arik[2] and A. M. Gavrilik[3]

[1]Department of Physics and Center for Quantum Research and Applications, Eskisehir Osmangazi University, Eskisehir, Turkey
[2]Department of Physics, Bogazici University, Bebek, Istanbul, Turkey
[3]Department of Mathematical Methods in Theoretical Physics, Bogolyubov Institute for Theoretical Physics (BITP), Kiev, Ukraine

Journal of Physics: Conference Series **766** (2016) 011001 doi:10.1088/1742-6596/766/1/011001

ICQSA-2016 Conference (May 25-27, 2016), ESOGU-Eskisehir, TURKEY

Chair: Hasan Gonen, Prof.Dr. (Rector of Eskisehir Osmangazi University)
Co-chair: Abdullah Algin, Prof.Dr. (Eskisehir Osmangazi University)

Committees:

Scientific Committee

M. Arik (Bogazici U., Istanbul, Turkey)
N. M. Atakishiyev (Universidad Nacional Auto-Noma de Mexico, Cuernavaca, Mexico)
D. Baleanu (Cankaya U., Ankara, Turkey)
N. Berker (Sabanci U., Istanbul, Turkey)
F. G. S. L. Brandao (Caltech, California, USA)
F. Buyukkilic (Ege U., Izmir, Turkey)
S. Catto (CUNY Graduate School and the Rockefeller University, New York, USA)
R. Chakrabarti (CMI, Chennai, India)
W. S. Chung (Gyeongsang National U., Jinju, South Korea)
D. Demirhan (Ege U., Izmir, Turkey)
S. Durukanoglu Feyiz (Sabanci U., Istanbul, Turkey)
A. M. Gavrilik (BITP, Kiev, Ukraine)
Mo-Lin Ge (Nankai U., Tianjin, China)
N. Imoto (Osaka U., Osaka, Japan)
M. Keskin (Erciyes U., Kayseri, Turkey)
M. R. Kibler (Universite de Lyon, Lyon, France)
A. Lavagno (INFN, Torino, Italy)
A. Mostafazadeh (Koc U., Istanbul, Turkey)
O. E. Mustecaplioglu (Koc U., Istanbul, Turkey)
R. Norte (Delft University of Technology, Delft, The Netherlands)
M. Ozer (YTU, Istanbul, Turkey)
O.K. Pashaev (IYTE, Izmir, Turkey)
R. Sever (METU, Ankara, Turkey)
I. Siap (YTU, Istanbul, Turkey)
U. Tirnakli (Ege U., Izmir, Turkey)
C. Tsallis (CBPF, Rio de Janeiro, Brazil)
M. R. Ubriaco (U. of Puerto Rico, San Juan, USA)

Plenary Speakers

M. Arik (Turkey)
D. Baleanu (Turkey)
S. Catto (USA)
W. S. Chung (South Korea)
S. Durukanoglu Feyiz (Turkey)
A. M. Gavrilik (Ukraine)
N. Imoto (Japan)
O. E. Mustecaplioglu (Turkey)
R. Norte (The Netherlands)
O.K. Pashaev (Turkey)
I. Siap (Turkey)
M. R. Ubriaco (USA)

Organizing Committee

Programme Committee
H.K. Aksoy (Eskisehir Osmangazi U., Eskisehir, Turkey)
A. Algin (Co-Chair, Eskisehir Osmangazi Univ., Eskisehir, Turkey)
E. Algin (Eskisehir Osmangazi U., Eskisehir, Turkey)
A. A. Altintas (Okan U., Istanbul, Turkey)
H. Anilan (Eskisehir Osmangazi U., Eskisehir, Turkey)
A.S. Arikan (Sakarya U., Sakarya, Turkey)
T.E. Cosan (Eskisehir Osmangazi U., Eskisehir, Turkey)
N. Degirmenci (Anadolu U., Eskisehir, Turkey)
R. Edizkan (Eskisehir Osmangazi U., Eskisehir, Turkey)
A. Odabas (Eskisehir Osmangazi U., Eskisehir, Turkey)
D. Peker (Eskisehir Osmangazi U., Eskisehir, Turkey)
E. Uslu (Eskisehir Osmangazi U., Eskisehir, Turkey)
A. Yazici (Eskisehir Osmangazi U., Eskisehir, Turkey)
C. Yuce (Anadolu U., Eskisehir, Turkey)

Local Organizing Committee
N. Akkurt (Eskisehir Osmangazi U., Eskisehir, Turkey)
C. Asici (Eskisehir Osmangazi U., Eskisehir, Turkey)
K.O. Ay (Eskisehir Osmangazi U., Eskisehir, Turkey)
H. Cetinkaya (Dumlupinar U., Kutahya, Turkey)
E. Dil (Sinop U., Sinop, Turkey)
M. Kaya (Eskisehir Osmangazi U., Eskisehir, Turkey)
A. Olkun (Eskisehir Osmangazi U., Eskisehir, Turkey)
M. Ozgur (Eskisehir Osmangazi U., Eskisehir, Turkey)
M. Senay (Eskisehir Osmangazi U., Eskisehir, Turkey)
E. Silik (Eskisehir Osmangazi U., Eskisehir, Turkey)
G. Topcu (Ankara, Turkey)

Peer review statement

All papers published in this volume of *Journal of Physics: Conference Series* have been peer reviewed through processes administered by the proceedings Editors. Reviews were conducted by expert referees to the professional and scientific standards expected of a proceedings journal published by IOP Publishing.

Journal of Physics: Conference Series **766** (2016) 012001 doi:10.1088/1742-6596/766/1/012001

Clifford Algebras and Their Decomposition into Conjugate Fermionic Heisenberg Algebras

Sultan Catto[1,2,†], Yasemin Gürcan[3], Amish Khalfan[4,††], Levent Kurt[5] and V. Kato La[6]

[1]Physics Department, The Graduate School, City University of New York, New York, NY 10016-4309
[2]Theoretical Physics Group, Rockefeller University, 1230 York Avenue, New York, NY 10021-6399
[3,5] Department of Science, Borough of Manhattan CC, The City University of NY, New York, NY 10007
[4] Physics Department, LaGuardia CC, The City University of New York, LIC, NY 11101
[6] Columbia University, New York, NY 10027

† Work supported in part by the PSC-CUNY Research Awards, and DOE contracts No. DE-AC-0276 ER 03074 and 03075; and NSF Grant No. DMS-8917754.
†† Work supported in part by the PSC-CUNY Research Award No:69379-00 47

Abstract.

We discuss a construction scheme for Clifford numbers of arbitrary dimension. The scheme is based upon performing direct products of the Pauli spin and identity matrices. Conjugate fermionic algebras can then be formed by considering linear combinations of the Clifford numbers and the Hermitian conjugates of such combinations. Fermionic algebras are important in investigating systems that follow Fermi-Dirac statistics. We will further comment on the applications of Clifford algebras to Fueter analyticity, twistors, color algebras, M-theory and Leech lattice as well as unification of ancient and modern geometries through them.

1. Clifford Algebras

These are associative hermitian matrix algebras[1] with basis γ_α ($\alpha = 1, \cdots, 2n+1$) that obey

$$\gamma_\alpha \gamma_\beta = \frac{1}{2}\{\gamma_\alpha, \gamma_\beta\} = \frac{1}{2}(\gamma_\alpha\gamma_\beta + \gamma_\beta\gamma_\alpha) = \delta_{\alpha\beta} I \tag{1}$$

where I is a unit matrix of the same dimension as γ_α. We have

$$\gamma_\alpha = \gamma_\alpha^\dagger \,, \quad \text{and} \quad \gamma_\alpha(\gamma_\beta\gamma_\rho) = (\gamma_\alpha\gamma_\beta)\gamma_\rho \tag{2}$$

The basis elements γ_α are linearly independent.

The Clifford relation means that the square of each γ_α is unity and two different γ_α anticommute.

(a) Suppose there are $2n$ such elements γ_α. Define

$$\gamma_{2n+1} = \epsilon\gamma_1\gamma_2\cdots\gamma_{2n-1}\gamma_{2n} \tag{3}$$

Then

$$\{\gamma_{2n+1}, \gamma_i\} = 0 \qquad (i = 1, \cdots, 2n) \tag{4}$$

Also

$$\gamma_{2n+1}^2 = \epsilon^2(-1)^n \tag{5}$$

so that if $\epsilon = (-1)^{\frac{n}{2}}$ we have $\gamma_{2n+1}^2 = 1$. Thus the basis is generated by $2n$ elements $\gamma_1, \cdots, \gamma_{2n}$.

(b) Construction of the algebra (Clifford, Dirac, Jordan and Wigner):

Let $n = 1$. The answer is given by the hermitian traceless 2×2 Pauli matrices τ_1, τ_2, τ_3. In this case $\tau_3 = -i\tau_1\tau_2$. we have

$$\tau_i\tau_j + \tau_j\tau_i = 2\delta_{ij}I \tag{6}$$

where I is a 2×2 unit matrix. The τ_i act on complex 2-dimensional vector space U. We can also write $\tau_j = ie_j$ where e_j are the 3 quaternion units.

Let $n = 2$. We take two commuting sets of Pauli operators $\sigma_i^{(1)}$ and $\sigma_j^{(2)}$ acting in the vector space $U \otimes U$. We have seen that

$$\sigma_i^{(1)} = \tau_i \times I \quad \text{and} \quad \sigma_j^{(2)} = I \times \tau_i \tag{7}$$

are 4×4 matrices with the properties

$$[\sigma_i^{(1)}, \sigma_j^{(2)}] = 0 \tag{8}$$

We define

$$\begin{aligned} \gamma_1 = \sigma_1^{(2)} \quad &, \qquad \gamma_3 = \sigma_2^{(2)} \\ \gamma_2 = \sigma_3^{(2)}\sigma_1^{(1)} \quad &, \qquad \gamma_4 = \sigma_3^{(2)}\sigma_2^{(1)} \end{aligned} \tag{9}$$

Then we find

$$\gamma_5 = \sigma_3^{(1)}\sigma_3^{(2)} \quad \text{and} \quad \{\gamma_\alpha, \gamma_\beta\} = 2\delta_{\alpha\beta} \tag{10}$$

Dirac's notation: $\sigma_j^{(1)} = \sigma_j, \quad \sigma_i^{(2)} = \rho_i$.

Then $\gamma_1 = \rho_1$, $\gamma_3 = \rho_2$, $\gamma_2 = \rho_3\gamma_1$, $\gamma_4 = \rho_3\sigma_2$, $\gamma_5 = \rho_3\sigma_3$.

Next we look at its generalization.

For arbitrary n we construct $2n + 1$ different $2^n \times 2^n$ matrices that act on the vector space

$$\underbrace{U \otimes U \otimes \cdots \otimes U}_{n \quad factors} \tag{11}$$

of dimension 2^n in terms of n commuting sets of Pauli matrices

$$\begin{aligned} \sigma_j^{(n)} &= \tau_j \otimes I \otimes I \otimes \cdots \otimes I \\ \sigma_j^{(n-1)} &= I \otimes \tau_j \otimes I \cdots \otimes I \\ \cdots\cdots & \\ \sigma_j^{(1)} &= I \otimes I \otimes \cdots \otimes \tau_j \end{aligned} \tag{12}$$

Then we construct the following $2n + 1$ hermitian matrices:

$$\begin{aligned} \gamma_1 = \sigma_1^{(n)} \quad &, \quad \gamma_{n+1} = \sigma_2^{(n)} \\ \gamma_2 = \sigma_3^{(n)}\sigma_1^{(n-1)} \quad &, \quad \gamma_{n+2} = \sigma_3^{(n)}\sigma_2^{(n-1)} \\ \gamma_3 = \sigma_3^{(n)}\sigma_3^{(n-1)}\sigma_1^{(n-2)} \quad &, \quad \gamma_{n+3} = \sigma_3^{(n)}\sigma_3^{(n-1)}\sigma_2^{(n-2)} \\ \cdots\cdots & \\ \gamma_{2n+1} &= \sigma_3^{(n)}\sigma_3^{(n-1)}\cdots\sigma_3^{(2)}\sigma_3^{(1)} \end{aligned} \tag{13}$$

They satisfy

$$\gamma_\alpha\gamma_\beta + \gamma_\beta\gamma_\alpha = 2\delta_{\alpha\beta}I \quad (\alpha, \beta = 1, \cdots, 2n + 1) \tag{14}$$

(c) General Representation of the Clifford Algebra:

The automorphism group of the Clifford algebra is the unitary group $U(2^n)$. Indeed, if

$$\gamma'_\alpha = U\gamma_\alpha U^\dagger \,, \qquad (UU^\dagger = I) \tag{15}$$

then

$$\begin{aligned} \gamma'_\alpha\gamma'_\beta + \gamma'_\beta\gamma'_\alpha &= U\gamma_\alpha U^\dagger U\gamma_\beta U^\dagger + U\gamma_\beta U^\dagger U\gamma_\alpha U^\dagger \\ &= 2U(\delta_{\alpha\beta}I)U^\dagger = \delta_{\alpha\beta}I \end{aligned} \tag{16}$$

For example the following different representations of the Dirac algebra:

$$\begin{aligned} \mathrm{I}: \qquad & \gamma_1 = \rho_2\sigma_1 \,, \quad \gamma_2 = \rho_2\sigma_2 \,, \\ & \gamma_3 = \rho_2\sigma_3 \,, \quad \gamma_4 = \rho_1 \,, \quad \gamma_5 = \rho_3 \qquad \text{Weyl rep.} \end{aligned} \tag{17}$$

$$\begin{aligned} \mathrm{II}: \qquad & \gamma_1 = \rho_2\sigma_1 \,, \quad \gamma_2 = \rho_2\sigma_2 \,, \\ & \gamma_3 = \rho_2\sigma_3 \,, \quad \gamma_4 = -\rho_3 \,, \quad \gamma_5 = \rho_1 \qquad \text{Dirac rep.} \end{aligned} \tag{18}$$

$$\begin{array}{lll} \text{III}: & \gamma_1 = \sigma_1\,, \quad \gamma_2 = \rho_2\sigma_2\,, & \\ & \gamma_3 = \sigma_3\,, \quad \gamma_4 = \sigma_2\rho_1\,, \quad \gamma_5 = \sigma_2\rho_3 & \text{Majorana rep.} \end{array} \tag{19}$$

are unitarily equivalent to the standard representation given above.

(d) Reality properties in the standard representation:

From the construction given above $n+1$ elements of the basis $(\gamma_1, \cdots, \gamma_n$ and $\gamma_{2n+1})$ are real and symmetric while the remaining n elements $(\gamma_{n+1}, \cdots, \gamma_{2n})$ are imaginary and antisymmetric.

(e) Generation of the $2^n \times 2^n$ matrix algebra from the Clifford algebra.

γ_α and I close under the symmetric Jordan product. They do not close under the matrix product since for instance

$$\begin{aligned} \gamma_\alpha\gamma_\beta &= \frac{1}{2}\{\gamma_\alpha\gamma_\beta\} + \frac{1}{2}[\gamma_\alpha, \gamma_\beta] \\ &= \frac{1}{2}\delta_{\alpha\beta}I + i\sigma_{\alpha\beta} \end{aligned} \tag{20}$$

where

$$\sigma_{\alpha\beta} = \frac{1}{2i}[\gamma_\alpha, \gamma_\beta] \tag{21}$$

(called the spin matrices) are hermitian and linearly independent of γ_α. Similarly $\gamma_\alpha\gamma_\beta\gamma_\rho$ will contain a part independent of γ_α and $\sigma_{\beta\rho}$ up to products of γ_α of degree n. The product of degree $2n+1$, namely

$$\gamma_1\gamma_2\cdots\gamma_{2n}\gamma_{2n+1} \tag{22}$$

is proportional to the unit matrix as we have already seen. Hence a linear combination of all the powers of γ_α from 0 to n span the whole $2^n \times 2^n$ matrix algebra. In the case of the Dirac algebra we have I, γ_α and $\sigma_{\alpha\beta}$ respectively 1, 5, and $\frac{5\times4}{2} = 10$ in number spanning the 16 independent elements of the 4×4 matrix algebra.

(f) Let $a = a_\alpha\gamma_\alpha$. We have $a^2 = a_\alpha a_\alpha I$.

Let us now show an application: Take 4 of the γ_α for $n = 2$: γ_μ $(\mu = 1, 2, 3, 4)$. Let $\partial = \gamma_\mu\frac{\partial}{\partial x_\mu}$. Let $\partial\psi = m\psi$ where ψ is a 4-column. Then

$$\partial^2\psi = \partial_\mu\partial_\mu\psi = \Box\psi = m^2\psi \tag{23}$$

This is the relativistic wave equation for ψ. $\partial\psi = m\psi$ is the Dirac equation.

(g) Rotation of $2n+1$ dimensional vectors.

Consider an infinitesimal transformation of the vector a_α

$$a'_\alpha = a_\alpha + \delta a_\alpha \tag{24}$$

This transformation is a rotation if

$$a'_\alpha a'_\alpha = (a_\alpha + \delta a_\alpha)(a_\alpha + \delta a_\alpha) = a_\alpha a_\alpha \tag{25}$$

This entails $a_\alpha \delta a_\alpha = 0$. Hence if

$$a = \gamma_\rho a_\rho\,, \qquad \delta a = \gamma_\rho \delta a_\rho \tag{26}$$

we must have

$$\{a, \delta a\} = 0 \tag{27}$$

Now let

$$\delta a = [\frac{i}{4}\sigma_{\alpha\beta}\omega_{\alpha\beta}, a] = [\frac{i}{2}\omega, a] \tag{28}$$

where $\omega = \frac{1}{2}\sigma_{\alpha\beta}\omega_{\alpha\beta}$. We have

$$\begin{aligned} \{a, \delta a\} &= \frac{i}{2}(a[\omega, a] + [\omega, a]a) \\ &= \frac{i}{2}(a\omega a - a^2\omega + \omega a^2 - a\omega a) = 0 \end{aligned} \tag{29}$$

Hence

$$a \to a + \frac{i}{2}[\omega, a] \tag{30}$$

represents a rotation and $\frac{1}{2}\sigma_{\alpha\beta}$ are called spin generators. In component form we write

$$a'_\alpha = a_\alpha + \omega_{\alpha\beta} a_\beta \tag{31}$$

In the case of the Dirac algebra ($n = 2$)

$$\frac{1}{2}\sigma_{\alpha\beta} \qquad (\alpha, \beta = 1, \cdots, 5) \tag{32}$$

generate 5 dimensional rotations, while

$$\frac{1}{2}\sigma_{\mu\nu} \qquad (\mu, \nu = 1, \cdots, 4) \tag{33}$$

generate 4 dimensional rotations. When $\omega_{23}, \omega_{31}, \omega_{12}$ are real and $\omega_{41}, \omega_{42}, \omega_{43}$ are imaginary

$$\delta a = [\frac{i}{2}\sigma_{\mu\nu}\omega_{\mu\nu}, a] \tag{34}$$

for the vector a with a_1, a_2, a_3 real and a_4 imaginary represents a Lorentz transformation in 3 space and 1 time directions (the Minkowski space-time). Under such a Lorentz transformation

$$\partial \to \partial + [\frac{i}{2}\omega, \partial] \tag{35}$$

while

$$\psi \to \psi + \frac{i}{2}\omega\psi \tag{36}$$

Thus ∂ transforms like a vector while ψ transforms like a spinor. Under such a Lorentz transformation the bilinear forms

$$\bar{\psi}\psi \quad \text{and} \quad i\bar{\psi}\gamma_5\psi \tag{37}$$

(where $\bar{\psi} = \psi^\dagger\gamma_4$) remains invariant, while

$$\bar{\psi}\gamma_\mu\psi \quad \text{and} \quad i\bar{\psi}\gamma_5\gamma_\mu\psi \tag{38}$$

transform like vectors.

2. Grassmann Algebra as Subalgebra of Clifford Algebra

Grassmann algebra and its conjugate Grassmann algebra

The Grassmann algebra is the algebra of associative anticommuting numbers. It is of order n if we can choose a basis b_j $(j = 1, \cdots, n)$ such that

$$b_i(b_jb_k) = (b_ib_j)b_k \tag{39}$$

and

$$\{b_ib_j\} = b_ib_j + b_jb_i = 0\,, \quad (\text{or} \quad b_i \cdot b_j = 0) \tag{40}$$

The associativity relation is automatically satisfied if b_i is chosen to be matrices and they are multiplied with the rules of matrix multiplication.

A standard set satisfying the anticommutativity relation can be obtained from the Clifford algebra of order $2n+1$ by choosing

$$b_j = \frac{1}{2}(\gamma_j + i\gamma_{n+j})\,, \qquad (j = 1, \cdots, n) \tag{41}$$

We then verify that $b_j^2 = 0$, and $b_jb_k = -b_kb_j$ for $j \neq k$. In the standard representation γ_j $(j \leq n)$ are real while γ_{n+j} are imaginary so that b_j are real non-hermitian $2^n \times 2^n$ matrices. Another set of Grassmann numbers b'_j of the same order can be obtained from b_j by a unitary transformation

$$b'_j = Ub_jU^\dagger \tag{42}$$

with $UU^\dagger = I$, where I is the $2^n \times 2^n$ unit matrix, or by the transformation to the conjugate set

$$b_j \to b_j^\dagger = \frac{1}{2}(\gamma_j - i\gamma_{n+j}) \tag{43}$$

which are also real and obey

$$b_j^\dagger b_k^\dagger + b_k^\dagger b_j^\dagger = 0 \tag{44}$$

Let us now give examples: For $n = 1$ we have

$$b = \frac{1}{2}(\sigma_1 + i\sigma_2) = \begin{pmatrix} 0 & 1 \\ 0 & 0 \end{pmatrix} \tag{45}$$

$$b^\dagger = \frac{1}{2}(\sigma_1 - i\sigma_2) = \begin{pmatrix} 0 & 0 \\ 1 & 0 \end{pmatrix} \tag{46}$$

$$b^2 = 0\,, \qquad b^{\dagger 2} = 0 \tag{47}$$

$$b' = UbU^\dagger\,, \quad \text{and} \quad b'^\dagger = Ub^\dagger U^\dagger \tag{48}$$

are not real in general.

For $n = 2$ (Dirac algebra) we can construct a Grassmann algebra of order 2 and its conjugate, briefly:

$$b_1 = \frac{\rho_1 + i\rho_2}{2} \tag{49}$$

$$b_2 = \rho_3 \frac{\sigma_1 + i\sigma_2}{2} \tag{50}$$

$$b_1^2 = b_2^2 = 0\,, \quad b_1 b_2 = -b_2 b_1 \tag{51}$$

$$b_1^\dagger = \frac{\rho_1 - i\rho_2}{2} \tag{52}$$

$$b_2^\dagger = \rho_3 \frac{\sigma_1 - i\sigma_2}{2} \tag{53}$$

$$b_1^{\dagger 2} = b_2^{\dagger 2} = 0\,, \quad b_1^\dagger b_2^\dagger = -b_2^\dagger b_1^\dagger \tag{54}$$

3. Fermionic Heisenberg Algebra

If the Grassmann algebra b_j and its conjugate $b_j^\dagger$ are taken together we obtain

$$b_j b_k^\dagger + b_k^\dagger b_j = \frac{1}{4}\{\gamma_j + i\gamma_{n+j}, \gamma_k - i\gamma_{n+k}\} = \delta_{jk} I \tag{55}$$

These relations are also invariant under the unitary transformation $b_j \to U b_j U^\dagger$. Thus from b_j and $b_k^\dagger$ we obtain the Fermionic Heisenberg algebra

$$\{b_j, b_k\} = 0\,, \quad \{b_j^\dagger, b_k^\dagger\} = 0\,, \quad \{b_j, b_k^\dagger\} = \delta_{jk} I \tag{56}$$

These have the same structure as Poisson bracket relations among canonical variables (coordinates and momenta) in classical mechanics. The difference is that the Poisson bracket is anticommutative while the bracket used for the Fermionic Heisenberg algebra is commutative since it is an anticommutator. In the bosonic Heisenberg algebra commutator brackets with same formal properties as Poisson brackets are used.

The $2n$ generators $\gamma_1, \cdots, \gamma_n$ of the Clifford algebra of order $2n + 1$ can be recovered from the two conjugate Grassmann algebras through

$$\gamma_j = b_j + b_j^\dagger\,, \quad \gamma_{n+j} = -i(b_j - b_{n+j}^\dagger) \tag{57}$$

The last element γ_{2n+1} of the Clifford basis can be expressed by means of $b_j, b_j^\dagger$ through the relation

$$\gamma_1 \gamma_{n+1} \gamma_2 \gamma_{n+2} \cdots \gamma_n \gamma_{2n} \gamma_{2n+1} = i^n I \tag{58}$$

Now

$$\gamma_j \gamma_{n+j} = i(I - 2b_j^\dagger b_j) \tag{59}$$

so that

$$\gamma_{2n+1} = \prod_{j=1}^{n} (I - 2b_j^\dagger b_j) \tag{60}$$

4. Further Remarks

The present authors earlier developed octonion and split octonion[2],[3] algebras that led into algebraic foundations of color symmetries in nature, into quark-antidiquark symmetries[4],[5], exotics and pentaquark formations. The formulations we put forth in this paper also have deeper applications in hadronic physics and in astrophysics. Further applications into Fueter analyticity, quaternion analyticity, twistor formalism, hyperbolic extensions of exceptional groups, M-theory and Leech lattices will be published in our forthcoming papers.

References

[1] S. Catto, Y. Gürcan, A. Khalfan and L. Kurt, *Journal of Physics,* **411** (2013) 012009
[2] S. Catto, Y. Gürcan, A. Khalfan and L. Kurt, *Journal of Physics,* **563** (2014) 0120096
[3] S. Catto, Y. Gürcan, A. Khalfan and L. Kurt, *Journal of Physics,* **670** (2016) 012016
[4] S. Catto and F. Gürsey, *Nuovo Cimento* **A 86** *(1985) 201*
[5] S. Catto and F. Gürsey, *Nuovo Cimento* **A99** *(1988) 685*

Acknowledgments:
One of us (SC) would like to thank professor Abdullah Algın for the invited talk given at ICQSA-2016 meeting in Eskişehir, Turkey, and for his kind hospitality. He is also grateful to late professor Feza Gürsey for useful discussions on many of the ideas presented in this coference. We would like to thank Professors Vladimir Akulov, Itzhak Bars and Francesco Iachello for many enlightening conversations.

Journal of Physics: Conference Series **766** (2016) 012002 doi:10.1088/1742-6596/766/1/012002

Relativistic Approximate Solutions for a Two-Term Potential: Riemann-Type Equation

Altug Arda

Department of Physics Education, Hacettepe University, 06800, Ankara, Turkey

E-mail: arda@hacettepe.edu.tr

Abstract. Approximate analytical solutions of a two-term potential are studied for the relativistic wave equations, namely, for the Klein-Gordon and Dirac equations. The results are obtained by solving of a Riemann-type equation whose solution can be written in terms of hypergeometric function ${}_2F_1(a, b; c; z)$. The energy eigenvalue equations and the corresponding normalized wave functions are given both for two wave equations. The results for some special cases including the Manning-Rosen potential, the Hulthén potential and the Coulomb potential are also discussed by setting the parameters as required.

1. Introduction

The investigation of the non-relativistic and relativistic bound/scattering state solutions of exponential-type potentials has become an important area within quantum mechanics. For last decades, many authors have made much efforts to obtain the analytical solutions of the exponential-type potentials, especially about the problems based on the Morse potential, Hulthén potential, and the Woods-Saxon potential, with the help of various methods [1-32]. Within this context, we deal with the following two-term exponential potential

$$V(r) = -V_0 \frac{e^{-\beta r}}{1 - qe^{-\beta r}} + V_1 \frac{e^{-2\beta r}}{(1 - qe^{-\beta r})^2}, \tag{1}$$

which firstly proposed in [33]. Jia *et. al.* have studied the bounded solutions of the Schrödinger equation for this potential by using SUSY-approach [34], and Benarami *et. al.* have investigated the same problem in terms of Green's function [35]. Arda and Sever have also presented the approximate, analytical solutions of the above potential within functional analysis method for $q \geq 1$ and $q = 0$ [36].

In the present work, we tend to give the approximate, analytical solutions of the Klein-Gordon and Dirac equations which will be written in the form of a Riemann-type equation. We write the eigenvalue equations and the corresponding "wave functions" both for the above equations with the two-term exponential potential. We present also briefly the results for the Manning-Rosen potential, the Hulthén potential and Coulomb potential by setting the potential parameters as required. The readers can find an application of the Riemann equation [37] within the quantum mechanics in [27] where the bounded and scattering state solutions of the Klein-Gordon and Dirac equations have been studied for an exponential potential.

2. Analytical Solutions

2.1. Klein-Gordon Equation

Time-independent part of the Klein-Gordon equation in the presence of scalar, $S(r)$, and vector, $V(r)$, potentials is written [37]

$$\left[\hbar^2 c^2 \vec{\nabla}^2 + (E - V(r))^2 - (m_0 c^2 + S(r))^2\right] \Psi(\vec{r}) = 0\,, \tag{2}$$

for a spinless particle with mass m_0 and energy E, where c is speed of light. Since the potentials in (2) are spherical symmetric, we write the radial part of (2) by taking $\Psi(\vec{r}) = R(r)Y(\theta, \phi)$

$$\frac{1}{r^2}\frac{d}{dr}\left(r^2 \frac{dR(r)}{dr}\right) + \frac{1}{\hbar^2 c^2}\left[(E - V(r))^2 - (m_0 c^2 + S(r))^2 - \frac{C}{r^2}\right] R(r) = 0\,, \tag{3}$$

where we have to write the parameter C in terms of the angular momentum quantum number ℓ as $C = \ell(\ell + 1)$ because of the mathematical restrictions [37].

For the case where the scalar potential is equal to the vector potential, we get from (3)

$$\left[\frac{d^2}{dr^2} - \frac{\ell(\ell+1)}{r^2} + Q^2(E^2 - m_0^2 c^4) - 2Q^2(E + m_0 c^2)V(r)\right] u(r) = 0\,. \tag{4}$$

which is obtained by writing $R(r) = \frac{u(r)}{r}$ with $Q^2 = 1/\hbar^2 c^2$.

By inserting two-term potential (1) into (4), using the approximation instead of the centrifugal term $\frac{1}{r^2} \sim \beta^2 \frac{e^{-\beta r}}{(1 - qe^{-\beta r})^2}$, and defining a new variable as $z/q = e^{-\beta r}$ in (4) ($1 \leq z \leq 0$ for $0 \leq r < \infty$ with $q \to 1$), we obtain

$$\frac{d^2 u(z)}{dz^2} + \frac{1}{z}\frac{du(z)}{dz} + \left[-\frac{A_1^2}{z} + A_2^2 - \frac{A_3^2}{q}\frac{1}{1 - z}\right]\frac{u(z)}{z(1 - z)} = 0\,, \tag{5}$$

where

$$A_1^2 = \frac{Q^2}{\beta^2}(m_0^2 c^4 - E^2)\,;\, A_2^2 = A_1^2 + \frac{2Q^2}{q\beta^2}(E + m_0 c^2)\left(V_0 + \frac{V_1}{q}\right)\,;\, A_3^2 = \ell(\ell+1) + \frac{2Q^2 V_1}{q\beta^2}(E + m_0 c^2)\,. \tag{6}$$

Eq. (5) is a special form of the Riemann equation having regular singular points at ρ, σ and τ [38, 39]

$$\begin{aligned}\frac{d^2 y(x)}{dx^2} &+ \left(\frac{1 - a - a'}{x - \rho} + \frac{1 - b - b'}{x - \sigma} + \frac{1 - c - c'}{x - \tau}\right)\frac{dy(x)}{dx} \\ &+ \left[\frac{aa'(\rho - \sigma)(\rho - \tau)}{x - \rho} + \frac{bb'(\sigma - \tau)(\sigma - \rho)}{x - \sigma} + \frac{cc'(\tau - \rho)(\tau - \sigma)}{x - \tau}\right] \\ &\times \frac{y(x)}{(x - \rho)(x - \sigma)(x - \tau)} = 0\,,\end{aligned} \tag{7}$$

which can be reduced to the following form [27]

$$\frac{d^2 y(x)}{dx^2} + \left(\frac{1 - a - a'}{x} - \frac{1 - c - c'}{1 - x}\right)\frac{dy(x)}{dx} + \left(\frac{aa'}{x} - bb' + \frac{cc'}{1 - x}\right)\frac{y(x)}{x(1 - x)} = 0\,. \tag{8}$$

where the parameters satisfy Fuch's relation as $a + a' + b + b' + c + c' - 1 = 0$ [39]. Since the hypergeometric differential equation with solutions ${}_2F_1(p', q'; r'; x)$ is a special case of Riemann's equation [39], the solution of (5) can be written in terms of hypergeometric function [27]

$$y(x) = x^a (1 - x)^c\, {}_2F_1(p', q'; r'; x)\,, \tag{9}$$

where

$$p' = a + b + c\,, \quad q' = a + b' + c\,, \quad r' = 1 + 2a\,. \tag{10}$$

By comparing (5) with (8), we can set the parameters as

$$a = +A_1\,, \quad b = -A_2\,, \quad c = 1 + D\,, \tag{11}$$

with $D = -1/2 + \sqrt{1/4 + A_3^2/q}$, and we obtain the solutions of (5)

$$u(z) \sim z^{A-1}(1-z)^{1+D}\,{}_2F_1(A_1 - A_2 + 1 + D, A_1 + A_2 + 1 + D; 1 + 2A_1; z)\,, \tag{12}$$

In order to have a physical solution, we must write $A_1 - A_2 + 1 + D = -n(n = 0, 1, 2, \ldots)$. This expression is the quantization condition of our system and gives the wave functions of the Klein-Gordon equation for the two-term potential

$$u(z) = N z^{A_1 - 1}(1-z)^{1+D}\,{}_2F_1(-n, -n + 2A_2; 1 + 2A_1; z)\,. \tag{13}$$

with the normalization constant N. By using (6) in quantization condition, we obtain the corresponding energy eigenvalue equation for the two-term potential

$$E^2 - m_0 c^2 = (E + m_0 c^2)\left(V_0 + \frac{V_1}{q}\right) - \left[\frac{\frac{Q}{q\beta}(E + m_0 c^2)\left(V_0 + \frac{V_1}{q}\right)}{n + 1 + D}\right]^2 + \left(\frac{n + 1 + D}{2Q/\beta}\right)^2. \tag{14}$$

The normalization condition $\int_0^\infty |u(r)|^2 dr = 1$ turns into $\frac{1}{\beta}\int_0^1 \frac{1}{z}|u(z)|^2 dz = 1$ which gives

$$\frac{1}{\beta}N^2 \int_0^1 z^{2A_1 - 1}(1-z)^{2+2D}[{}_2F_1(-n, -n + 2A_2; 1 + 2A_1; z)]^2 dz = 1\,,$$

With the help of the following expression

$$\int_0^1 z^{P'}(1-z)^{Q'}[{}_2F_1(-n, n + P' + Q' - 1; P' + 2; z)]^2 dz = \frac{n!(n + Q'/2)\Gamma(n + Q')\Gamma(P' + 1)\Gamma(P' + 2)}{(n + 2Q' + P' + 1)\Gamma(n + P' + 2)\Gamma(n + P' + Q' + 1)}\,, \tag{15}$$

with $P' = 2p'' - 1$, $Q' = 2q'' + 2$ and $p'' > 0, q'' > -3/2$, we write the normalization constant

$$N = \left[\frac{\beta(n + A_1 + D + 1)\Gamma(n + 2A_1 + 1)\Gamma(2A_2)}{n!(n + 1 + D)\Gamma(n + 2 + 2D)\Gamma(2A_1)\Gamma(1 + 2A_1)}\right]^{1/2}. \tag{16}$$

for $q \to 1$.

We discuss briefly now the results for the special cases of the two-term potential (1) by starting with the Manning-Rosen potential. If we write the parameters in (1) as

$$V_0 = \frac{A}{2b^2}\,, \; V_1 = \frac{\alpha(\alpha - 1)}{2b^2}\,, \; \beta = \frac{1}{b}\,, \; q = 1\,, \tag{17}$$

we obtain the above potential, and (14) gives the energy eigenvalue equation for the Klein-Gordon equation with Manninng-Rosen potential ($\hbar = c = 1$)

$$E^2 - m_0^2 = \frac{1}{2b^2}(E + m_0)[A + \alpha(\alpha - 1)] - \left[\frac{\frac{1}{2b}(E + m_0)[A + \alpha(\alpha - 1)]}{\Gamma}\right]^2 + \frac{b^2\Gamma^2}{4}. \tag{18}$$

with $\Gamma = n + 1/2 + \sqrt{1/4 + \ell(\ell+1) + (E+m)\alpha(\alpha-1)}$.

In order to obtain the energy eigenvalue equation for the Hulthén potential, we set the parameters in (1) as

$$V_1 = 0\,, \;\; q = 1\,, \tag{19}$$

and (14) gives the following expression ($\hbar = c = 1$)

$$E^2 - m_0^2 = (E+m_0)V_0 - \left[\frac{\frac{1}{\beta}(E+m_0)V_0}{n+\ell+1}\right]^2 + \frac{\beta^2}{4}(n+\ell+1)^2\,. \tag{20}$$

The Coulomb potential could be obtained from the Hulthén potential by taking $\beta r \ll 1$

$$V(r) = -\frac{Ze^2}{r}\,,$$

where we set $V_0 = Ze^2\beta$. We write the energy eigenvalues of the Klein-Gordon equation for the Coulomb potential from (14) ($\hbar = c = 1$)

$$\frac{E-m_0}{E+m_0} = -\frac{Z^2e^4}{(n+\ell+1)^2}\,. \tag{21}$$

2.2. Dirac Equation

For the case of time-independent vector and scalar potentials, the Dirac equation is given [37]

$$\{c\hat{\alpha}.\hat{p} + \hat{\beta}[m_0 + S(r)]\}\Psi(\vec{r}) = [E - V(r)]\Psi(\vec{r})\,, \tag{22}$$

where m_0 is the mass of a spin-$\frac{1}{2}$ particle, E is the relativistic energy, $\hat{p}$ is the linear momentum operator and $\hat{\alpha}$ and $\hat{\beta}$ are 4×4 matrices given by

$$\hat{\alpha} = \begin{pmatrix} 0 & \hat{\sigma} \\ \hat{\sigma} & 0 \end{pmatrix}\,, \quad \hat{\beta} = \begin{pmatrix} \mathbb{1} & 0 \\ 0 & -\mathbb{1} \end{pmatrix}\,, \tag{23}$$

The Dirac spinor $\Psi(\vec{r})$ on the basis of complete set $[\hat{H}, \hat{\kappa}, \hat{\mathbf{J}}^2, \hat{\mathbf{J}}_z]$ is defined as [37]

$$\Psi(\vec{r}) = \frac{1}{r}\begin{pmatrix} f_{n\kappa}(r)\phi_{jm} \\ ig_{n\kappa}(r)\chi_{jm} \end{pmatrix} \text{ for } \kappa = j + \frac{1}{2}\,, \tag{24a}$$

$$\Psi(\vec{r}) = \frac{1}{r}\begin{pmatrix} f_{n\kappa}(r)\chi_{jm} \\ ig_{n\kappa}(r)\phi_{jm} \end{pmatrix} \text{ for } \kappa = -(j + \frac{1}{2})\,, \tag{24b}$$

with

$$\phi_{jm} = \begin{pmatrix} \sqrt{\frac{j+m}{2j}}\,Y_{j-1/2,m-1/2} \\ \sqrt{\frac{j-m}{2j}}\,Y_{j-1/2,m+1/2} \end{pmatrix}\,, \quad \chi_{jm} = \begin{pmatrix} \sqrt{-\frac{j-m+1}{2j+2}}\,Y_{j+1/2,m-1/2} \\ \sqrt{\frac{j+m+1}{2j+2}}\,Y_{j+1/2,m+1/2} \end{pmatrix}\,, \tag{25}$$

where $Y_{j,m}$ are the normalized spherical harmonics [37]. In the complete set $[\hat{H}, \hat{\kappa}, \hat{\mathbf{J}}^2, \hat{\mathbf{J}}_z]$ of the conservative quantities, the operator $\hat{H}$ denotes the Hamiltonian of the system under consideration, $\hat{\kappa}$ describes the spin-orbit coupling operator, $\hat{\mathbf{J}}$ and $\hat{\mathbf{J}}_z$ are the total angular momentum and it's z-component operators, respectively [37]. We notice that we use the functions given in Eqs. (24a), (24b) and Eq. (25) without the subscripts in the rest of computation.

By using Eqs. (24a) and (24b) and with the help of Eq. (22) we have two couple differential equations of first order

$$\hbar c\left(\frac{d}{dr}-\frac{\kappa}{r}\right)f(r)=[m_0c^2+E+S(r)-V(r)]g(r)\,, \tag{26a}$$

$$\hbar c\left(\frac{d}{dr}+\frac{\kappa}{r}\right)g(r)=[m_0c^2-E+S(r)+V(r)]f(r)\,, \tag{26b}$$

which gives us the followings for $S(r)=V(r)$

$$f(r)=\frac{\hbar c}{m_0c^2-E+2V(r)}\left(\frac{d}{dr}+\frac{\kappa}{r}\right)g(r)\,, \tag{27a}$$

$$g(r)=\frac{\hbar c}{m_0c^2+E}\left(\frac{d}{dr}-\frac{\kappa}{r}\right)f(r)\,, \tag{27b}$$

Inserting $g(r)$ in (27b) into (27a) gives us the equation

$$\left[\frac{d^2}{dr^2}-\frac{\kappa(\kappa-1)}{r^2}+Q^2(E^2-m_0^2c^4)-2Q^2(E+m_0c^2)V(r)\right]f(r)=0\,. \tag{28}$$

which is similar to (4) with a change $\ell(\ell+1)\to\kappa(\kappa-1)$. By following the same steps in above subsection, we write the Dirac equation for the two-term potential as

$$\frac{d^2f(z)}{dz^2}+\frac{1}{z}\frac{df(z)}{dz}+\left[-\frac{A_1^2}{z}+A_2^2-\frac{(A_3^{\mp})^2}{q}\frac{1}{1-z}\right]\frac{f(z)}{z(1-z)}=0\,, \tag{29}$$

with

$$(A_3^-)^2=\kappa(\kappa-1)+\frac{2Q^2V_1}{q\beta^2}(E+m_0c^2)\,. \tag{30}$$

and $(A_3^+)^2=(A_3^-)^2(\kappa\to\kappa+1)$. The upper indices $\mp$ of the constant A_3^2 is related with the upper component $f(r)$, and $g(r)$, respectively.

It is easy to obtain the energy eigenvalue equation for the Dirac equation with the two-term potential from (14), and we write

$$E^2-E\left(V_0+\frac{V_1}{q}\right)-m_0c^2\left(V_0+\frac{V_1}{q}+1\right)+\left[\frac{\frac{Q}{q\beta}(E+m_0c^2)\left(V_0+\frac{V_1}{q}\right)}{n+\frac{1}{2}+\sqrt{\frac{1}{4}+\frac{(A_3^{\mp})^2}{q}}}\right]^2-\left(\frac{n+\frac{1}{2}+\sqrt{\frac{1}{4}+\frac{(A_3^{\mp})^2}{q}}}{2Q/\beta}\right)^2=0\,. \tag{31}$$

The corresponding upper component of Dirac spinor from (13) is given as

$$f(z)=Nz^{A_1-1}(1-z)^{1+D^{\mp}}\,{}_2F_1(-n,-n+2A_2;1+2A_1;z)\,, \tag{32}$$

with the normalization constant

$$N=\left[\frac{\beta(n+A_1+D^{\mp}+1)\Gamma(n+2A_1+1)\Gamma(2A_2)}{n!(n+1+D^{\mp})\Gamma(n+2+2D^{\mp})\Gamma(2A_1)\Gamma(1+2A_1)}\right]^{1/2}. \tag{33}$$

with $D^{\mp}=-\frac{1}{2}+\sqrt{\frac{1}{4}+\frac{(A_3^{\mp})^2}{q}}$.

3. Conclusion

We have studied the approximate, analytical bound state solutions of the Klein-Gordon and Dirac equations for the two-term potential. We have obtained the energy eigenvalue equations, and the corresponding normalized "wave functions" by converting the wave equations to a Riemann-type equation. We have presented the results with the help of an approximation instead of the centrifugal term, and discussed the cases including the Manning-Rosen potential, the Hulthén potential and the Coulomb potential.

4. Acknowledgments

This research was partially supported by Hacettepe University Scientific Research Coordination Unit, Project Code: FBB-2016-9394.

References

[1] Chen G 2005 *Phys. Lett.* A **339** 300
[2] de Castro, and Hott M 2005 *Phys. Lett.* A **342** 53
[3] Jia C S, Li Y, Sun Y, Liu J Y, and Sun L T 2003 *Phys. Lett.* A **311** 115
[4] Egrifes H, Demirhan D, and Buyukkılıc F 2000 *Phys. Lett.* A **275** 229
[5] Lai C S 1982 *Phys. Rev.* A **26** 2245
[6] Dong S, Garcia-Ravelo J and Dong S H 2007 *Phys. Scr.* **76** 393
[7] Ikhdair S M, and Sever R 2007 *J. Math. Chem.* **41** 329
[8] Jia C S, Diao Y F, Li M, Yang Q B, Sun L T, and Huang R Y 2004 *J. Phys.* A **37** 11275
[9] Sever R, and Tezcan C 1987 *Phys. Rev.* A **35** 2725
[10] Roy B, *Phys. Rev.* A **34** 5108
[11] Diao Y F, Yi L Z, and Jia C S, *Phys. Lett.* A **332** 157
[12] Taşkın F, Boztosun I, and Bayrak O 2008 *Int. J. Theor. Phys.* **47** 1612
[13] Berkdemir C 2006 *Nucl. Phys.* A **770** 32
[14] Bayrak O, and Boztosun I 2006 *J. Phys.* A **39** 6955
[15] Yu J, Dong S H, and Sun G H 2004 *Phys. Lett.* A **322** 290
[16] Qiang W C, Zhou R S, and Gao Y 2007 *J. Phys.* A **40** 1677
[17] Berkdemir C, and Han J 2005 *Chem. Phys. Lett.* **409** 203
[18] Duru I H 1983 *Phys. Rev.* D **28** 2689
[19] Filho E D, and Ricotta R M 2000 *Phys. Lett.* A **269** 269
[20] Morales D A 2004 *Chem. Phys. Lett.* **394** 68
[21] Ikhdair S M 2010 *J. Math. Phys.* **51** 023525
[22] Ikhdair S M, and Sever R 2006 *J. Math. Chem.* **42** 461
[23] Filho E D, and Ricotta R M 1995 *Mod. Phys. Lett.* A **10** 1613
[24] Sogut K, and Havare A 2010 *J. Phys.* A **43** 225204
[25] Jia C S, Wang J Y, He S, and Sun L T 2000 *J. Phys.* A **33** 6993
[26] Gonul B, Ozer O, Cancelik Y, and Kocak M 2000 *Phys. Lett.* A **275** 238
[27] Haouat S, and Chetouani L 2008 *Phys. Scr.* **77** 025005
[28] Dominguez-Adame G 1989 *Phys. Lett.* A **136** 175
[29] Rojas C, and Villalba V M 2005 *Phys. Rev.* A **71** 052101
[30] Maghsoodi E, Hassanabadi H, Zarrinkamar S, and Rahimov H 2012 *Phys. Scr.* **85** 055007
[31] Feizi H, Shojaei M R, and Rajabi A A 2012 *Eur. Phys. J Plus* **127** 41
[32] Zarrinkamar S, Rajabi A A, and Hassanabadi H 2010 *Ann. Phys.* **325** 2522
[33] Sun J X 1999 *Acta Phys. Sin.* **48** 1992
[34] Jia C S, Wang J Y, He S, and Sun L T 2000 *J. Phys.* A **33** 6993
[35] Benamira F, Guechi L, Mameri S, and Sadoun M A 2007 *J. Math. Phys.* **48** 032102
[36] Arda A, and Sever R 2012 *J. Math. Chem.* **50** 1920
[37] Greiner W 2000 *Relativistic Quantum Mechanics-Wave Equations* (Berlin: Springer)
[38] Gradshteyn I S, and Ryzhik I M 2007 *Table of Integrals, Series, and Products* (New York: Academic Press)
[39] Kimura T 1969 *Funk. Ekvacioj* **12** 269

Exact quantization of Cauchy-Euler type forced parametric oscillator

Şirin A. Büyükaşık and Zehra Çayiç
Department of Mathematics, Izmir Institute of Technology, Gulbahce Campus 35430 Urla, Izmir, Turkey
E-mail: sirinatilgan@iyte.edu.tr, zehracayic@iyte.edu.tr

Abstract. Driven and damped parametric quantum oscillator is solved by Wei-Norman Lie algebraic approach, which gives the exact form of the evolution operator. This allows us to obtain explicitly the probability densities, time-evolution of initially Glauber coherent states, expectation values and uncertainty relations. Then, as an exactly solvable model, we introduce the driven Cauchy-Euler type quantum parametric oscillator, which appears as self-adjoint quantization of the classical Cauchy-Euler differential equation. We discuss some typical behavior of this oscillator under the influence of external terms and give a concrete example.

1. Introduction

Quantum harmonic oscillator with explicitly time-dependent Hamiltonian is a fundamental model, which appears in many physical branches such as quantum optics, quantum fluid, ion-traps, and cosmology. It is a useful model also in quantum information and quantum computation.

In the present work, we first solve the evolution problem for quantum harmonic oscillator

$$\begin{cases} i\hbar\frac{\partial}{\partial t}\Psi(q,t) = \hat{H}(t)\Psi(q,t), \\ \Psi(q,t_0) = \Psi_0(q), \quad -\infty < q < \infty, \end{cases} \tag{1}$$

with generalized quadratic Hamiltonian

$$\hat{H}(t) = \frac{\hat{p}^2}{2\mu(t)} + \frac{\mu(t)\omega^2(t)}{2}\hat{q}^2 + \frac{B(t)}{2}(\hat{q}\hat{p} + \hat{p}\hat{q}) + D(t)\hat{q}, \tag{2}$$

where $\mu(t) > 0$, $\omega^2(t)$, $B(t)$ and $D(t)$ are sufficiently smooth, real-valued parameters depending on time. To understand better the behavior of such quantum systems it is always important to have exactly solvable models. The best known one is the Caldirola-Kanai oscillator with an exponentially increasing mass, [1, 2], which is widely used to study dissipation in quantum mechanics. In this article, we introduce the Cauchy-Euler type quantum parametric oscillator as another exactly solvable model, and briefly discuss some of its typical properties.

2. Quantization of the harmonic oscillator

2.1. The evolution operator and wave functions

As known, the evolution operator for system (1) can be found by the Wei-Norman Lie algebraic approach. Indeed, the Hamiltonian (2) can be written as time-dependent linear combination of Lie algebra generators as

$$\hat{H}(t) = -i\left(\frac{-\hbar}{\mu(t)}\hat{K}_- + \mu(t)\omega^2(t)\hat{K}_+ + 2\hbar B(t)\hat{K}_0 + D(t)\hat{E}_1\right), \tag{3}$$

where $\hat{E}_1 = iq$, $\hat{E}_2 = \partial/\partial q$, $\hat{E}_3 = i\hat{I}$ are generators of the Heisenberg-Weyl algebra, and $\hat{K}_- = -(i/2)(\partial^2/\partial q^2)$, $\hat{K}_+ = (i/2)q^2$, $\hat{K}_0 = (q(\partial/\partial q) + 1/2)/2$ are generators of the $su(1,1)$ algebra. Then, the evolution operator for the general oscillator can be written as product of exponential operators

$$\hat{U}(t,t_0) = e^{c(t)\hat{E}_3}e^{\frac{a(t)}{\hbar}\hat{E}_1}e^{-b(t)\hat{E}_2}e^{f(t)\hat{K}_+}e^{2h(t)\hat{K}_0}e^{g(t)\hat{K}_-}, \tag{4}$$

where $f(t), g(t), h(t), a(t), b(t), c(t)$ are unknown real-valued functions to be determined from the IVP defining the unitary operator $\hat{U}$, that is $i\hbar(d\hat{U}(t,t_0)/dt) = \hat{H}(t)\hat{U}(t,t_0)$, $\hat{U}(t_0,t_0) = \hat{I}$. Plugging (3) and (4) in that problem, after some algebra we obtain the exact form of the evolution operator

$$\begin{aligned}\hat{U}(t,t_0) \quad = \quad & \exp\left(\frac{i}{-\hbar}\int_{t_0}^{t}\left[\frac{-1}{2\mu(s)}p_p^2(s) + \frac{\mu(s)\omega^2(s)}{2}x_p^2(s)\right]ds\right) \\ & \times\exp\left(ip_p(t)q\right)\times\exp\left(-x_p(t)\frac{\partial}{\partial q}\right)\times\exp\left(i\frac{\mu(t)}{2\hbar}\left(\frac{\dot{x}_1(t)}{x_1(t)} - B(t)\right)q^2\right) \\ & \times\exp\left(\ln\left|\frac{x_1(t_0)}{x_1(t)}\right|\left(q\frac{\partial}{\partial q} + \frac{1}{2}\right)\right)\times\exp\left(\frac{i}{2}{}^{-}\hbar x_1^2(t_0)\left(\frac{x_2(t)}{x_1(t)}\right)\frac{\partial^2}{\partial q^2}\right),\end{aligned} \tag{5}$$

where the functions $x_1(t)$ and $x_2(t)$ are two linearly independent homogeneous solutions of the classical equation of motion,

$$\ddot{x} + \frac{\dot{\mu}}{\mu}\dot{x} + \left(\omega^2(t) - \left(\dot{B} + B^2 + \frac{\dot{\mu}}{\mu}B\right)\right)x = -\frac{1}{\mu}D, \tag{6}$$

satisfying the IC's $x_1(t_0) = x_0 \neq 0$, $\dot{x}_1(t_0) = x_0B(t_0)$, and $x_2(t_0) = 0$, $\dot{x}_2(t_0) = 1/\mu(t_0)x_1(t_0)$, and $x_p(t)$ is the particular solution of (6) satisfying the IC's $x_p(t_0) = \dot{x}_p(t_0) = 0$. Also $p_p(t)$ is the particular solution to the classical equation of motion in momentum space

$$\ddot{p} - \frac{(\dot{\mu\omega^2})}{\mu\omega^2}\dot{p} + \left(\omega^2(t) + \left(\dot{B} - B^2 - \frac{(\dot{\mu\omega^2})}{\mu\omega^2}B\right)\right)p = -\dot{D} + \left(\frac{(\dot{\mu\omega^2})}{\mu\omega^2} + B\right)D, \tag{7}$$

satisfying the IC's $p_p(t_0) = 0$, $\dot{p}_p(t_0) = -D(t_0)$. Next, applying the evolution operator (5) to the normalized eigenstates of the standard Hamiltonian

$$\varphi_k(q) = N_k e^{-\frac{m\omega_0}{2\hbar}q^2}H_k\left(\sqrt{\frac{m\omega_0}{-\hbar}}q\right), \quad k = 0,1,2,..., \tag{8}$$

we obtain the exact wave function in the form

$$\Psi_k(q,t) = N_k\sqrt{R_B(t)}\times\exp\left[i\left(k+\frac{1}{2}\right)\arctan\left(-m\omega_0x_1^2(t_0)\left(\frac{x_2(t)}{x_1(t)}\right)\right)\right]$$

$$\times \exp\left\{\frac{i}{2\hbar}\left[\left(\mu(t)\left(\frac{\dot{x}_1(t)}{x_1(t)} - B(t)\right) - (m\omega_0 x_1(t_0))^2\frac{x_2(t)}{x_1(t)}R_B^2(t)\right)(q - x_p(t))^2 + 2p_p(t)q\right]\right\}$$
$$\times \exp\left[\frac{i}{\hbar}\int_{t_0}^{t}\left(\frac{-1}{2\mu(s)}p_p^2(s) + \frac{\mu(s)\omega^2(s)}{2}x_p^2(s)\right)ds\right]$$
$$\times \exp\left[-\frac{1}{2}\left(\sqrt{\frac{m\omega_0}{\hbar}}R_B(t)\,(q - x_p(t))\right)^2\right] \times H_k\left(\sqrt{\frac{m\omega_0}{\hbar}}R_B(t)\,(q - x_p(t))\right). \tag{9}$$

Therefore, the probability density is

$$\rho_k(q,t) = N_k^2 R_B(t)\exp\left(-\left(\sqrt{\frac{m\omega_0}{\hbar}}R_B(t)(q - x_p(t))\right)^2\right) \times H_k^2\left(\sqrt{\frac{m\omega_0}{\hbar}}R_B(t)\,(q - x_p(t))\right), \tag{10}$$

where

$$R_B(t) = \sqrt{\frac{{x_0}^2}{x_1^2(t) + (m\omega_0 {x_0}^2 x_2(t))^2}}. \tag{11}$$

We see that, in (10) $R_B(t)$ is the squeezing (or spreading) coefficient and $x_p(t)$ acts as displacement of the wave packet. Clearly, $R_B(t)$ depends on $B(t)$, but does not depend on $D(t)$. On the other hand, $x_p(t)$, clearly depends on all parameters of the Hamiltonian (2).

2.2. *Coherent states, expectation values and uncertainty relation*

Coherent states of the standard harmonic oscillator (Glauber CS's), can be written in the form

$$\phi_\alpha(q) = \left(\frac{m\omega_0}{\pi\hbar}\right)^{1/4} e^{\frac{i}{\hbar}\langle\hat{p}\rangle_\alpha q} e^{-\frac{m\omega_0}{2\hbar}(q - \langle\hat{q}\rangle_\alpha)^2}, \quad \alpha \in \mathcal{C}. \tag{12}$$

Using (5) we can find the time-evolution of initially Glauber CS's, that is $\Phi_\alpha(q,t) = \hat{U}_g(t,t_0)\phi_\alpha(q)$, so that the corresponding probability density $\rho_\alpha(q,t) = |\Phi_\alpha(q,t)|^2$ becomes

$$\begin{aligned}\rho_\alpha(q,t) &= \sqrt{\frac{m\omega_0}{\pi\hbar}}R_B^2(t) \times \exp\left\{2\left[((m\omega_0 x_0)x_2(t)R_B(t))^2(\alpha_1^2 - \alpha_2^2)\right.\right.\\ &\quad \left.\left. -2(m\omega_0)x_1(t)x_2(t)R_B^2(t)\alpha_1\alpha_2 - \alpha_1^2\right]\right\}\\ &\quad \times \exp\left(2\sqrt{\frac{2m\omega_0}{\hbar}}R_B^2(t)\left(\alpha_1\frac{x_1(t)}{x_0} + \alpha_2(m\omega_0 x_0)x_2(t)\right)(q - x_p(t))\right)\\ &\quad \times \exp\left(-\left(\frac{m\omega_0}{\hbar}\right)R_B^2(t)(q - x_p(t))^2\right).\end{aligned} \tag{13}$$

Using again (5), we compute the position and momentum operators in Heisenberg picture as

$$\hat{q}_H(t) \equiv \hat{U}_g^\dagger(t,t_0)\hat{q}_S\hat{U}_g(t,t_0) = \frac{1}{x_0}x_1(t)\hat{q}_H(t_0) + x_0x_2(t)\hat{p}_H(t_0) + x_p(t), \tag{14}$$

$$\hat{p}_H(t) \equiv \hat{U}_g^\dagger(t,t_0)\hat{p}_S\hat{U}_g(t,t_0) = \frac{1}{x_0}\mu(t)p_1(t)\hat{q}_H(t_0) + x_0\mu(t)p_2(t)\hat{p}_H(t_0) + p_p(t). \tag{15}$$

Since $\langle\hat{q}\rangle_\alpha(t) \equiv \langle\Phi_\alpha(q,t)|\hat{q}|\Phi_\alpha(q,t)\rangle = \langle\phi_\alpha(q,t_0)|\hat{q}_H(t)|\phi_\alpha(q,t_0)\rangle$, the expectation value of position at coherent state $\Phi_\alpha(q,t)$ is

$$\langle\hat{q}\rangle_\alpha(t) = \sqrt{\frac{2\hbar}{m\omega_0}}\left(\frac{\alpha_1}{x_0}x_1(t) + \alpha_2(m\omega_0x_0)x_2(t)\right) + x_p(t). \tag{16}$$

Similarly, using that $\langle\hat{p}\rangle_\alpha(t) = \langle\phi_\alpha(q,t_0)|\hat{p}_H(t)|\phi_\alpha(q,t_0)\rangle$, the expectation value of the momentum becomes

$$\langle\hat{p}\rangle_\alpha(t) = \sqrt{\frac{2\hbar}{m\omega_0}}\mu(t)\Big(\frac{\alpha_1}{x_0}p_1(t) + \alpha_2(m\omega_0 x_0)p_2(t)\Big) + p_p(t), \tag{17}$$

where $p_1(t) = \mu(t)(\dot{x}_1(t) - B(t)x_1(t))$ and $p_2(t) = \mu(t)(\dot{x}_2(t) - B(t)x_2(t))$ are two linearly independent solutions of the homogeneous part of the classical oscillator (7) in momentum space. This explicitly shows that, the expectation values at coherent states satisfy the classical equations of motion. After some calculations, we obtain also the uncertainties

$$(\Delta\hat{q})_\alpha(t) = \sqrt{\frac{\bar{h}}{2m\omega_0}\frac{1}{R_B(t)}}, \tag{18}$$

$$(\Delta\hat{p})_\alpha(t) = \sqrt{\frac{m\omega_0\bar{h}}{2}R_B(t)}\sqrt{1 + \frac{\mu^2(t)}{(m\omega_0 R_B^2(t))^2}\left(\frac{\dot{R}_B(t)}{R_B(t)} + B(t)\right)^2}. \tag{19}$$

We note that the expectation values depend on all parameters of the Hamiltonian, however the uncertainties does not depend on the external linear term, which contributes only to the displacement of the wave packet. Finally, the uncertainty relation for the generalized harmonic oscillator with time dependent parameters is found

$$(\Delta\hat{q})_\alpha(\Delta\hat{p})_\alpha(t) = \frac{\bar{h}}{2}\sqrt{1 + \frac{\mu^2(t)}{(m\omega_0 R_B^2(t))^2}\left(\frac{\dot{R}_B(t)}{R_B(t)} + B(t)\right)^2}, \tag{20}$$

where clearly $(\Delta\hat{q})_\alpha(\Delta\hat{p})_\alpha \geq \bar{h}/2$ Summarizing, coherent states of the time-dependent parametric oscillator are displaced Gaussian wave packets, see (13), and follow the classical trajectory as can be seen from (16) and (17). However, they are spreading or squeezing in time, since $(\Delta\hat{q})_\alpha(t)$ given by (18) depends explicitly on time, and are no longer minimum uncertainty states, as we can see from (20).

3. Cauchy-Euler type quantum parametric oscillator

We introduce the Cauchy-Euler type quantum parametric oscillator and study its behavior under the influence of the mixed and external terms. Precisely, we consider the Hamiltonian

$$\hat{H}(t) = \frac{1}{2}t^{-\gamma}\hat{p}^2 + \frac{1}{2}w_0^2 t^{\gamma-2}\hat{q}^2 + B(t)\frac{(\hat{q}\hat{p}+\hat{p}\hat{q})}{2} + D(t)\hat{q}, \quad t > 0, \tag{21}$$

with variable mass $\mu(t) = t^\gamma$, variable frequency $\omega^2(t) = w_0^2/t^2$, $\gamma > 0$, $w_0^2 > 0$, mixed term parameter $B(t)$ and linear driving $D(t)$. When $B(t) = D(t) = 0$, the corresponding classical oscillator is the Cauchy-Euler type ordinary differential equation

$$\ddot{x} + (\gamma/t)\dot{x} + (w_0^2/t^2)x = 0, \;\; t > 0,$$

which solutions depend on the sign of $\Omega_0^2 = w_0^2 - (\gamma-1)^2/4$. Accordingly, we have the cases: (i) $\Omega_0^2 = 0$ critical damping, (ii) $\Omega_0^2 < 0$ over damping, and (iii) $\Omega_0^2 > 0$ under damping. Then, we choose $B(t) = \dot{y}(t)/y(t)$, where $y(t)$ satisfies the Cauchy-Euler equation $\ddot{y}+(\gamma/t)\dot{y}+(w_1^2/t^2)y = 0$, with the same γ, but w_1 possibly different than w_0. This special choice of $B(t)$ modifies the original frequency w_0^2/t^2, but does not change the Cauchy-Euler type structure of the oscillator. Indeed, in that case the classical oscillator corresponding to the Hamiltonian (21) takes the form

$$\ddot{x} + \frac{\gamma}{t}\dot{x} + \frac{(w_0^2 + w_1^2)}{t^2}x = -\frac{1}{t^\gamma}D(t), \;\; t > 0, \tag{22}$$

with time-variable damping $\Gamma(t) = \gamma/t$, $\gamma > 0$, modified frequency $w_g^2(t) = (w_0^2 + w_1^2)/t^2$, and driving force. In this short article, some typical behavior of the quantum oscillator is illustrated by an example.

Example: We take the Hamiltonian

$$\hat{H}(t) = \frac{t^{-4}}{2}\hat{p}^2 + \frac{t^2}{2}\hat{q}^2 + B(t)(\hat{q}\hat{p} + \hat{p}\hat{q}) + D(t)\hat{q}, \quad , \quad (\gamma = 4,\ w_0 = 1), \tag{23}$$

with mixed term parameter $B(t) = -|\Omega_1|[\tan(|\Omega_1|\ln t - \alpha) + \tan\alpha]/t$, where $\Omega_1^2 = 247/4$, $\alpha = \tan^{-1}(3/\sqrt{247})$, and external parameter $D(t) = \sqrt{t}$. The corresponding classical equation of motion for (23) becomes

$$\ddot{x} + (4/t)\dot{x} + (65/t^2)x = t^{-7/2}, \quad t > 0. \tag{24}$$

Note that, when $B(t) = 0$, the oscillator is in over damping case, but with above choice of $B(t)$ we have $w_1 = 8$, so that the modified frequency is $w_g^2(t) = (w_0^2 + w_1^2)/t^2 = 65/t^2$, and then oscillator becomes in under damping case. For $t_0 = 1$, two linearly independent solutions of (24) satisfying the initial conditions $x_1(1) = 1$, $\dot{x}_1(1) = 0$, and $x_2(1) = 0$, $\dot{x}_2(1) = 1$, are

$$x_1(t) = \frac{\sqrt{65}}{|\Omega_g|}t^{-3/2}\cos(|\Omega_g|\ln t - \beta), \quad x_2(t) = \frac{1}{|\Omega_g|}t^{-3/2}\sin(|\Omega_g|\ln t), \quad t > 0,$$

where $\beta = \tan^{-1}(3/\sqrt{251})$ and $\Omega_g^2 = 251/4$. The particular solution satisfying the initial conditions $x_p(1) = \dot{x}_p(1) = 0$, is $x_p(t) = -t^{-3/2}(1 - \cos(|\Omega_g|\ln t))/\Omega_g^2$. Thus, we obtain

$$R_B(t) = |\Omega_g|\sqrt{\frac{t^3}{65\cos^2(|\Omega_g|\ln t - \beta) + 25\sin^2(|\Omega_g|\ln t)}}, \quad t > 0, \tag{25}$$

which is smooth, shows oscillatory behavior, and $R_B(t) \to \infty$ as $t \to \infty$. Then the probability density becomes

$$\begin{aligned} \rho_k(q,t) &= N_k^2 R_B(t)\exp\left(-\left(5R_B(t)\left(q + \frac{1}{\Omega_g^2}t^{-(\frac{\gamma-1}{2})}(1 - \cos(|\Omega_g|\ln t))\right)\right)^2\right) \\ &\quad \times H_k^2\left(5R_B(t)\left(q + \frac{1}{\Omega_g^2}t^{-(\frac{\gamma-1}{2})}(1 - \cos(|\Omega_g|\ln t))\right)\right), \end{aligned} \tag{26}$$

and in Fig. 1 we plot it for $k = 1, 2$ respectively. We observe that, probability density amplitudes increase without bound, and wave packets are squeezing as time increases, since $R_B(t) \to \infty$ as $t \to \infty$. Also, probability is oscillatory in time, and is localized near $q = 0$ as $t \to \infty$, since $x_p(t) \to 0$. Knowing (25), the uncertainties and uncertainty relation take the form

$$(\Delta\hat{q})_\alpha(t) = \sqrt{\frac{1}{10}\frac{1}{R_B(t)}}, \tag{27}$$

$$(\Delta\hat{p})_\alpha(t) = \sqrt{\frac{5}{2}R_B(t)}\sqrt{1 + \frac{t^8}{25R_B^4(t)}\left(\frac{\dot{R}_B(t)}{R_B(t)} - \frac{|\Omega_1|}{t}\Big(\tan(|\Omega_1|\ln t - \alpha) + \tan\alpha\Big)\right)^2}, \tag{28}$$

$$(\Delta\hat{q})_\alpha(\Delta\hat{p})_\alpha(t) = \frac{1}{2}\sqrt{1 + \frac{t^8}{25R_B^4(t)}\left(\frac{\dot{R}_B(t)}{R_B(t)} - \frac{|\Omega_1|}{t}\Big(\tan(|\Omega_1|\ln t - \alpha) + \tan\alpha\Big)\right)^2}. \tag{29}$$

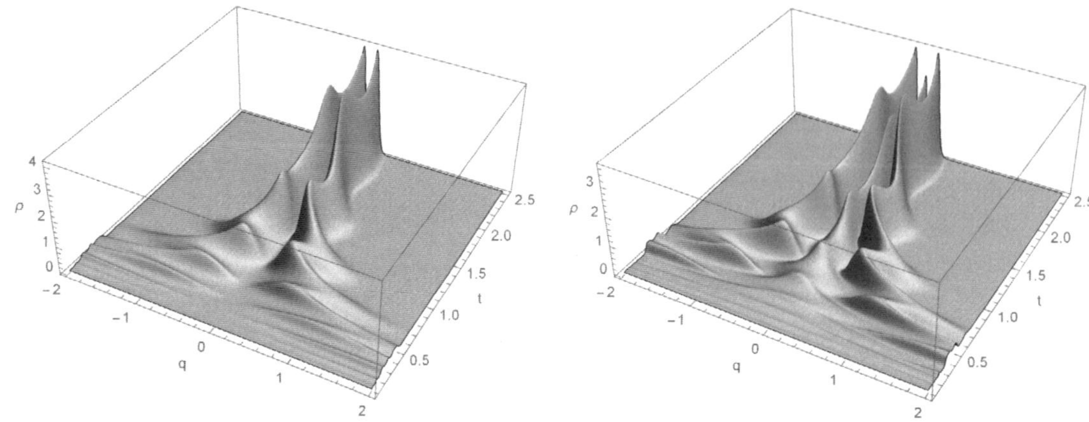

Figure 1. Probability density $\rho_k(q,t) = |\Psi_k(q,t)|^2$ for Cauchy-Euler type generalized oscillator when $B(t) = -|\Omega_1|[\tan(|\Omega_1|\ln t - \alpha) + \tan\alpha]/t$, $D(t) = \sqrt{t}$ $w_0 = 1, w_1 = 8, \gamma = 4$, and $k = 1, 2$, respectively.

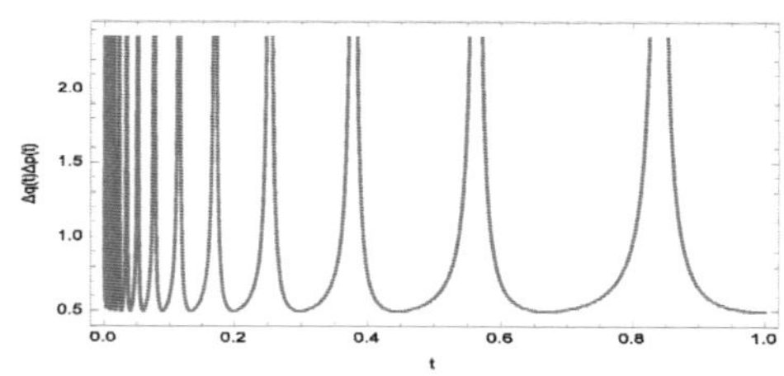

Figure 2. Uncertainty relation for driven Cauchy-Euler oscillator, $w_0 = 1, w_1 = 8, \gamma = 4$.

As a result, we see that by a special choice of the mixed term parameter $B(t)$, which modifies the original frequency, we are able to preserve the structure of the original oscillator. However, this parameter $B(t)$ has finite time singularities. In that case, the probability densities and expectations of position remain smooth, but singularities of $B(t)$ are reflected in the expectations of momentum and the uncertainty relation. Precisely, momentum is not defined at these points, and when time approaches these singularities, uncertainty relation tends to infinity, as one can see in Fig. (2).

References

[1] Caldirola P., 1941: Nouovo Cimento, **18**, 393 (1941).
[2] Kanai E., 1948: *Prog.Theo.Physics*, **3**, 440.
[3] Wei, J. and Norman, E., 1963: Lie Algebraic Solution of Linear Differential Equations. *J. Math. Phys.*, **4**, 575.
[4] Perelemov, A.M., 1986: *Generalized Coherent States and Their Applications*, (Berlin: Springer).
[5] Dattoli, G., Richetta, M. and Torre, A., 1988: Operator Disentanglement. *Phys. Rev. A*, **37**, 2007-2011.
[6] Lo, C.F., 1991: Quantum Mechanics of a General Driven Time-Dependent Oscillator. *Il Nuovo Cimento*, **13D**, 1279-1292.
[7] Büyükaşık, Ş.A., Pashaev, O.K. and Ulaş-Tigrak, E., 2009: Exactly Solvable Quantum Sturm-Liouville Problems. *J. Math. Phys.*, **50**, 072102
[8] Büyükaşık Ş. A. and Pashaev, O.K., 2010: *J. Math. Phys.* **51**, 122108.

Journal of Physics: Conference Series **766** (2016) 012004 doi:10.1088/1742-6596/766/1/012004

On the deformed Einstein equations and quantum black holes

E Dil[1], E Kolay[2] and C C Ersanli[1]

[1]Department of Physics, Sinop University, Korucuk, 57000-Sinop, Turkey
[2]Department of Statistics, Sinop University, Korucuk, 57000-Sinop, Turkey

Abstract. Recently q-deformed Einstein equations have been studied for extremal quantum black holes which have been proposed to obey deformed statistics by Strominger. In this study, we give the solutions of deformed Einstein equations by considering these equations for the charged black holes. Also we present the implications of the solutions, such as the deformation parameters lead the charged black holes to have a smaller mass than the classical Reissner-Nordström black holes. The reduction in mass of a classical black hole can be viewed as a transition from classical to quantum black hole regime.

1. Introduction

Recently, we have studied the q-deformed Einstein equations [1] for describing the gravitational fields of the Strominger's extremal quantum black holes which obey the deformed statistics [2]. Here, Verlinde's entropic gravity approach [3] have been used to obtain the q-deformed Eintein equations which are the gravitational field equations of charged quantum black holes with possible minimal mass. A charged black hole loses its mass via Hawking radiation and reaches a minimum mass proportional to the charge. The resulting black hole with a minimal mass is considered to be a micro black hole. Since these black holes can be considered as deformed bosons or deformed fermions, the micro black holes defined by q-deformed Einstein equations are assumed be the Ubriaco's q-deformed bosons [4].

We first give a brief summary of obtaining the q-deformed Einstein equations then obtain the solutions of the deformed Einstein equations for charged black holes. Since the solutions of standard Einstein equations for charged black holes are the Reissner-Nordström solutions in classical gravity, the solutions of the deformed Einstein equations for charged black holes can be considered in quantum gravity. Lastly the implications of the solutions are represented. These are that the deformation parameters lead the charged black holes to have a smaller mass than the usual Reissner-Nordström black holes. This reduction in mass of a usual black hole can be considered as a transition from classical to quantum black hole regime.

2. q-Deformed Einstein equations

We obtain the q-deformed Einstein equations which describe the gravitational fields quantum black holes by using the entropy of the Ubriaco's q-deformed Bose gas models in Verlinde's entropic gravity approach. The quantum algebraic structure of this q-deformed Bose gas model is given by the q-deformed boson algebra [4]

$$a_2 a_2^* - q^2 a_2^* a_2 = 1, \tag{1}$$

$$a_1 a_1^* - q^2 a_1^* a_1 = 1 + (q^2 - 1) a_2^* a_2 , \tag{2}$$

$$a_2 a_1 = q a_1 a_2 , \tag{3}$$

$$a_2 a_1^* = q a_1^* a_2 . \tag{4}$$

Here a and a^* represents the deformed annihilation and creation operators, respectively. q is also a real deformation parameter with $0 \le q < \infty$. The grand partition function of the q-deformed boson model is [4]

$$Z = \prod_k \sum_{m=0}^{\infty} (m+1) e^{-\beta \varepsilon_k \{m\}} z^m , \tag{5}$$

where $\beta = 1/kT$ and k is the Boltzmann constant, $z = e^{\beta\mu}$ is the fugacity, ε_k is the energy of the single-particle state, m is the occupation number of the single-particle state, and $\{m\}$ is the deformed occupation number and given by

$$\{m\} = \frac{1 - q^{2m}}{1 - q^2} . \tag{6}$$

The deformed entropy of the model is also given as

$$S = \frac{4\pi V (2m)^{3/2}}{h^3 T} E^{5/2} \left[\frac{5\sqrt{\pi}}{4} z + \frac{5\sqrt{\pi}}{2} \delta(q) z^2 - \frac{\sqrt{\pi}}{2} z \ln z - 2\sqrt{\pi}\delta(q) z^2 \ln z + \cdots \right], \tag{7}$$

where $E = kT$ is the average energy of single particle, V is the volume enclosed by the deformed bosons, m is the mass of deformed bosons, T is the temperature of the model and $\delta(q) = (1/4).\{[3/(1+q^2)^{3/2}] - (1/\sqrt{2})\}$ [4]. The deformed entropy in (7) is used to obtain the one-parameter deformed or equivalently the q-deformed Einstein equations for q-deformed bosons.

We apply the Verlinde's proposal to the q-deformed Bose gas models, in order to construct the q-deformed Einstein equations from the entropy in (7). The fundamental notion needed to derive the gravity is information in the Verlinde's proposal. It is formally the amount of information associated with the matter and its location, measured in terms of entropy. When matter is displaced in space due to a reason, the result is a change in the entropy and this change causes a reaction force. This force is the gravity being an entropic force as an inertial reaction against the force causing the increase of the entropy [3].

The source of gravity is energy or matter and it is distributed evenly over the degrees of freedom in spacetime. The existence of energy or matter in spacetime causes a temperature in the spacetime. The product of the change of entropy during the displacement of source and the temperature is in fact the work and this work is originally led by the force which is known to be gravity [3].

By using the Verlinde's idea, q-deformed Einstein equations have recently been derived from the deformed entropy (7) of the q-deformed Bose gas model [1]. Eventually the q-deformed Einstein equations is given, as [1]

$$\frac{10\pi V (2mE)^{3/2}}{h^3} g(z,q) \left(R_{\mu\nu} - \frac{1}{2} g_{\mu\nu} R \right) = 8\pi G T_{\mu\nu} , \tag{8}$$

where

$$g(z,q) = \left[\frac{5\sqrt{\pi}}{4} z + \frac{5\sqrt{\pi}}{2} \delta(q) z^2 - \frac{\sqrt{\pi}}{2} z \ln z - 2\sqrt{\pi}\delta(q) z^2 \ln z + \cdots \right]. \tag{9}$$

By rewriting (8), we get

$$\Psi^q \left(R_{\mu\nu} - \frac{1}{2} g_{\mu\nu} R \right) = 8\pi G T_{\mu\nu} , \tag{10}$$

where $\Psi^q = 10\pi V(2mE)^{3/2} g(z,q)/h^3$.

The equation (10) is the q-deformed Einstein equations, and it is assumed to describe the gravitational fields generated by the extremal quantum black holes which obey the statistics of deformed particles in accordance with the Strominger's proposal.

In the next section, we solve the deformed Einstein equations for a charged extremal black hole, and investigate the implications of the solutions.

3. Solution of the q-deformed Einstein equations

Deformed Einstein field equations will be used to describe the geometry of the spacetime around a charged and spherically symmetric quantum black hole. Therefore we need to solve the deformed Einstein-Maxwell equations due to the charge of the quantum black holes. Also, because of the spherical symmetry the metric has to be in the form for 4-dimension is [5]

$$ds^2 = -e^{2\alpha(r,t)}dt^2 + e^{2\beta(r,t)}dr^2 + r^2 d\theta^2 + r^2 \sin^2\theta\, d\phi^2 . \tag{11}$$

The energy-momentum tensor $T_{\mu\nu}$ in (10) is

$$T_{\mu\nu} = F_{\mu\rho}F_{\nu}^{\rho} - \frac{1}{4} g_{\mu\nu}F_{\rho\sigma}F^{\rho\sigma} , \tag{12}$$

where $F_{\mu\nu}$ is the electromagnetic field strength tensor and the trace of $T_{\mu\nu}$ is zero due to the electromagnetic strength tensor. Taking the trace of (30) leads to

$$\Psi^q R_{\mu\nu} = 8\pi\, GT_{\mu\nu} . \tag{13}$$

Spherical symmetry and absence of magnetic charge for our quantum black hole gives the electromagnetic field strength tensor has only radial electric field components $E_r = F_{tr} = -F_{rt} = f(r,t)$.

By equating the non-zero components of the Ricci tensor for the metric (11) and the corresponding non-zero components of the energy-momentum tensor obtained by (12), we find $\alpha(r,t) = \alpha(r) = -\beta(r)$. The solution of the Maxwell equations $g^{\mu\nu}\nabla_\mu F_{\nu\sigma} = 0$ and $\nabla_{[\mu}F_{\nu\rho]} = 0$ gives $f(r,t) = f(r) = Q/\sqrt{4\pi}r^2$. Using this electromagnetic field strength tensor for one of the components of the Ricci tensor and the corresponding energy-momentum tensor in (13) gives the solutions of the deformed Einstein equations:

$$ds^2 = \Delta dt^2 + \Delta^{-1}dr^2 + r^2 d\theta^2 + r^2 \sin^2\theta\, d\phi^2 , \tag{14}$$

where

$$\Delta = 1 - \frac{2Gm}{r} + \frac{1}{\Psi^q}\frac{GQ^2}{r^2} . \tag{15}$$

The singularities and the event horizons for these black holes are determined by the function Δ and the radius r. There is a true curvature singularity at $r = 0$, since the metric goes to infinity for this value. The coordinate singularity also occurs at $\Delta = 0$ and the conditions giving this singularity occur from the solution of $\Delta = 0$, such as

$$r_\pm = Gm \pm \sqrt{G^2m^2 - \frac{GQ^2}{\Psi^q}} . \tag{16}$$

The case $Gm^2 = Q^2/\Psi^q$ stands for the quantum charged black hole case, since the mass of the black hole decreases to the minimum value from the case $Gm^2 > Q^2/\Psi^q$. A minimum mass solution for the quantum black holes is reached by the Hawking radiation and remains stationary. $Gm^2 = Q^2/\Psi^q$ case makes $\Delta = 0$ at a single radius $r_\pm = Gm$ and refers to a single event horizon. This deformed case solution $Gm^2 = Q^2 / \Psi^q$ is the analogue of classical Reissner-Nordström solution $m = Q/\sqrt{G}$ which is often examined in the studies of quantum gravity. For $Gm^2 > Q^2 / \Psi^q$ case, the mass of the black hole is allowed to be in very large classical scales due to the ability of getting bigger values than the charge. Whereas the mass of the deformed black hole is allowed to decrease very small values which could fall into the quantum regime, because the decrease of the mass is governed by a very small term $1/\Psi^q$ being order of $h^{6/7}$.

Mass reduction with respect to the classical Reissner-Nordström case can be compared for q-deformed cases. Using $1/\Psi^q$ from (10) gives the mass of the extremal quantum black hole in case of $Gm^2 = Q^2 / \Psi^q$, such as

$$m^q = \left(\frac{h^3}{10\pi V(2E)^{3/2}} \frac{1}{g(z,q)} \right)^{\frac{2}{7}} \left(\frac{Q^2}{G} \right)^{\frac{2}{7}}, \tag{17}$$

The comparison of the deformed black hole mass and classical Reissner-Nordström black hole mass $m = Q/\sqrt{G}$ reads as

$$m^q = \left(\frac{1}{10\pi V(2E)^{3/2} Q^{3/2}} \right)^{\frac{2}{7}} \left(\frac{h^3 G^{3/4}}{g(z,q)} \right)^{\frac{2}{7}} m, \tag{18}$$

The equation in (18) implies that the mass of the charged extremal black hole in the deformed quantum case can decrease to a smaller value than that of the classical Reissner-Nordström case. We figure out the decrease in the mass by examining the behaviours of the factor $(h^3 G^{3/4} / g(z,q))^{2/7}$ in the classical mass of the charged black hole in (18). We illustrate the behaviour of $(h^3 G^{3/4} / g(z,q))^{2/7}$ with respect to z and q in Figure 1 and Figure 2, for $q < 1$ and $1 < q$, respectively.

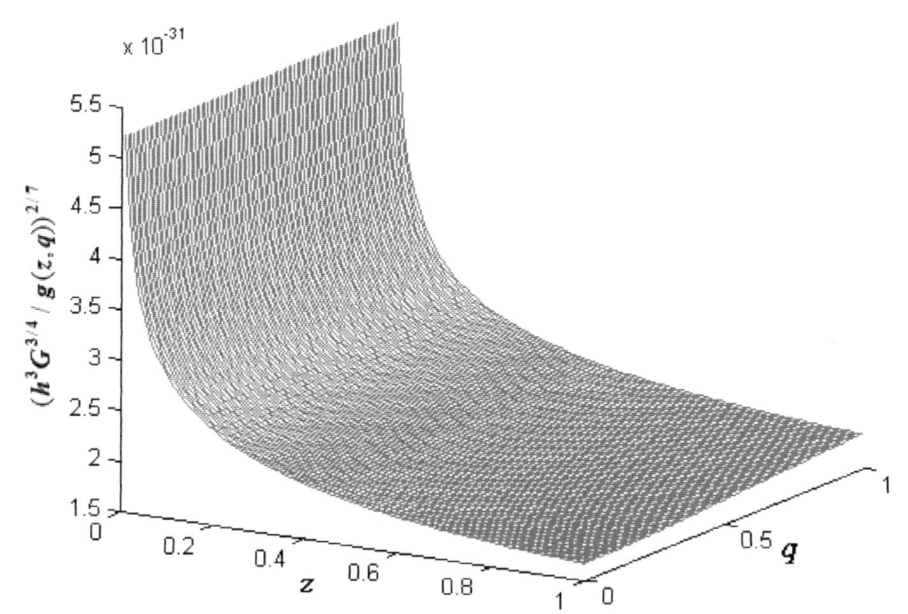

Figure 1. Mass reduction factor $(h^3G^{3/4} / g(z,q))^{2/7}$ for various values of the deformation parameter $q<1$.

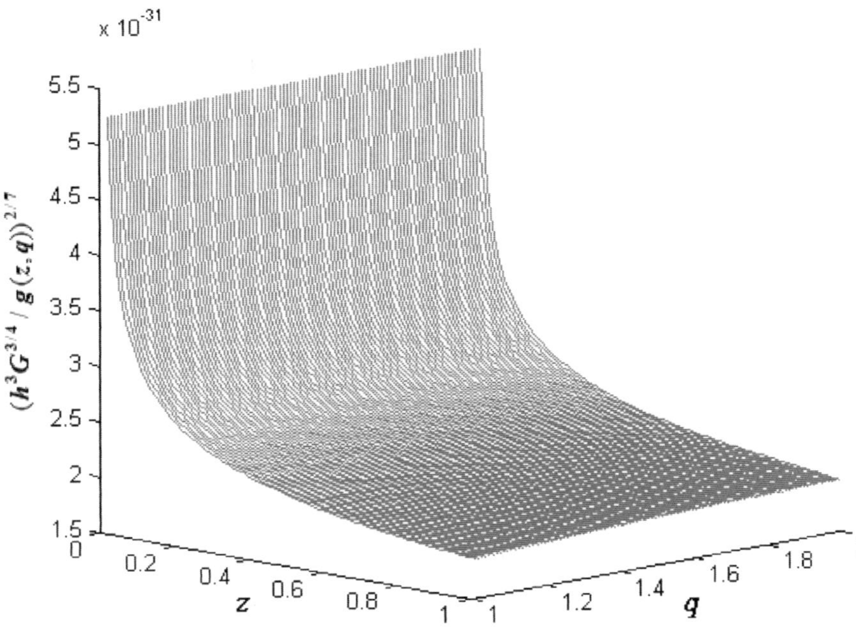

Figure 2. Mass reduction factor $(h^3G^{3/4} / g(z,q))^{2/7}$ for various values of the deformation parameter $q>1$.

4. Conclusions

In this study, we review the q-deformed statistics of bosons and obtained the q-deformed Einstein equations, which is based on the Strominger's idea, such that the micro quantum black holes obey the deformed statistics. We then obtain the solutions of q-deformed Einstein equations for the charged extremal quantum black holes. We then interpret the solutions for q-deformed Einstein equations.

There occur two singularities from the solutions of deformed Einstein equations, such that the true and coordinate singularities. The possible decrease in mass via Hawking radiation to the minimum value which is determined by the charge of quantum black hole is investigated and the difference between the classical black holes and quantum black holes is compared from the equations (18). We represent the decrease in quantum mass m^q in Figure 1 and Figure 2. According to the Figures 1 and 2, the mass of the quantum black hole m^q in (18) is at least 10^{-30} times smaller than the classical black hole mass m, in the deformed case, with the inverse of the volume, charge and energy factors, mass m^q gets smaller than $10^{-30}m$.

Three independent studies, Strominger's micro black holes obeying the deformed statistics, Verlinde's entropic gravity approach and the thermostatistics of deformed Bose gas model, seem to be consistent with each other, since the theoretical possibility of concentrating a mass into its reduced Planck mass leads to a quantum black hole whose dynamics described by the deformed Einstein equations and the mass reduction to a smaller value than that of the classical black holes.

5. References

[1] Dil E 2015 *Canad. J. Phys.* **93** 1274
[2] Strominger A 1993 *Phys. Rev. Lett.* **71** 3397
[3] Verlinde E 2011 *J. High Energy Phys.* JHEP04(2011)029
[4] Ubriaco M R 1997 *Phys. Rev. E* **55(1)** 291
[5] Wald R M 1984 *General Relativity* (Chicago: The University of Chicago Press)

Journal of Physics: Conference Series **766** (2016) 012005 doi:10.1088/1742-6596/766/1/012005

SVZ Sum Rules in Quantum Mechanics

Halil Mutuk, Hasan Gümüş

Ondokuz Mayis University, Faculty of Arts and Sciences, Department of Physics, 55139, Samsun, Turkey

halil.mutuk@omu.edu.tr

Abstract. In this work we obtained ground state energy level of well known harmonic oscillator in quantum mechanics by using SVZ sum rules. We calculated energy eigenvalues with and without an external electric field.

1. Sum Rules

Sum rules were an important tool in the early days of quantum mechanics. A sum rule is simply (not many times) a formula for transitions between energy levels, in which the sum of the transition strengths is expressed in a formula. Sum rules have been used in many areas of physics such as atomic, molecular, solid state, nuclear and mostly in particle physics.

Quantum mechanical sum rules (Thomas-Reiche-Kuhn (TRK) sum rule [1-4] and Bethe sum rule [4-6]) were widely used in many branches of physics as stated above. TRK sum rule for energy eigenstates,

$$\sum_k (E_k - E_n)\, |\langle n|x|k\rangle|^2 = \frac{\hbar^2}{2m} \qquad 1.1$$

was used to study the physics of electric-dipole interactions with atoms [4]. Another sum rule was formulated by Bethe as

$$\sum_k (E_k - E_n)\, \left|\langle n|e^{iqx}|k\rangle\right|^2 = \frac{\hbar^2 q^2}{2m}. \qquad 1.2$$

This was used to study for energy loss mechanisms of charged particles in matter. Sum rules are derived mostly from Heisenberg's view of quantum mechanics and renowned uncertainty principle [4]. Although these sum rules are derived in the early days of quantum mechanics, there can be found modern applications in [7-12].

Shifman-Vainshtein-Zakharov (SVZ) sum rule (mostly known as QCD (Quantumchromo Dynamics) Sum Rule) was derived in [13] and is being used widely in hadron physics. It is a semi-phenomenological way of determining the characteristics of hadrons. It is a perturbative method. The ground state energy and wave function can be found with an accuracy of 10%-20%. Since it was formulated, it has been applied to numerous problems in high energy physics especially properties of hadrons such as masses, decay and coupling constants, form factors, magnetic moments etc. In this study we applied SVZ method to linear harmonic oscillator in an electric field. We obtained energy levels for this system. A general discussion about SVZ sum rules for bound states can be found in [14]. We compared our results with that study.

2. SVZ sum rules in quantum mechanics

The following consideration is so close to [14]. We used SVZ sum rules to determine ground state energy of harmonic oscillator in quantum mechanics. For this we write the Euclidean Green function

$$M(\tau) = G(\mathbf{0}, -i\tau; \mathbf{0}, 0) \tag{2.1}$$

where

$$G(\boldsymbol{x_2}, t_2; \boldsymbol{x_1}, t_1) = \sum_{k=0}^{\infty} \psi_k(\boldsymbol{x_2})\, \psi_k^*(\boldsymbol{x_1}) e^{-iE_k(t_2-t_1)} \tag{2.2}$$

is the Green function for time-dependent Schrödinger equation and τ is the time parameter. The sum rule for Euclidean Green functions is

$$M(\tau) \equiv |\psi_0(\mathbf{0})|^2 e^{-E_0\tau} + M_c(\tau) \tag{2.3}$$

where $M_c(\tau)$ shows continuum contribution to $M(\tau)$ (contribution of states lying above the ground state). It is given by

$$M_c(\tau) = \sum_{k=1}^{\infty} |\psi_k(\mathbf{0})|^2 e^{-E_k\tau}. \tag{2.4}$$

L.h.s. and r.h.s of (2.3) can be approximately equal if there exist a perturbative expansion, $M_{pert}(\tau)$ which is the first few terms of the Born series for $M(\tau)$ (small Euclidean time, $\tau < \tau_B$) and if $M(\tau)$ is dominated by ground state contribution (large Euclidean times, $\tau > \tau_C$). Given τ as $\tau_C < \tau < \tau_B$ the ground state energy is given by

$$E_0 = -\frac{d}{d\tau}\ln\left(M_{pert}(\tau) - M_c(\tau)\right). \tag{2.5}$$

2.1. Linear harmonic oscillator in an external electric field

In this part we study a generic example of quantum mechanics: energy levels of linear harmonic oscillator. The Hamiltonian in an electric field for 1D is given by

$$H = -\frac{1}{2m}\frac{d^2}{dx^2} + \frac{1}{2}m\omega^2x^2 - eEx. \tag{2.6}$$

The energy eigenvalues are given as

$$E_k = \left(k + \frac{1}{2} - \varepsilon\right)\omega \text{ where } \varepsilon \equiv \frac{(eE)^2}{2mw^3};\ k = 0,1,2,\cdots \tag{2.7}$$

and eigenfunctions are given as

$$\psi_k(x) = \phi_k(x - x_0);\ x_0 = \frac{eE}{m\omega^2} \tag{2.8}$$

where $\phi_k(x)$ are eigenfunctions of 1D linear harmonic oscillator without electric field. The Euclidean Green function for this example is

$$M(\tau) = e^{\varepsilon\omega\tau} \sum_{k=0}^{\infty} \phi_k(-x_0)\phi_k^*(-x_0)e^{-\left(k+\frac{1}{2}\right)\omega\tau}. \tag{2.9}$$

Now let's focus on the potential $V(x) = \frac{1}{2}m\omega^2x^2 - eEx$ and its Born series

$$M_{pert}(\tau) = M_0(\tau)\left(1 - \frac{(\omega\tau)^2}{12} + \varepsilon\frac{(\omega\tau)^3}{12} + \frac{(\omega\tau)^4}{160} + \cdots.\right) \tag{2.10}$$

Here $M_0(\tau) = \sqrt{\frac{m}{2\pi\tau}}$ is the Euclidean Green function of a free particle. Finally we can write down the lowest field-induced correction to $M(\tau)$ as

$$M_{pert}(\tau) = M_0(\tau)\varepsilon\frac{(\omega\tau)^3}{12}. \tag{2.11}$$

So we have the first part for r.h.s. of (2.5). To have the second part i.e. $M_c(\tau)$, we use free motion approximation such as

$$M_0(\tau) = \int_0^{\infty} dE\rho_0(E)e^{-E\tau} \qquad 2.12$$

where $\rho_0(E) = \frac{1}{\pi}\sqrt{\frac{m}{2E}}$. Under this assumption we have

$$M_c(\tau) = \int_{E_c}^{\infty} dE\rho_0(E)e^{-E\tau}. \qquad 2.13$$

Here E_c is determined with the ground state parameters [14]. The ground state energy can be obtained using (2.11) and (2.12) in (2.5). For very small fields we use equation (2.5). For normal fields we use

$$\Delta E_0 = -\frac{d}{d\tau}\ln\left(1 + \frac{\Delta M_{pert}(\tau) - \Delta M_c(\tau)}{|\psi_0(\mathbf{0})|^2 e^{-E_0\tau}}\right) \qquad 2.14$$

where $\Delta M_{pert}(\tau) = M'_{pert}(\tau) - M_{pert}(\tau)$ and $\Delta M_c(\tau) = M'_c(\tau) - M_c(\tau)$. Primed (unprimed) quantities refer to the case with (without) electric field.

3. Numerical results

For very small fields we used equation (2.5). For normal fields it is better to use (2.14). Table 1 shows with the results using (2.5) and Table 2 shows with the results using (2.14).

Table 1. Ground state energy level shift (in units of ω) using (2.5)

$\varepsilon \equiv \frac{(eE)^2}{2mw^3}$	E_c	$E_0\ (this\ work)$	$E_0\ (ref.14)$	$\Delta E_0 (this\ work)$	$\Delta E_0\ (ref.14)$	$\Delta E_0\ (exact)$
0.	0.90	0.397	0.405	0.	0	0.
0.01	0.81	0.387	0.387	-0.01	-0.018	-0.01
0.02	0.72	0.375	0.368	-0.022	-0.037	-0.02
0.03	0.63	0.364	0.349	-0.033	-0.056	-0.03

Table 2. Ground state energy level shift using (2.14)

$\varepsilon \equiv \frac{(eE)^2}{2mw^3}$	E_c	$\Delta E_0\ (this\ work)$	$\Delta E_0\ (ref.14)$	$\Delta E_0\ (exact)$
0.01	1.43	-0.011	-0.011	-0.01
0.02	1.36	-0.021	-0.023	-0.02
0.03	1.29	-0.035	-0.036	-0.03
0.04	1.22	-0.047	?	-0.04
0.05	1.15	-0.058	-0.061	-0.05

We can see from Table 1. for a zero external field the ground state energy is $E_0 = 0.397$ (in units of ω) where the exact value is 0.5. In a non-zero electric field, the ground state energy level shift i.e. $\Delta E_0 = E_0(\varepsilon) - E_0(0)$ was calculated where $E_0(\varepsilon)$ is the ground state energy with a non-zero field. The energy level shift values of our work well agrees with the exact values.

From Table 2 we can see that by using refined formula, energy level shift values of our work is consistent well with exact values also with the values of [14]. The question mark in the fourth column means that authors of [14] didn't have ΔE_0 value for $\varepsilon = 0.04$ in their paper.

To sum up, we obtained ground state energy of harmonic oscillator with and without electric field by using SVZ sum rules. The key point is that we didn't use directly Schrödinger equation to find energy eigenvalues.

Acknowledgement

The corresponding author is so grateful to S. Olejnik for his various contribution.

Journal of Physics: Conference Series **766** (2016) 012005 doi:10.1088/1742-6596/766/1/012005

References

[1] Thomas W 1925 *Naturwissenschaften* **13** 627
[2] Kuhn W 1925 *Z. Phys.* **33** 408
[3] Reiche F, Thomas W 1925 *Z. Phys.* **34** 510
[4] Belloni M, Robinett RW 2008 *American Journal of Physics* **76** 798
[5] Bethe H 1930 *Ann. Phys.* **5** 325
[6] Bransden BH, Joachin CJ 2003 *Physics of Atoms and Molecules*, 2nd edition, Prentice Hall
[7] Jackiw R 1967 *Phys. Rev.* **157** 1220
[8] Wang S 1999 *Phys. Rev. A* **60** 262
[9] Franta D, Necas D, Zajickova L 2013 *Thin Solid Films* **534**, 432
[10] Barnett SM, Loudon R 2012 *Phys. Rev. Lett.* **108**
[11] Kamohara M et al. 2008 *Radiat. Phys. Chem.* **77** 1153
[12] Aucar GA, Oddershede J, Sabin JR 1995 *Phys. Rev. A* **52**, 1054
[13] Shifman MA, Vainshtein AI, Zakharov VI 1979 *Nuc. Phys. B* **147**, 385, 448, 519
[14] Novak I, Olejnik S 1985 *Czech. J. Phys.* **77** 1153

Comparison of Finite Differences and WKB approximation Methods for PT symmetric complex potentials

Leila Naceri, Meziane Chekkal, Amine B. Hammou

Universit des Sciences et de la Technologie d'Oran- Mohamed Boudiaf, Faculté de Physique, BP 1505 El M'naouer, Oran 31000, Algeria

E-mail: naceri.lila25@gmail.com, rile89@yahoo.com, amine.hammou@univ-usto.dz, hbamine@gmail.com

Abstract. We consider the one dimensional schrödinger eigenvalue problem on a finite domain (Strum-Liouville problem) for several PT-symmetric complex potentials, studied by Bender and Jones using the WKB approximation method. We make a comparison between the solutions of theses PT-symmetric complex potentials using both the finite difference method (FDM) and the WKB approximation method and show quantitative and qualitative agreement between the two methods.

1. Introduction

Recently, there has been growing interest in non-hermitian Hamiltonians having real energy spectra [1], these complex potentials extend quantum mechanics to several open fields of theoretical, experimental as well as technological physics [2]. In conventional formulation of quantum mechanics the Hamiltonian must be Hermitian $H = H^+$ (mathematical axiom) in order to ensure the reality of the energy spectrum (the eigenvalues of H). In alternative formulation the mathematical axiom of Hermiticity is replaced by the physically transparent axiom called space-time reflection symmetry (PT-symmetry). In this light, Bender and his collaborators [3] [4] and later others [5] have studied and solved non-Hermitian Hamiltonians, who leads to real eigenvalues by using numerical and analytical techniques see [9]-[10] and references there in.

The goal of this paper is to investigate a class of PT-symmetric Hamiltonian with complex potentials, for which we compare between WKB approximation and finite difference method and systematically show the qualitative and quantitative agreement between the two.

2. WKB Analysis of PT -Symmetric Sturm-Liouville problems

Most problems encountered in PT-symmetric QM cannot be solved exactly. The solution of the schrödinger eigenvalue problem exist only for an infinite domain, one must resort to approximate methods, a variety of such methods have been developed and each has its own area of applicability. One of these methods is the Wentzel-Kramers-Brillouin approximation used by Bender and Jones [6]-[7] to investigate PT-symmetric systems defined on finite domain whose

Journal of Physics: Conference Series **766** (2016) 012006 doi:10.1088/1742-6596/766/1/012006

eneregy eigenvalues grow like n^2 for large n. Another future of the PT eigenvalue problem on a finite domain is the showing up of a sequence of critical points at which pairs of eigenvalues cease to be real and become complex conjugates of one another. For the potentials considered here this sequence of critical points is associated with a turning point on the imaginary axis in the complex plane. In the next section we will briefly review the WKB analysis of [6][7], where the authors focus on the complex potentials to calculate the real eigenvalues of energy and plotted as function of the coupling parameter λ.

2.1. *WKB calculation of eigenvalues for "one turning point "*

In their paper [6] Bender and Jones considered the one -dimensional time independent schrödinger equation :

$$-\Psi^{''}(x) + V(ix)\Psi(x) = \lambda\Psi(x), \tag{1}$$

where λ is the energy eigenvalue, and the following boundary condition $\Psi(+1) = 0$, $\Psi(-1) = 0$ are imposed. In a preparatory step, they treat both the eigenvalues λ and g (coupling parameter) as large and proportional $\lambda = a \cdot g$, thus substituting in (1)one obtains:

$$-\Psi^{''}(x) = -g \cdot Q(x) \cdot \Psi(x), \tag{2}$$

where $Q(x) = V(ix) + a$. We must notice that the asymptotic solution of (2) is controlled by a turning point (see FIG[1]) on the imaginary X axis, satisfying $V(-b) + a = 0$.
We will here summarize the methodology and skip the details of the calculation presented in [6], where we refer the interested reader.

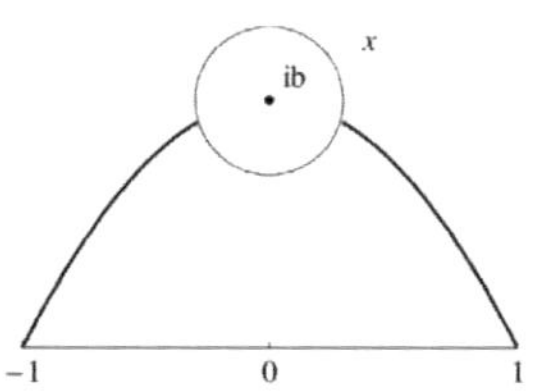

Figure 1. schematic path with end point $x = \pm 1$ of one turning point $x = ib$.

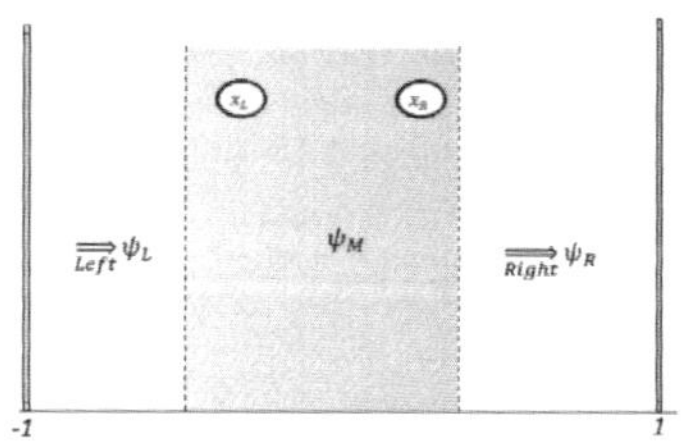

Figure 2. schematic the path of integration of WKB function for two turning point.

Part1: firstly we approximate our schrödinger equation near the turning point, we start by $x = ib + y \cdot c$, where $y \ll 1$ then $Q(x) = iycV^{'}(-b)$, substitution of this in (2) gives

$$\frac{\partial^2}{\partial y^2}\Psi(y) = -igc^3V^{'}(-b)y\Psi(y), \tag{3}$$

the approximated schrödinger equation have a similar form as the standard Airy differential equation, we then convert (2) as

$$\frac{\partial^2}{\partial y^2}\Psi(y) = y\Psi(y), \qquad with \qquad \begin{cases} y &= \frac{x-ib}{\gamma}e^{-i\pi/6} \\ c &= \gamma e^{i\pi/6} \\ \gamma &= g[V^{'}(-b)]^{-1/3} \end{cases} \tag{4}$$

For the sake of convenience the solution of the standard Airy differential equation is well known, where the solution of the schrödinger equation will take the following form:

$$\Psi(y) = K_1 Ai(y) + K_2 Ai(\omega y) \Leftrightarrow \Psi(y) = K_1 Ai(\frac{x - ib}{\gamma} e^{-i\pi/6}) + K_2 Ai(\omega \frac{x - ib}{\gamma} e^{-i\pi/6}) \quad (5)$$

Part2: They approximate the schrödinger equation away from the turning point on both sides using WKB function and invokes the appropriate boundary condition, it will be done as:

$$\Re[x] < 0 \Rightarrow \Psi_L(x) \sim \frac{L_1}{[Q(x)]^{\frac{1}{4}}} \exp[i \int_x^{ib} ds\sqrt{gQ(s)}] + \frac{L_2}{[Q(x)]^{\frac{1}{4}}} \exp[-i \int_x^{ib} ds\sqrt{gQ(s)}] \quad (6)$$

$$\Psi_L(-1) = 0 \Rightarrow 0 = L_1 \exp[i \int_{-1}^{ib} ds\sqrt{gQ(s)}] + L_2 \exp[-i \int_{-1}^{ib} ds\sqrt{gQ(s)}] \quad (7)$$

$$\Re[x] > 0 \Rightarrow \Psi_R(x) \sim \frac{R_1}{[Q(x)]^{\frac{1}{4}}} \exp[i \int_{ib}^{x} ds\sqrt{gQ(s)}] + \frac{R_2}{[Q(x)]^{\frac{1}{4}}} \exp[-i \int_{ib}^{x} ds\sqrt{gQ(s)}] \quad (8)$$

$$\Psi_R(1) = 0 \Rightarrow R_1 \exp[i \int_{ib}^{1} ds\sqrt{gQ(s)}] + R_2 \exp[-i \int_{ib}^{1} ds\sqrt{gQ(s)}] \quad (9)$$

At this point, we will have 6 arbitrary unknown constants. Obviously our key technical step to determine them by the asymptotic matching between the solution of Airy function near the turning point and WKB function away from the turning point, where they must be identical and can be expressed as:

$$\begin{cases} \frac{K_1}{2\sqrt{\pi}} = L_2 g^{\frac{1}{4}} \gamma^{\frac{1}{2}} e^{-i\pi/6} \\ \frac{1}{2\sqrt{\pi}}(-K_1 e^{2i\pi/3} + K_2) = L_1 g^{\frac{1}{4}} \gamma^{\frac{1}{2}} \end{cases} \quad \begin{cases} \frac{K_1}{2\sqrt{\pi}} = R_1 g^{\frac{1}{4}} \gamma^{\frac{1}{2}} e^{-i\pi/6} \\ \frac{K_2}{2\sqrt{\pi}} = R_2 g^{\frac{1}{4}} \gamma^{\frac{1}{2}} \end{cases} \quad (10)$$

Then it is more advantageous to combine it, when the calculation can be easily carried out to a secular equation that determines the real energy eigenvalues:

$$\sin[\int_{-1}^{1} ds\sqrt{gQ(s)}] + \frac{1}{2} \exp[i \int_{-1}^{ib} ds\sqrt{gQ(s)} - i \int_{ib}^{1} ds\sqrt{gQ(s)}] = 0 \quad (11)$$

2.2. WKB calculation of eigenvalues for " More than one turning point "

Let us mention here that the previous secular equation (11) fails to reproduce the expected energy eigenvalues for more than one turning point. Therefore the authors in a subsequent paper [7] readapted their treatment with some modification to the case the path of integration passes through two turning points x_L and x_R as schematized in FIG[2]. The WKB approximation of the wave function Ψ_R (8) and Ψ_L (6) are respectively the wave functions on the left region of the x_L and the right region of x_R turning points. Nevertheless, there is additional basic conditions requiring the wave function in the region between the two turning points Ψ_M to be PT-symmetric, to finally yield the new following secular equation:

$$\sin[\int_{-1}^{1} ds\sqrt{gQ(s)}] + e^{\Delta} \cos[i \int_{x_L}^{x_R} ds\sqrt{gQ(s)}] = 0, \quad where \quad \Delta \equiv 2ImI_L, I_L \equiv [\int_{-1}^{x_L} ds\sqrt{gQ(s)}] \quad (12)$$

In fact this constraint can be viewed as a compact way of setting a secular equation of the WKB analysis involving the pertinent boundary condition and also for complex potential which had more then one turning point, this is what makes the method so effective.

3. Finite Difference Method

Finite difference method is a numerical method based on mathematical discretization of differential equation of boundary problems. The basic idea of FDM (Finite Difference Method) is to replace each term of the partial derivative $\frac{df(x)}{dx}$ by a finite difference approximation $\frac{\Delta f(x)}{\Delta x}$ obtained by Taylor's expansion near the point of interests and neighboring points as schematized in FIG[3].

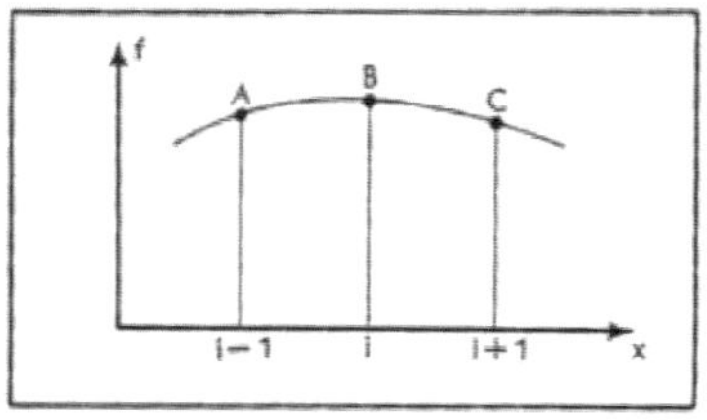

Figure 3. describe the three area of function $f(x)$ presented in finite difference scheme.

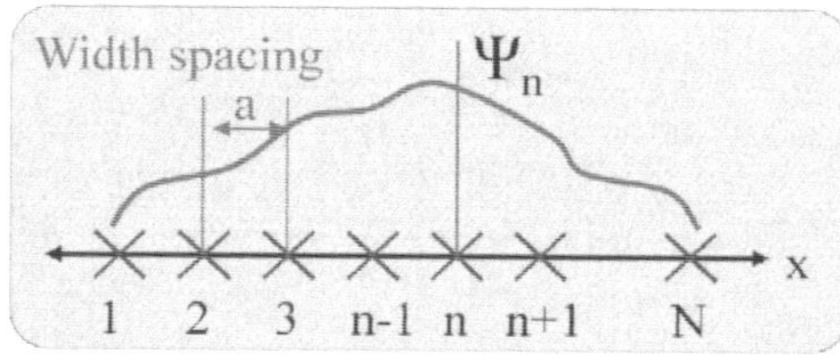

Figure 4. the describtion of the wave function in discretization space with the boundary condition.

Let us summarize the FDM to solve the time independent schrödinger equation in a close form, where we must notice that we will consider a complex PT-symmetric potential. This leads to discretize the geometric domain of the X-space into grid points x_n, $n = 1, 2.....N$ (see FIG[4]) and at each point the stationary 1D schrödinger equation can be written as:

$$H\Psi = E\Psi \Leftrightarrow H_n\Psi(x_n) = E\Psi(x_n) \Leftrightarrow -\Psi''(x_n) + V(x_n)\Psi(x_n) = E\Psi(x_n). \tag{13}$$

We now have a system of N equations, producing N solutions. The next step is to approximate the second derivative of the wave function $\Psi''(x)$ using finite difference scheme [8] such that:

$$\begin{cases} [\frac{d\Psi}{dx}]_{n+1} = \frac{\Psi_{n+1}-\Psi_n}{\Delta x} \\ [\frac{d\Psi}{dx}]_n = \frac{\Psi_n-\Psi_{n-1}}{\Delta x} \end{cases} \qquad [\frac{d^2\Psi}{dx^2}]_n = \frac{[\frac{d\Psi}{dx}]_{n+1} - [\frac{d\Psi}{dx}]_n}{\Delta x} = \frac{\Psi_{n+1} - 2\Psi_n + \Psi_{n-1}}{(\Delta x)^2}. \tag{14}$$

Where $\Delta x_{n+1} = \Delta x_n = \Delta x = a.$

The system (13) becomes:

$$E\Psi_n = V(x_n)\Psi_n - t_0[\Psi_{n+1} - 2\Psi_n + \Psi_{n-1}] \qquad t_0 \equiv \frac{1}{(\Delta x)^2}. \tag{15}$$

After using an appropriate FD scheme to approximate the schrödinger equation, we must impose the boundary condition (see FIG[4]) as a consequence of the consideration of finite domain $\Psi(x_1) = \Psi(x_N) = 0$. In fact the system of equations (15) can be easily written in matrix notation, where we must represent the wave functions in vector and the Hamiltonian in matrix forms who includes the potential as:

$$E \cdot \begin{bmatrix} \Psi_1 \\ \Psi_2 \\ \vdots \\ \Psi_n \\ \vdots \\ \Psi_N \end{bmatrix} = \begin{bmatrix} 2t_0 + V(x_1) & -t_0 & 0 & 0 & 0 & 0 \\ -t_0 & 2t_0 + V(x_2) & -t_0 & 0 & 0 & 0 \\ 0 & -t_0 & \ddots & \ddots & 0 & 0 \\ 0 & 0 & \ddots & \ddots & \ddots & 0 \\ 0 & 0 & 0 & \ddots & \ddots & -t_0 \\ 0 & 0 & 0 & 0 & -t_0 & 2t_0 + V(x_N) \end{bmatrix} \cdot \begin{bmatrix} \Psi_1 \\ \Psi_2 \\ \vdots \\ \Psi_n \\ \vdots \\ \Psi_N \end{bmatrix} \tag{16}$$

Journal of Physics: Conference Series **766** (2016) 012006 doi:10.1088/1742-6596/766/1/012006

4. Numerical Results and Discussion

We give numerical results showing the agreement between both methods. In order to make a comparison between the FDM and the WKB approximation for non-hermetian Hamiltonians we choose a class of complex PT-symmetric potentials studied by Bender and Jones [6]-[7]. For the FDM we have developed a code for solving the eigenvalue problem of the Hamiltonian (16) with boundary condition similar to those considered using WKB analysis [6]-[7]. We have taken into account additional considerations, as variation of the dimension of our Hamiltonian matrix, which also connect to the number of the grid points as well as the width of spacing distance between them (see FIG[4]).

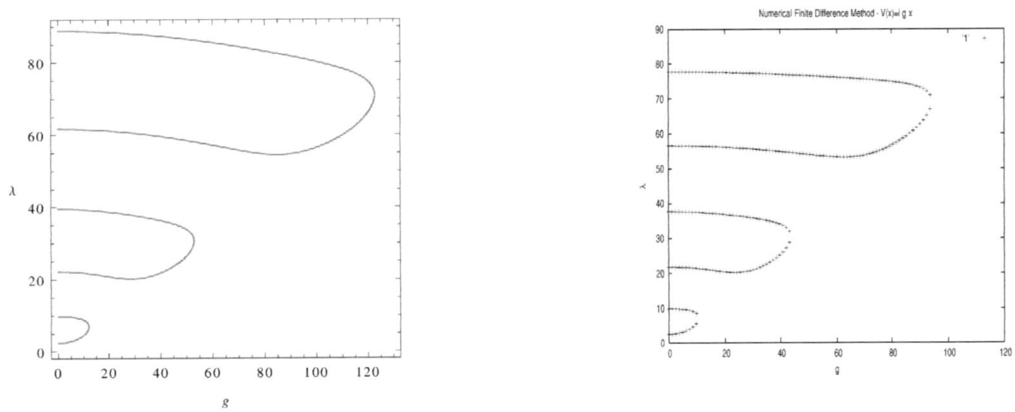

Figure 5. Real energy levels for WKB analysis(Right) and FD method(Left) of $V(x) = igx$.

We start our comparison between the two methods for potentials which have one turning point $V(x) = igx$, the results are represented in FIG[5], where we do find that FD method reproduces the one obtained from WKB analysis. The same result is obtained for the sinusoidal potential $V(x) = ig\sin(2x)$, shown in FIG[6].

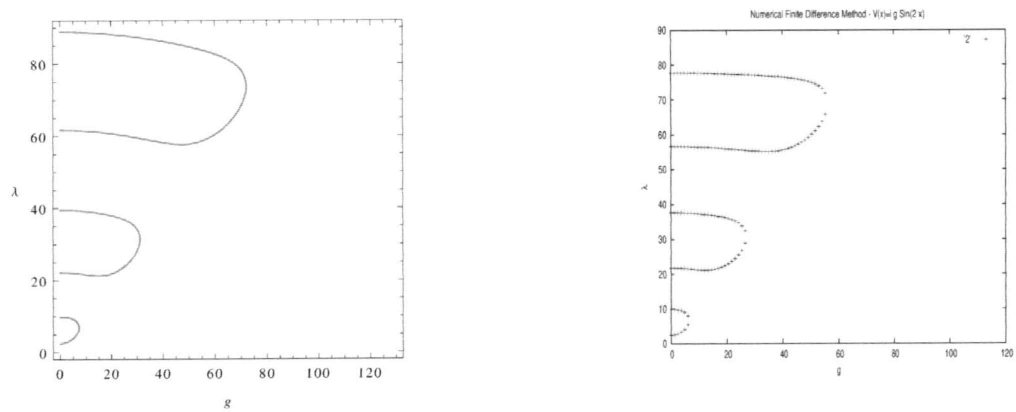

Figure 6. Real energy levels for WKB analysis (Right) and FD method (Left) of $V(x) = ig\sin(2x)$.

Briefly, we must note that in WKB analysis the energy eigenvalues λ and the coupling parameter g was treated as large and proportional, in contrary the range of validity of FD method is for small values of both (λ, g) nevertheless we note the excellent agreement between the two results.
We end this paper by showing numerical results for potentials having more than one turning point using the potential $V(x) = igx^3$ as an example. In FIG[7] one sees that FD method approximately reproduces the structure of the eigenvalues obtained by the WKB analysis. This is a consequence of truncation errors involved in the numerical computations due to the fact that in FD representation of derivatives with Taylor's series expansion the higher order terms are neglected. In other words it means that the truncation error identifies the difference between the exact solution of a differential equation and its finite difference solution.

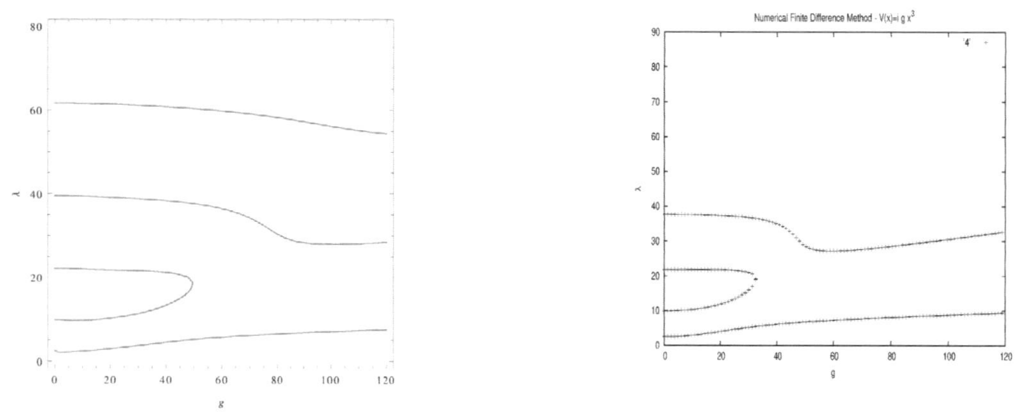

Figure 7. Real energy levels for WKB analysis (Right) and FD method (Left) of $V(x) = igx^3$.

5. Conclusion

Summarizing the results presented in this paper: The FD method can effectively offer an easy way to solve the one-dimensional schrödinger complex PT-symmetric eigenvalues problem.
The developed methodology of FDM provides a significant advantage over conventional analytical method like WKB analysis. Further study may include the high-order terms of Taylor's series expansion to obtain more accuracy (diminish the truncation error) especially for the complex potentials (having more then one turning points). Which will be the subject of future investigations.

[1] C.M.Bender and S. Boettcher, Phys. Rev. Lett. 80, 5243 (1998).
[2] C. M. Bender, Europhys. News 47, 17 (2016).
[3] C. M. Bender, Rep. Prog. Phys. 70, 947 (2007).
[4] Z. Ahmed, C. M. Bender and M. V. Berry, J. Phys. A: Math. Gen. 38, L627 (2005).
[5] G. Lvai, M. Znojil: "Systematic search for PT-symmetric potentials with real energy spectra"Journal of Physics A 33 (2000) 7165
[6] C.M.Bender and H.F.Jones,journal of physics A-mathematical and theoretical, Vol: 45, ISSN: 1751-8113 (2012).
[7] C.M.Bender and H.F.Jones,journal of physics A-mathematical and theoretical, Vol: 85, ISSN: 1050-2947 (2012).
[8] G.E.Forsythe and W.R.Wasow,Finite Difference Methods for Partial Differential Equations,Wiley,New York,(1960).
[9] A. Mostafazadeh, J. Math. Phys., 44, 974 (2003); LANL ArXiv: math-ph/0209018.
[10] B. Bagchi, F. Cannata and C. Quesne, Phys. Lett. A269, 79 (2000).

International Conference on Quantum Science and Applications (ICQSA-2016) IOP Publishing
Journal of Physics: Conference Series **766** (2016) 012007 doi:10.1088/1742-6596/766/1/012007

The role of curvature in quantum statistical mechanics

M R Ubriaco

Laboratory of Theoretical Physics, Department of Physics, University of Puerto Rico, Río Piedras Campus, San Juan, PR 00936, USA

E-mail: ubriaco@ltp.uprrp.edu

Abstract. In this manuscript, we calculate the scalar curvature of a two-dimensional thermodynamic space to study the properties of two thermodynamic systems. In particular, we study the stability and possible anyonic behavior of quantum group invariant systems and systems with fractal distribution functions.

1. Introduction

During the last five decades there has been a lot of development in using geometry to study some properties of thermodynamic systems [1]-[5]. In particular, it consists in defining a metric in a two dimensional parameter space and calculate the corresponding scalar curvature as a measure of the correlations strength of the system [7]-[15], with applications to classical and quantum gases [6][12][16][17], magnetic systems [18]-[21], non-extensive statistical mechanics [22]-[24], anyon gas, fractional statistics and deformed boson and fermion systems [25], systems with fractal distribution functions [26], quantum group invariant systems [27], systems with M-statistics [28], and those related to Dunkl differential-difference operators [29]. Some of the basic results of these approaches include the relationships between the metric with the correlations of the stochastic variables, and the scalar curvature R with the stability of the system, and the facts that the scalar curvature R vanishes for the classical ideal gas, $R > 0 (R < 0)$ for a boson (fermion) ideal gas, and it is singular at a critical point. In Section 2 we give a general but brief introduction to the calculation of the thermodynamic metric and the corresponding scalar curvature. In Sections 3 and 4 we apply the formalism to study the stability and possible anyonic behavior for the cases of quantum group invariant systems and systems with fractal distribution functions, respectively. In constrast to the quantum group case the fractal density distribution is not exponential.

2. Thermodynamic curvature

In order to make the discussion as general as possible, we start with a general probability density ρ which is not necessarily exponential. Therefore, let us define

$$\rho = \frac{1}{Z}\prod_{l=0} F_l. \tag{1}$$

Starting with the relative entropy for two close densities $\rho(\beta^\alpha)$ and $\rho(\beta_\alpha + d\beta_\alpha)$, where $\beta_1 = \beta$ and $\beta_2 = -\beta\mu$, (two-dimensional coordinate space). The metric is defined [7] as the second order term in the expansions of the information distance

$$I = Tr\rho\left(\ln\rho(\beta^\alpha) - \ln\rho(\beta^\alpha + d\beta^\alpha)\right). \tag{2}$$

Expanding, at second order we obtain the metric

$$g_{\alpha\gamma} = \frac{\partial^2 \ln Z}{\partial\beta_\alpha\partial\beta_\gamma} - Tr\left(\rho\sum_{l=0}\frac{\partial^2 \ln F_l}{\partial\beta_\alpha\partial\beta_\gamma}\right), \tag{3}$$

where the second term vanishes for exponential distributions with an exponent linear in the inverse temperature β. The scalar curvature R is given by

$$R = \frac{2}{detg}R_{1212}, \tag{4}$$

where the Riemann tensor reduces to

$$R_{\alpha\beta\gamma\delta} = g^{\eta\theta}(\Gamma_{\eta\gamma\lambda}\Gamma_{\theta\beta\gamma} - \Gamma_{\eta\alpha\gamma}\Gamma_{\theta\beta\lambda}) \tag{5}$$

The calculation of R simply reduces to solve the following determinant

$$R = \frac{1}{2(detg)^2}\begin{vmatrix} g_{11} & g_{22} & g_{12} \\ g_{11,1} & g_{22,1} & g_{21,1} \\ g_{11,2} & g_{22,2} & g_{21,2}\end{vmatrix}, \tag{6}$$

where $g_{\alpha\beta,\lambda} = \partial g_{\alpha\beta}/\partial\beta_\lambda$. The scalar curvature values give us the following information: Classical case: $R = 0$, Bosons: $R > 0$, Fermions :$R < 0$, at a phase transition: $R \to \infty$, and $R \approx 0$ implies a more stable system.

3. Geometry of Quantum Group invariant systems

In the last twenty years there has been considerable interest on applications of quantum groups [30][31], in addition to the theory of integrable models, to diverse areas of theoretical physics. The vast published literature on this subject includes formulations of quantum group versions of Lorentz and Poincaré algebras [32], its use as internal quantum symmetries in quantum mechanics and field theories [33], molecular and nuclear physics [34], and the formulation and study on the implications of imposing quantum group invariance in thermodynamic systems [35]. In particular, it has been shown [36] that quantum group gases exhibit anyonic behavior in two and three dimensions.

More recent applications include gravity theories wherein the discreteness and non-commutative properties of space-time at the Planck scale are approached mathematically by replacing the local symmetry by a quantum group symmetry [37] and applications to the phenomena of entanglement [38]. In particular,we focuses on the quantum group $SU_q(2)$ which consists of the set of matrices $T = \begin{pmatrix} a & b \\ c & d \end{pmatrix}$ with elements $\{a, b, c, d\}$ generating the algebra

$$ab = q^{-1}ba, \quad ac = q^{-1}ca, \quad bc = cb, \quad dc = qcd, \quad db = qbd,$$
$$da - ad = (q - q^{-1})bc, \quad det_q T \equiv ad - q^{-1}bc = 1,$$

with the unitary conditions [39] $\overline{a} = d, \overline{b} = q^{-1}c$ and $q \in \mathbf{R}$. Hereafter, we take $0 < q < \infty$.

3.1. $SU_q(2)$-bosons

The $SU_q(2)$-invariant bosonic algebra is given by the following relations:

$$\Phi_2\overline{\Phi}_2 - q^2\overline{\Phi}_2\Phi_2 = 1, \quad \Phi_1\overline{\Phi}_1 - q^2\overline{\Phi}_1\Phi_1 = 1 + (q^2 - 1)\overline{\Phi}_2\Phi_2,$$
$$\Phi_2\Phi_1 = q\Phi_1\Phi_2, \quad \Phi_2\overline{\Phi}_1 = q\overline{\Phi}_1\Phi_2.$$

The simplest Hamiltonian written in terms of the operators Φ_j is simply written as

$$\mathcal{H}_B = \sum_{\kappa} \varepsilon_{\kappa}(\mathcal{N}_{1,\kappa} + \mathcal{N}_{2,\kappa}), \tag{7}$$

where $[\overline{\Phi}_{i,\kappa}, \Phi_{\kappa',j}] = 0$ for $\kappa \neq \kappa'$ and $\mathcal{N}_{i,\kappa} = \overline{\Phi}_{i,\kappa}\Phi_{i,\kappa}$. For a given κ the $SU_q(2)$ bosons are written in terms of boson operators ϕ_i and $\phi_i^\dagger$ with usual commutation relations: $[\phi_i, \phi_j^\dagger] = \delta_{ij}$ as follows

$$\Phi_j = (\phi_j^\dagger)^{-1}\{N_j\}q^{N_{j+1}}, \quad \overline{\Phi}_j = \phi_j^\dagger q^{N_{j+1}}, \quad j = 1, 2$$

leading to the interacting boson Hamiltonian

$$\mathcal{H}_B = \sum_{\kappa} \frac{\varepsilon_{\kappa}}{q^2 - 1} \sum_{m=1}^{\infty} \frac{2^m \ln^m q}{m!} (N_{1,\kappa} + N_{2,\kappa})^m, \tag{8}$$

where $N_{i,\kappa}$ is the ordinary boson number operator and the q-number $\{n\} = \frac{1-q^{2n}}{1-q}$. Equation (7) becomes the standard free boson hamiltonian at $q = 1$.

The grand partition function $\mathcal{Z}_B$ is given by

$$\begin{aligned} \mathcal{Z}_B &= Tr\, e^{-\beta \sum_{\kappa} \varepsilon_{\kappa}(\mathcal{N}_{1,\kappa} + \mathcal{N}_{2,\kappa})} \\ &= \prod_{\kappa} \sum_{n=0}^{\infty} \sum_{m=0}^{\infty} e^{-\beta\varepsilon_{\kappa}\{n+m\}} e^{\beta\mu(n+m)}, \end{aligned} \tag{9}$$

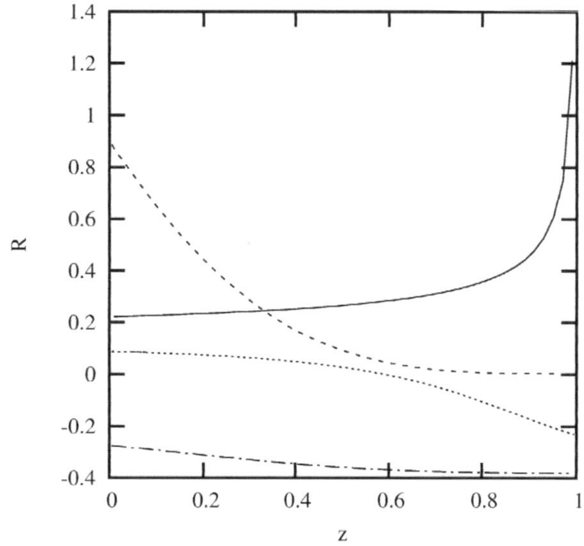

Figure 1. The scalar curvature R, in units of $\lambda^3 V^{-1}$, as a function of the fugacity z for quantum group bosons at $D = 3$ and constant β for the cases of $q = 1$ (solid line), $q = 0.5$ (dashed line) and $q = 1.15$ (dotted line) and $q = 2$ (dashed-dotted line).

which after rearrangement simplifies to the Equation

$$\mathcal{Z}_B = \prod_{\kappa} \sum_{m=0}^{\infty} (m+1) e^{-\beta \varepsilon_\kappa \{m\}} z^m, \tag{10}$$

where $z = e^{\beta\mu}$ is the fugacity. Since the density distribution is exponential, the second term in Equation (3) vanishes. The scalar curvature for $D = 2, 3$ is given by the Equation

$$R_D = \frac{(D+4\nu)\pi^\nu \lambda^D}{V_D} \left(\frac{c_\nu b_\nu^2 + a_\nu d_\nu b_\nu - 2a_\nu c_\nu^2}{[(D+4\nu)a_\nu c_\nu - (D+2\nu-1)b_\nu^2]^2} \right), \tag{11}$$

where $\nu = \frac{D-2}{2}$ and for example

$$a_\nu = \int_0^\infty x^\nu \ln\left(1 + \sum_{n=1} (n+1) e^{-x\{n\}} z^n \right) dx, \tag{12}$$

where b_ν, c_ν and d_μ are obtained by taking the corresponding derivatives with respect to β_1 and β_2. Figures 1 displays the results of a numerical calculation of Equation (11) for the scalar curvature of a QGB system in $D = 3$. Figure 1 shows that for $0 < q \leq 1$ the curvature is positive for all z values, and therefore the system is bosonic. In addition, the system becomes more stable than the standard Bose-Einstein case at higher values of z. For $1 < q \approx 1.2$ the system exhibits anyonic behavior as a function of z and becomes purely fermionic for all values of z and $q > 1.2$. The calculation of the Equation (11) for $D = 2$ shows that for $0 < q < 1$ the system is bosonic and more stable at $z \approx 1$ than the standard case $q = 1$. For $q > 1$ the system goes from bosonic to fermionic and becomes bosonic becoming more unstable as z approaches the value of 1.

Journal of Physics: Conference Series **766** (2016) 012007 doi:10.1088/1742-6596/766/1/012007

3.2. $SU_q(2)$-fermions

The $SU_q(2)$ invariant fermionic algebra is defined by the set of relations

$$\{\Psi_2, \overline{\Psi}_2\} = 1, \quad \{\Psi_1, \overline{\Psi}_1\} = 1 - (1 - q^{-2})\overline{\Psi}_2\Psi_2$$
$$\Psi_1\Psi_2 = -q\Psi_2\Psi_1, \quad \overline{\Psi}_1\Psi_2 = -q\Psi_2\overline{\Psi}_1$$
$$\{\Psi_1, \Psi_1\} = 0 = \{\Psi_2, \Psi_2\}.$$

The corresponding Hamiltonian in terms of the operators Ψ_i is simply

$$\mathcal{H}_F = \sum_\kappa \varepsilon_\kappa(\mathcal{M}_{1,\kappa} + \mathcal{M}_{2,\kappa}), \tag{13}$$

where $\mathcal{M}_{i\kappa} = \overline{\Psi}_{i,\kappa}\Psi_{i,\kappa}$ and $\{\overline{\Psi}_{\kappa,i}, \Psi_{\kappa',j}\} = 0$ for $\kappa \neq \kappa'$. The occupation numbers are restricted to $m = 0, 1$ and therefore $SU_q(N)$ fermions satisfy the Pauli exclusion principle. A simple check shows that the $SU_q(2)$ fermionic algebra is consistent with the following representation of Ψ_i operators in terms of fermion operators ψ_j

$$\Psi_2 = \psi_2, \quad \overline{\Psi}_2 = \psi_2^\dagger$$
$$\Psi_1 = \psi_1\left(1 + (q^{-1} - 1)M_2\right), \quad \overline{\Psi}_1 = \psi_1^\dagger\left(1 + (q^{-1} - 1)M_2\right)$$

where $M_2 = \psi_2^\dagger\psi_2$.

From Equation (13) we see that the original Hamiltonian becomes the interacting Hamiltonian

$$H_F = \sum_\kappa \varepsilon_\kappa\left(M_{1,\kappa} + M_{2,\kappa} + (q^{-2} - 1)M_{1,\kappa}M_{2,\kappa}\right). \tag{14}$$

Therefore the parameter $q \neq 1$ mixes the two degrees of freedom in a nontrivial way through a quartic term in the Hamiltonian.

The grand partition function $\mathcal{Z}_F$ is written

$$\begin{aligned}\mathcal{Z}_F &= \prod_\kappa \sum_{n=0}^{1}\sum_{m=0}^{1} e^{-\beta\varepsilon_\kappa(n+m-(1-q^{-2})mn}e^{\beta\mu(n+m)} \\ &= \prod_\kappa\left(1 + 2e^{-\beta(\varepsilon_\kappa-\mu)} + e^{-\beta\left(\varepsilon_\kappa(q^{-2}+1)-2\mu\right)}\right)\end{aligned} \tag{15}$$

which for $q = 1$ becomes the square of a single fermion type grand partition function. Figure 2 is a graph of R vs. q for the values $z = 0.1, 0.5, 2, 10$. For $z = 0.1$ the system becomes bosonic for $q \geq 2.3$. This QGF system is more stable than the $q = 1$ system when $q < 1$ and it is more unstable than the $q = 1$ system when $q > 1$. On the other hand, quantum group, $SU_q(2)$, fermions do not exhibit anyonic behavior at $D = 2$. However, a calculation of the virial coefficients for the two parameter case reveals that $SU_{q/p}(2)$ fermions have anyonic behavior [40].

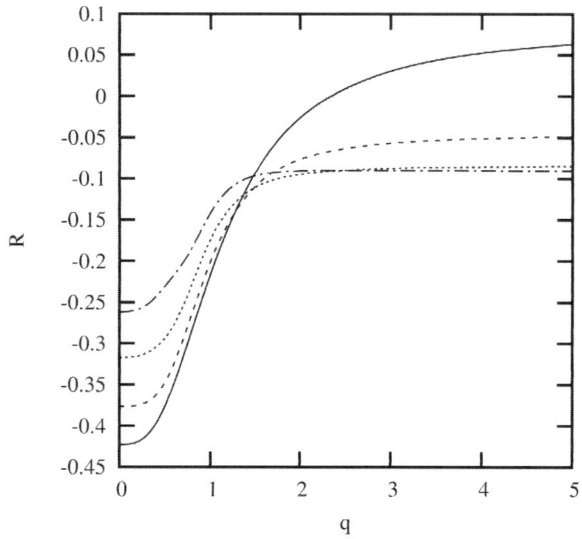

Figure 2. The scalar curvature R for quantum group fermions at $D = 3$, in units of $\lambda^3 V^{-1}$, as a function of the parameter q and values for the fugacity $z = 0.1$ (solid line), $z = 0.5$ (dashed line), $z = 2$ (dotted line) and $z = 10$ (dashed-dotted line).

4. Geometry of systems with fractal distribution functions

From the theory of fractals [41] we learned that given a statistical weight $\Omega(q,\delta)$ of a system with order parameter q and resolution δ, the fractal dimension is defined as the exponent $d = D_q$ which will make the product $lim_{\delta\to 0}\Omega(q,\delta)\delta^d$ finite. With use of the definition of the Boltzmann entropy $S(q,\delta) = \ln\Omega(q,\delta)$, the relation between the entropy and the fractal dimension D_q is given by

$$D_q = -\lim_{\delta\to 0}\frac{S(q,\delta)}{\ln\delta}. \tag{16}$$

Based on these definitions and with use of the Boltzmann's H theorem, the generalized entropy and distribution functions for classical and quantum gases were calculated in Ref.[42]. The average number of particles with energy ϵ was shown to be given by

$$< n(\epsilon) >= \frac{1}{\rho_j + a}, \tag{17}$$

where $\rho_j = [1 + \beta(q-1)(\epsilon - \mu)]^{1/(q-1)}$, $a = 0$ for the classical case, and the values $a = -1$ and $a = 1$ correspond to Bose-Einstein and Fermi-Dirac cases, respectively. For $q = 1$, Equation (17) becomes the standard textbook result for classical and quantum ideal gases. The distribution functions in Equation (17) were also obtained in Ref.[43] by considering a dilute gas approximation to the partition function of a non-extensive statistical mechanics originally proposed in Ref.[44]. In this Section we calculate the scalar curvature of systems with average particle number according to Equation (17). The probability densities that leads to Equation (17) were obtain in Ref. [45] and they are given by

$$\rho_{MB} = \frac{1}{Z_{MB}}\prod_{j=0}\frac{1}{n_j!}\rho^{-n_j}, \tag{18}$$

Journal of Physics: Conference Series **766** (2016) 012007 doi:10.1088/1742-6596/766/1/012007

for the classical case, and

$$\hat{\rho} = \frac{1}{Z} \prod_{j=0} \rho^{-\hat{n}_j}, \tag{19}$$

for the Bose-Einstein and Fermi-Dirac cases. Since the density distribution is not exponential, the second term in Equation (3) does not vanish and the metric for the three cases can be summarized in the general formula [26]

$$g_{\alpha\gamma} = \sum_{l=0} \frac{< n_l >^2}{\rho_l} \frac{\partial \rho_l}{\partial \beta^\alpha} \frac{\partial \rho_l}{\partial \beta^\gamma}. \tag{20}$$

The curvature is given by the expression

$$R = \frac{\lambda^3}{4V(detg)^2} \left(5h_{1/2}h_{3/2}^2 - 6h_{1/2}^2 h_{5/2} + h_{3/2}h_{-1/2}h_{5/2} \right), \tag{21}$$

where the function

$$h_\lambda = \frac{2}{\sqrt{\pi}(q-1)^{(\lambda+1)}} \frac{\Gamma(\lambda+1)\Gamma(\frac{q}{q-1} - \lambda)}{\Gamma(\frac{2q-1}{q-1})} \frac{1}{[1+(q-1)\gamma]^{\frac{q}{q-1}-\lambda}}. \tag{22}$$

After replacement of the definition of the function h_λ we get that the scalar curvature for the classical fractal case is identically equal to zero. Therefore, in this case the parameter q does not play any role as far as correlations are concerned.

For the bosonic and fermionic cases the functions h_λ are replaced in Equation (21) by the functions

$$G_\lambda^\pm = \frac{2}{\sqrt{\pi}} \int_0^\infty \frac{x^\lambda \Omega^{\frac{3-2q}{q-1}}}{(\Omega^{\frac{1}{q-1}} \pm 1)^2} dx, \tag{23}$$

where the $(+)$ sign is for fermions and the $(-)$ sign for bosons, and the function $\Omega = 1 + (q-1)(x - \beta\mu)$. Numerical calculations for the boson and fermion systems show that the corresponding values of R as a function of the fugacity z are closer to zero than those in the $q = 1$ case, implying that the departure from the value $q = 1$ makes the systems more stable. Therefore, for $q \neq 1$ bosons will be less attractive and fermions less repulsive that their standard counterparts. Our results also show that the sign of R remains unchanged as a function of z implying that these systems do not exhibit anyonic behavior, a fact that looks impossible to check by performing an expansion for $z \approx 0$ to obtain the second virial coefficient because the partition function is a function of $\ln z$.

We can summarize that the two interesting systems are two examples of the usefulness of the calculation of the thermodynamic curvature. These calculations give us important information about the stability and possible anyonic behavior of the system under study as a function of the temperature and the particular additional parameter involved.

Acknowledgment
I am very grateful to Professor Abdullah Algin and Osmangazi University Administration for their hospitality and giving me the opportunity of visiting their beautiful country. I also extend my gratitude to the physics graduate students Hakan Cetinkaya and Erbil Silik.

References
[1] Tisza L 1966 *Generalized Thermodynamics* (MIT, Cambridge).
[2] Griffiths R and Wheeler J 1970 *Phys. Rev.* A **2** 1047.
[3] Weinhold F 1975 *J. Chem. Phys.* **63** 2479.
[4] Amari S 1985 *Differential-Geometrical Methods in Statistics* (Springer-Verlag, Berlin).
[5] Amari S and Nagaoka H 2000 *Methods of Information Geometry* (AMS, Rhode Island).
[6] Ruppeiner G 1979 *Phys. Rev.* A **20** 1608.
[7] Ingarden R, Janyszek H, Kossakowski A and Kawaguchi 1982 *Tensor N.S.* **37** 105.
[8] Wootters W 1981 *Phys. Rev.* D **23** 357.
[9] Gilmore R 1984 *Phys. Rev.* A **30** 1994.
[10] Ruppeiner G 1985 *Phys. Rev.* A **32** 3141.
[11] Gilmore R 1985 *Phys. Rev.* A **32** 3144.
[12] Nulton J and Salamon P 1985*Phys. Rev.* A **31** 2520.
[13] Janyszek H 1986 *Rep. Math. Phys.* **24** 1; *Rep. Math. Phys.* **24** 11.
[14] Janiszek H and Mrugala R 1989*Rep. Math. Phys.* **27** 145.
[15] Ruppeiner G 2010 *Am. J. Phys.* **78** 1170, and references therein.
[16] Janiszek H and Mrugala R 1990 *J. Phys. A: Math. Theor.* **23** 467.
[17] Brody D and Hook D 2009 *J. Phys. A: Math. Theor.* **42** 023001.
[18] Janiszek H and Mrugala R 1989 *Phys. Rev.*A **39** (1989) 6515.
[19] Janyszek H 1990 *J. Phys. A:Math.* **23** 477.
[20] D. Brody and N. Rivier 1995 *Phys. Rev.*E **51** 1006.
[21] Janke W, Johnston D and Kenna R 2004 *Physica* A **336** 181.
[22] Trasarti-Battistoni R cond-mat/0203536.
[23] Portesi M, Plastino A and Pennini F 2006 *Physica* A **365** 173.
[24] Ohara A 2007 *Phys. Lett.* A **370** 184.
[25] Mirza B and Mohammadzadeh H 2008 *Phys. Rev.* E **79** 021127;2009 *Phys. Rev.* E **80** 011132; 2010 *Phys. Rev.* E **82** 031137; 2011 *J. Phys. A: Math. Theor.* **44** 475003.
[26] Ubriaco M R 2012 *Phys. Lett.* A **376** 2899.
[27] Ubriaco 2012 *Phys. Lett.* A **376** 3581.
[28] Ubriaco 2013 *Physica* A **392** 4868.
[29] Ubriaco 2014 *Physica* A **414** 128.
[30] See, for example, Jimbo M ed. 1990 *Yang-Baxter equation in integrable systems*, Advanced series in Mathematical Physics V.10 (World Scientific).
[31] Chari V and Pressley A 1994 *A Guide to Quantum Groups*, (Cambridge Univ. Press).
[32] See, for example, Carow-Watamura U, Schlieker M, Scholl M and Watamura S 1990 *Z. Phys.* C **48** 150 ; Ogievetsky O, Schmidke W, Wess J and Zumino B 1990 *Int. J. Mod. Phys.* A **6** 3081 .
[33] Aref'eva I and Volovich I 1991 *Phys. Lett* B **264** 62; Kempf A 1994 *J. Math. Phys.* **35** 4483 ; Brzezinski T and Majid S 1993 *Phys. Lett.* B **298** 339 ; Castellani L 1994 *Mod. Phys. Lett.* A **9** 2835 ; Ubriaco M R 1993 *Mod. Phys. Lett.* A **8** 2213 ; 1995 **10** 2223(E) ; Ubriaco M R 1994 *Mod. Phys. Lett.* A **9** 1121 ; Sudbery A 1996 *Phys. Lett.* B **375** 75.
[34] Iwao S 1990 *Prog. Theor. Phys.* **83** 363 . Bonatsos D, Argyres E and Raychev P 1991 *J. Phys.* A **24** L403; Capps R 1994 *Prog. Theor. Phys.* **91** 835.
[35] Ubriaco M R 1996 *Phys. Lett.* A **219** 205 ; 1998 *Phys. Rev.*E **57** 179 ; 1998 *Phys. Rev.*E **58** 4191 .
[36] Ubriaco M R 1997 *Phys. Rev.*E **55** 291 .
[37] Meusburger C and Noui K 2010 *Adv. Theor. Math. Phys.* **14** 6 1651, and references therein.
[38] Korbicz J, Wehr J and Lewenstein M 2009 *J. Math. Phys.* **50** 062104.
[39] Vokos S, Zumino B and Wess J 1989 *Symmetry in Nature* (Scuola Normale Superiore Publ., Pisa, Italy).

[40] Algin A, Arik M and Arikan A 2002 *Phys. Rev.*E **65** 026140 .
[41] Feder J 1988 *Fractals* (Plenum, New York).
[42] Büyükkiliç F and Demirhan D 1993 *Phys. Lett.* A **181** 24.
[43] Büyükkiliç F, Demirhan D and Güleç A 1995 *Phys. Lett.* A **197** 209.
[44] Tsallis C 1988 *J. Stat. Phys.* **52** 479.
[45] Ubriaco M R 1999 *Phys. Rev.*E **60** 165.

Journal of Physics: Conference Series **766** (2016) 012008 doi:10.1088/1742-6596/766/1/012008

General thermostatistical properties of a q-deformed fermion gas in two dimensions

Abdullah Algin[1] and Mustafa Senay[2]

[1]Department of Physics and Centre for Quantum Research and Applications, Eskisehir Osmangazi University, Meselik, 26480 Eskisehir, Turkey

[2]Graduate School of Sciences, Eskisehir Osmangazi University, Meselik, 26480 Eskisehir, Turkey

E-mail: aalgin@ogu.edu.tr; mustafasenay86@hotmail.com

Abstract. Starting with a deformed fermionic grand partition function, we study the high and low temperature thermostatistical properties of a special q-deformed fermion gas in two spatial dimensions. Many of the deformed thermostatistical functions such as the specific heat and the entropy are derived in terms of the real deformation parameter q for the range $q < 1$. For high temperatures, we specifically focus on the behavior of both the entropy function and the deformed virial coefficients in the equation of state for the q-fermion gas in two dimensions. Possible physical applications of the present q-fermion gas are briefly discussed.

1. Introduction

In recent years, quantum algebras [1-3] have attracted a great deal of interest for applications in natural and applied sciences. The statistical and thermodynamical properties of such boson and fermion systems have been extensively investigated [4-16]. As a matter of fact, interacting quantum systems exhibit many interesting phenomena not only for three spatial dimensions, but also in low-dimensional spaces. When confined to two dimensions, fermionic systems show phenomenon such as the fractional quantum Hall effect [17] and the behavior of an anyon gas [18]. In the meantime, some approaches with deformed fermionic and bosonic oscillator algebras to analyze the thermostatistical properties of such physical systems in two dimensions have been carried out [19,20]. However, for two dimensions, both quantum mechanical and thermostatistical properties of deformed oscillator systems have been less studied and less applied to the up-to-date scientific researches of different fields.

Motivated by the above considerations, in the present work, we aim to study the high and low temperature thermodynamical and statistical properties of a special q-deformed fermion gas model in two spatial dimensions. The q-deformed fermionic oscillator algebra, which we are interested in here, has been first introduced by Viswanathan *et al.* [21] and some of its statistical properties were also discussed by Chaichian *et al.* [8]. Therefore, we will use these fermion oscillators to construct a deformed gas of q-fermions describing a q-fermion gas model. Due to the historical reasons, we will call this model as the VPJC-type q-fermion gas. We should mention that high and low temperature thermostatistical properties of the VPJC q-fermion gas for three spatial dimensions have been elaborately investigated in [22,23]. Here, we continue to study further on the same model, and consider its quantum statistical behavior in two dimensions. In this framework, we will derive many

Journal of Physics: Conference Series **766** (2016) 012008 doi:10.1088/1742-6596/766/1/012008

of the important thermostatistical functions of the model depending on a function of the deformation parameter q. We specifically focus on the behavior of both deformed entropy function and deformed virial expansion of the equation of state for the VPJC q-fermion gas confined in two spatial dimensions. The effect of fermionic q-deformation on the thermodynamical quantities of the model in two dimensions and possible physical applications of the results are finally discussed.

2. The VPJC-type q-fermion gas model

The model containing the VPJC-type q-deformed fermion oscillators is based on the following deformed anti-commutation and commutation relations [21] for the deformed fermionic creation and annihilation operators c^*, c and the number operator $\hat{N}$:

$$cc^* + q c^* c = 1 \ , \qquad (1)$$
$$[\hat{N}, c^*] = c^* \ , \qquad [\hat{N}, c] = -c \ ,$$

where q is the real positive deformation parameter having the range $q < 1$. It has different properties in the sense of its quantum mechanical and thermostatistical aspects from the other fermion algebras in the literature [24-26]. Also, the deformed fermion number operator for the VPJC-fermion model is defined as

$$c^* c = [\hat{N}] = \frac{1 - (-1)^{\hat{N}} q^{\hat{N}}}{1+q} \ , \qquad (2)$$

whose spectrum is

$$[n] = \frac{1 - (-1)^n q^n}{1+q} \ . \qquad (3)$$

This is called the q-fermionic basic number for the VPJC-fermion model. By using the relations in Eqs. (1)-(3), it is possible to construct the representations of fermion operators in the Fock space, and therefore one can see that the VPJC-fermion model does not satisfy the Pauli exclusion principle for the interval $0 < q < 1$. However, it reduces to the standard fermion relations in the limit $q = 1$. Moreover, it has the following fermionic Jackson derivative (JD) operator $\hat{D}_x^{(q)}$ [26]:

$$\hat{D}_x^{(q)} f(x) = \frac{1}{x}\left[\frac{f(x) - f(-qx)}{1+q}\right] \ , \qquad (4)$$

which is a fermionic extension of the usual JD operator [27,28]. We will see below that this fermionic JD operator plays an important role for examining thermostatistics of the VPJC-type q-fermion gas even in two spatial dimensions.

3. Thermostatistical properties of the VPJC q-fermion gas

We consider a free deformed gas of q-fermions satisfying the relations in Eqs. (1)-(4) confined in a two-dimensional area (A). We can assume the Hamiltonian $\hat{H}$ for such a q-fermion gas model as $\hat{H} = \sum_i (\varepsilon_i - \mu)\hat{N}_i$, where μ is the chemical potential and ε_i is the kinetic energy in the state i related to the number operator $\hat{N}_i$. Hence, we can determine the mean value of the occupation number $n_{i,q}$ for the VPJC-type q-fermion gas using the relation $[n_i] = Tr(e^{-\beta\hat{H}} c_i^* c_i)/Z$, where $Z = Tr(e^{-\beta\hat{H}})$ is the partition function and $\beta = 1/kT$ with the Boltzmann constant k and the temperature T. From the above relations along with the use of the cyclic property of the trace [5,7], we obtain the following result for the mean occupation number $n_{i,q}$ of the VPJC q-fermion gas:

$$n_{i,q} = \frac{1}{|\ln q|}\left|\ln\left(\frac{\left|z^{-1}e^{\beta\varepsilon_i} - 1\right|}{z^{-1}e^{\beta\varepsilon_i} + q}\right)\right| \ , \qquad (5)$$

where $z = e^{\beta\mu}$ is the fugacity and $0 < q < 1$. Using Eqs. (4) and (5), one can deduce the following logarithm of the grand partition function:

$$\ln Z^{(VPJC)} = \left(\frac{1+q}{|\ln q|}\right)\sum_i \left|\ln|(1-ze^{-\beta\varepsilon_i})|\right| , \tag{6}$$

which satisfies the modified thermodynamic expression $N = z\hat{D}_z^{(q)} \ln Z^{(VPJC)} \equiv \sum_i n_{i,q}$, where $n_{i,q}$ is given in Eq. (5) and $\hat{D}_z^{(q)}$ has the same form as in Eq. (4) with the variable z instead of x. In order to find the high temperature thermostatistical functions of the VPJC q-fermion gas, we can replace the sums over states by integrals for a large area and a large number of particles as in the standard procedure [18,29]. From the thermodynamic relations $PA/kT = \ln Z^{(VPJC)}$ and $N = \sum_i n_{i,q}$, we obtain the following results for a two-dimonsional system by using the fermionic JD operator in Eq. (4) :

$$\frac{P^{(q)}}{kT} = \frac{1}{\lambda^2} f_2(z,q) , \qquad \frac{N^{(q)}}{A} = \frac{1}{\lambda^2} f_1(z,q) , \tag{7}$$

where $\lambda = h/\sqrt{2\pi mkT}$ is the thermal wavelength and the q-deformed function $f_n(z,q)$ is

$$\begin{aligned} f_n(z,q) &= \frac{1}{\Gamma(n)} \int_0^\infty \frac{x^{n-1}}{|\ln q|} \left|\ln\left(\frac{|1-ze^{-x}|}{1+qze^{-x}}\right)\right| dx \\ &= \frac{1}{|\ln q|}\left[\sum_{l=1}^\infty (-1)^{(l-1)} \frac{(zq)^l}{l^{n+1}} - \sum_{l=1}^\infty \frac{z^l}{l^{n+1}}\right] , \end{aligned} \tag{8}$$

where $x = \beta\varepsilon$. These functions are different from both the $h_n^\kappa(z_\kappa, q)$ functions of Ref. [20] and the standard $f_n(z)$ functions for an undeformed fermion gas in two dimensions [18]. Also, the internal energy $U^{(q)}$ is obtained as

$$U = -\left(\frac{\partial \ln Z^{(VPJC)}}{\partial\beta}\right)_{z,A} \quad \Rightarrow \quad \frac{U^{(q)}}{A} = \frac{1}{\lambda^2} kT\, f_2(z,q) . \tag{9}$$

It is noteworthy that from Eqs. (7) and (9), the relation $P^{(q)} = U^{(q)}/A$ is satisfied for the VPJC q-fermion gas. The specific heat of our model can be derived from both the thermodynamic definition $C_A = (\partial U/\partial T)_{N,A}$ and Eq. (4) as

$$\frac{C_A^{(q)}\lambda^2}{kA} = 2(z\hat{D}_z^{(q)} f_3(z,q)) - \frac{(z\hat{D}_z^{(q)} f_2(z,q))^2}{(z\hat{D}_z^{(q)} f_1(z,q))} , \tag{10}$$

where the function $f_n(z,q)$ is given in Eq. (8). The Helmholtz free energy $F = \mu N - PA$ can be found from Eqs. (7) and (9) as

$$F^{(q)} = N^{(q)} kT \left[\ln z - \left(\frac{f_2(z,q)}{f_1(z,q)}\right)\right] , \tag{11}$$

and the q-deformed entropy function $S^{(q)}/kN^{(q)}$ for the VPJC q-fermion gas model for high temperatures is derived from the relation $S = (U-F)/T$ as

$$\frac{S^{(q)}}{kN^{(q)}} = \left[2\left(\frac{f_2(z,q)}{f_1(z,q)}\right) - \ln z\right] . \tag{12}$$

For comparison, in figure 1, we show the plots of the q-deformed entropy function $S^{(q)}/kN^{(q)}$ and the entropy function S/kN of an undeformed fermion gas as a function of z for values of the deformation parameter q for the cases $q < 1$ and $q = 1$, respectively. For $q < 1$ and at the same fugacity, the entropy values of the VPJC q-fermion gas in two dimensions decrease with the values of q. At the same fugacity, it is also lower than the result of an undeformed fermion gas in two dimensions as shown in figure 1.

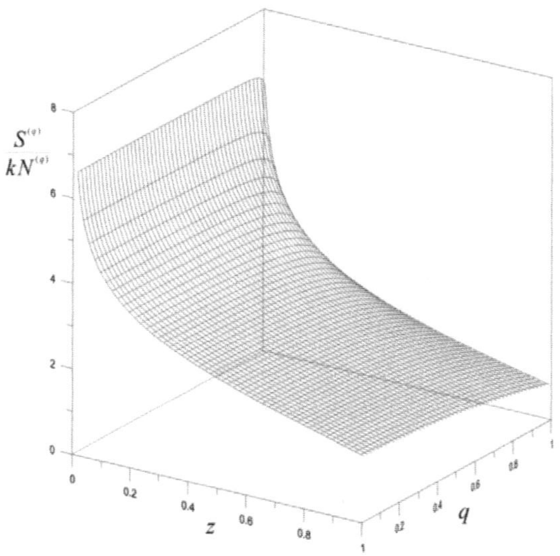

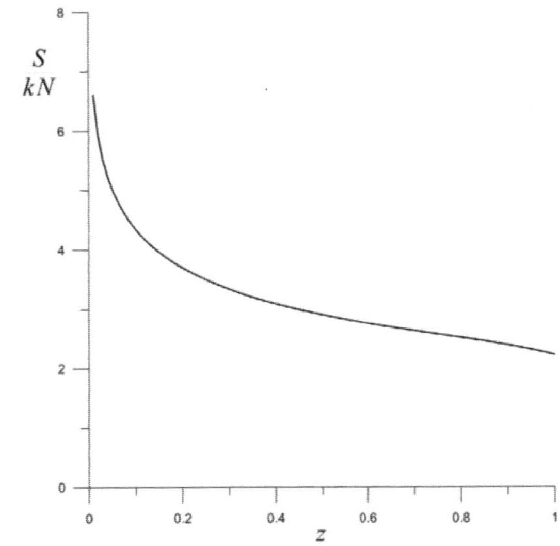

Figure 1. The entropy functions $S^{(q)}/kN^{(q)}$ and S/kN as a function of z for a two-dimensional system for the cases $q < 1$ (**left**) and $q = 1$ (**right**).

Furthermore, for high temperatures, the equation of state for the VPJC q-fermion gas model can be calculated from Eqs. (7) as a deformed virial expansion in the two-dimensional space as follows:

$$\frac{P^{(q)}A}{N^{(q)}kT} = a_1(q) + a_2(q)\left(\frac{N^{(q)}\lambda^2}{A}\right) + a_3(q)\left(\frac{N^{(q)}\lambda^2}{A}\right)^2 + a_4(q)\left(\frac{N^{(q)}\lambda^2}{A}\right)^3 + a_5(q)\left(\frac{N^{(q)}\lambda^2}{A}\right)^4 + \dots , \tag{13}$$

where the first five deformed virial coefficients are

$$\begin{aligned}
a_1(q) &= 1 , \\
a_2(q) &= \frac{1}{2^3}\frac{|\ln q|}{(q-1)^2}(q^2+1) , \\
a_3(q) &= \frac{1}{2^4}\frac{|\ln q|^2}{(q-1)^4}(q^2+1)^2 - \frac{2}{3^3}\frac{|\ln q|^2}{(q-1)^3}(q^3-1) , \\
a_4(q) &= \frac{5}{2^7}\frac{|\ln q|^3}{(q-1)^6}(q^2+1)^3 - \frac{1}{3.2^2}\frac{|\ln q|^3}{(q-1)^5}(q^2+1)(q^3-1) + \frac{3}{2^6}\frac{|\ln q|^3}{(q-1)^4}(q^4+1) , \\
a_5(q) &= \frac{7}{2^8}\frac{|\ln q|^4}{(q-1)^8}(q^2+1)^4 - \frac{1}{12}\frac{|\ln q|^4}{(q-1)^7}(q^2+1)^2(q^3-1) + \frac{1}{2^4}\frac{|\ln q|^4}{(q-1)^6}(q^2+1)(q^4+1) + \frac{2}{3^4}\frac{|\ln q|^4}{(q-1)^6}(q^3-1)^2 \\
&\quad - \frac{4}{5^3}\frac{|\ln q|^4}{(q-1)^5}(q^5-1) ,
\end{aligned} \tag{14}$$

where all the virial coefficients are q-dependent with the interval $0 < q < 1$, except that the first one. In order to visualize the behavior of these virial coefficients, we present the plots of $a_2(q)$, $a_3(q)$, $a_4(q)$, $a_5(q)$ as a function of q for a two-dimensional system for the case $q < 1$ in figure 2. As is shown in figure 2, we have numerically analyzed the behavior of these virial coefficients and have found that only the third virial coefficient $a_3(q)$ has both positive and negative values. The other virial coefficients have always positive values in the interval $0 < q < 1$. Thus, the high-temperature behavior of the VPJC q-fermion gas depends on the parameter q even in two dimensions.

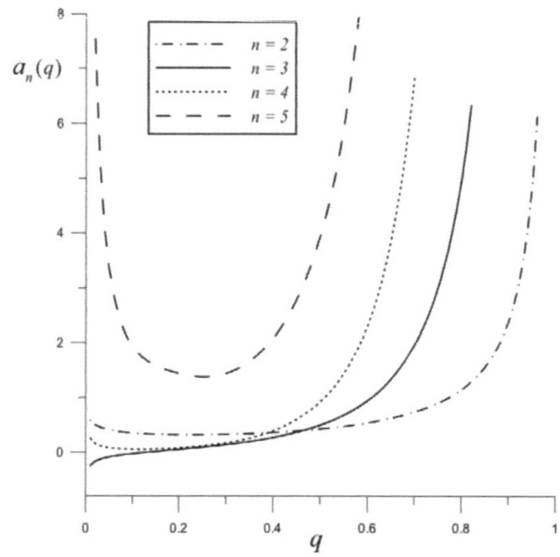

Figure 2. The virial coefficients $a_n(q)$ as a function of q for a two-dimensional system for the case $q < 1$.

We now consider some of the low-temperature thermostatistical properties of our model in two dimensions. From the relations $N^{(q)}(T)=\int_0^\infty g(\varepsilon)n(\varepsilon,T,q)d\varepsilon$ and $U^{(q)}(T)=\int_0^\infty \varepsilon g(\varepsilon)n(\varepsilon,T,q)d\varepsilon$, where $g(\varepsilon)=2\pi mA/h^2$, we derive the following relations for the chemical potential and the total energy of the VPJC-fermion model by using both Eq. (5) and the Sommerfeld expansion method [30,31] for low temperatures:

$$\mu=\frac{\hbar^2}{2m}\left(\frac{4\pi N^{(q)}(0)}{A}\right)\equiv\varepsilon_F\,,\qquad U^{(q)}(T)=\frac{(N^{(q)}(0))\,\varepsilon_F}{2}\left[1-\frac{I^{(VPJC)}(q)}{4}\left(\frac{kT}{\varepsilon_F}\right)^2\right],\tag{15}$$

where ε_F is the Fermi energy and $I^{(VPJC)}(q)$ is defined as

$$I^{(VPJC)}(q)=\int_{-\infty}^{\infty}dx\frac{x^2}{\ln q}\left[\frac{1}{(1-e^{-x})}-\frac{1}{(1+qe^{-x})}\right],\tag{16}$$

where $x=\beta(\varepsilon-\mu)$. This function has always negative values for the interval $0<q<1$. The Helmholtz free energy and the entropy function for our model in two spatial dimensions can be calculated for low temperatures as

$$\frac{F^{(q)}(T)}{N^{(q)}(0)}=\frac{\varepsilon_F}{2}\left[1+\frac{I^{(VPJC)}(q)}{4}\left(\frac{kT}{\varepsilon_F}\right)^2\right],\qquad \frac{S^{(q)}(T)}{kN^{(q)}(0)}=-\frac{I^{(VPJC)}(q)}{4}\left(\frac{kT}{\varepsilon_F}\right),\tag{17}$$

where $I^{(VPJC)}(q)$ is given in Eq. (16).

4. Conclusions

Following our previous works in [22,23], we further investigated the high and low temperature thermostatistical properties of the VPJC-type q-fermion gas model in two dimensions. According to figure 1, it should be emphasized that an enhancement of the values of q results in a reduction of the values of entropy function $S^{(q)}/kN^{(q)}$ for the case of two-dimensional system. This leads to a less chaoticity in the VPJC q-fermion gas. This model also describes an intermediate-statistics behavior via Eq. (5), and from the results obtained above, the deformation parameter q with the range $q<1$ is responsible for the behavior of the VPJC q-fermion gas as a whole. We hope that our results peculiar to two spatial dimensions could be useful for studies on several quantum systems such as strongly interacting electron gases and semiconductor quantum wells.

Acknowledgements

This work is supported by the Scientific and Technological Research Council of Turkey (TÜBİTAK) under Project No. 113F226.

References

[1] Arik M and Coon D D 1976 *J. Math. Phys.* **17** 524
[2] Biedenharn L C 1989 *J. Phys.* A **22** L873
[3] Macfarlane A J 1989 *J. Phys.* A **22** 4581
[4] Ng Y J 1990 *J. Phys.* A **23** 1023
[5] Lee C R and Yu J P 1990 *Phys. Lett.* A **150** 63
Lee C R and Yu J P 1992 *Phys. Lett.* A **164** 164
[6] Song H S, Ding S X and An I 1993 *J. Phys.* A **26** 5197
[7] Tuszynski J A, Rubin J L, Meyer J and Kibler M 1993 *Phys. Lett.* A **175** 173
[8] Chaichian M, Felipe R G and Montonen C 1993 *J. Phys.* A **26** 4017
[9] Cai S, Su G and Chen J 2010 *Int. J. Mod. Phys.* B **24** 3323
Cai S, Su G and Chen J 2007 *J. Phys.* A **40** 11245
[10] Tristant D and Brito F A 2014 *Physica* A **407** 276
[11] Dai W S and Xie M 2013 *Ann. Phys.* (NY) **332** 166
[12] Lavagno A and Narayana Swamy P 2002 *Phys. Rev.* E **65** 036101
[13] Lavagno A and Narayana Swamy P 2002 *Physica* A **305** 310
[14] Lavagno A and Narayana Swamy P 2010 *Found. Phys.* **40** 814
[15] Gavrilik A M, Kachurik I I and Mishchenko Yu A 2011 *J. Phys.* A **44** 475303
Gavrilik A M and Mishchenko Yu A 2015 arXiv:quant-ph/1504.00529v1
[16] Brito F A and Marinho A A 2011 *Physica* A **390** 2497
Marinho A A, Brito F A and Chesman C 2012 *Physica* A **391** 3424
[17] Wilczek F (Ed.) 1990 *Fractional Statistics and Anyon Superconductivity* (Singapore: World Scientific)
[18] Khare A 2005 *Fractional Statistics and Quantum Theory* (Singapore: World Scientific)
[19] Ubriaco M R 1998 *Phys. Lett.* A **241** 01
Ubriaco M R 1996 *Phys. Lett.* A **219** 205
Ubriaco M R 1996 *Mod. Phys. Lett.* A **11** 2325
[20] Lavagno A and Narayana Swamy P 2009 *Int. J. Mod. Phys.* B **23** 235
[21] Viswanathan K S, Parthasarathy R and Jagannathan R 1992 *J. Phys.* A **25** L335
[22] Algin A and Senay M 2012 *Phys. Rev.* E **85** 041123
[23] Algin A and Senay M 2016 *Physica* A **447** 232
[24] Parthasarathy R and Viswanathan K S 1991 *J. Phys.* A **24** 613
Parthasarathy R and Viswanathan K S 2016 arXiv: hep-ph/1605.08524v3
[25] Narayana Swamy P 2006 *Eur. Phys. J.* B **50** 291
[26] Algin A 2011 *Int. J. Theor. Phys.* **50** 1554
[27] F. Jackson 1909 *Messenger Math.* **38** 57
[28] Schork M 2005 *Russ. J. Math. Phys.* **12** 394
[29] Pathria R K and Beale P D 2011 *Statistical Mechanics (3rd ed.)* (London: Butterworth-Heinemann)
[30] Ashcroft N W and Mermin N D 1987 *Solid State Physics* (Philadelphia: Sounders Col. Pub.)
[31] Kittel C 2005 *Introduction to Solid State Physics* (New York: Wiley)

Journal of Physics: Conference Series **766** (2016) 012009 doi:10.1088/1742-6596/766/1/012009

Parametric investigation of a thermally driven QCD Deconfining Phase Transition in a finite volume at zero chemical potential

S. Bensalem and A. Ait El Djoudi

Laboratoire de Physique des Particules et Physique Statistique,
Ecole Normale Supérieure-Kouba, B.P. 92, 16050, Vieux-Kouba, Algiers, Algeria

E-mail: benlarbise@gmail.com, aiteldjoudi@ens-kouba.dz

Abstract. This work deals with a statistical description of a thermally driven deconfining phase transition (DPT) from a hadronic gas consisting of massless pions to a color-singlet Quark-Gluon Plasma (QGP), in a finite volume. The thermodynamical approach, within a coexistence model is used to investigate the Quantum Chromo-Dynamics DPT occurring between the two phases, at vanishing chemical potential. Considering the color singletness condition for the QGP phase, with massless up and down quarks, the exact total partition function of the studied system is obtained and then employed to calculate mean values of physical quantities, well characterizing the system near the transition. The finite-size effects on the DPT have been investigated through the study of the thermal behavior of the order parameter, the susceptibility and the second cumulant of the probability density. The similarity between the susceptibility and the second cumulant representing the variance is probed for the studied DPT and a parameterization of the variance is proposed for the first time.

1. Introduction

According to the state of the art in particle physics, the quark-lepton level is considered to be the fundamental level. Quarks interact with each other via the strong interaction, where the mediator bosons are the gluons [1]. The strong interaction is described through the Quantum Chromo-Dynamics (QCD) theory, which confirms that free quarks are never observed as free particles, but are confined inside hadrons such as the proton and the neutron. However, at high temperatures and/or densities a hadronic matter (HM) can turn into a new state of matter called Quark-Gluon Plasma (henceforth QGP), which is composed of the elementary particles; quarks and gluons [2]. It is worth mentioning that the quarks and gluons are still confined inside the plasma, but not inside hadrons, whence, the nomination "Deconfinement Phase Transition"(DPT) from a HM phase to a QGP phase.

The study of the finite-size effects on the DPT from the strongly interacting Hadronic Gas (HG) to the called QGP is possible through the study of several thermodynamic quantities, called response functions, among which we cite: the order parameter, its derivatives with respect to temperature and the cumulants of the probability density. Such a study has been performed by one of us in previous contributions [3, 4], allowing to note a remarkable qualitative similarity between the variations as function of temperature and volume of the order parameter derivatives and those of their homologue cumulants of the probability density, namely: between the first derivative of the order parameter and the variance representing the second cumulant, as well as between the second derivative of the order parameter and the skewness, which is related to the third cumulant.

In this work, we present a numerical study of a thermally driven QCD-DPT in order to investigate this similarity, which does not appear to have been previously developed, focusing on the following main objectives: a numerical study of the thermal QCD-DPT in a finite volume, an investigation of the similarity between the first derivative of the order parameter with respect to temperature, i.e., the susceptibility, and its homologue cumulant, i.e., the variance, and a parametric study aiming to provide an authentic approach to estimate the effective transition temperature for the studied DPT.

2. Statistical description of the Deconfinement Phase Transition in finite volumes

In order to conceive the QCD deconfinement phase transition in a finite volume, the coexistence model reported in [5] is a powerful tool. This model considers a mixed Hadronic Gas-Quark Gluon Plasma (HG-QGP) system, which exhibits a finite volume V. The volumes of the HG and QGP phases, noted V_{HG} and V_{QGP} respectively, form together the total volume V, and the fraction of volume occupied by the HG phase is designated by the parameter h, such that:

$$\begin{cases} V = V_{HG} + V_{QGP} \\ V_{HG} = hV. \end{cases} \tag{1}$$

The system is treated as a grand canonical ensemble, i.e., the thermodynamic variables are: the temperature (T), the volume (V) and the chemical potential (μ). In the framework of the coexistence model, the probability of finding the system in the state h is defined by [5]:

$$p(h,T,V,\mu) = \frac{Z(h,T,V,\mu)}{\int_0^1 Z(h,T,V,\mu)dh}, \tag{2}$$

here, $Z(h,T,V,\mu)$ is the total partition function of the system. The mean value of a given thermodynamic quantity is defined as [5]:

$$\langle X(T,V,\mu)\rangle = \int_0^1 X(h,T,V,\mu)p(h,T,V,\mu)dh = \frac{\int_0^1 X(h,T,V,\mu)Z(h)dh}{\int_0^1 Z(h)dh}, \tag{3}$$

where, $X(h,T,V,\mu)$ is the total thermodynamic quantity of the system in the state h.

Taking into account the color-charge confinement property, requiring that the QGP phase must be a colorless object in the color space [6], and under the assumption of non-interacting phases, the total partition function of the system can be expressed as follows:

$$Z(h) = \frac{4}{9\pi^2}\exp\left[\frac{\pi^2}{30}VT^3\right]\int_{-\pi}^{\pi}\int_{-\pi}^{\pi} d\varphi d\psi M(\varphi,\psi)\exp\left[(1-h)VT^3\left(g(\varphi,\psi)-\frac{\pi^2}{30}-\frac{B}{T^4}\right)\right], \tag{4}$$

where, $M(\varphi,\psi)$ represents the weight function (Haar measure) having the following expression:

$$M(\varphi,\psi) = \left[\sin\left(\frac{1}{2}\left(\psi+\frac{\varphi}{2}\right)\right)\sin\left(\frac{\varphi}{2}\right)\sin\left(\frac{1}{2}\left(\psi-\frac{\varphi}{2}\right)\right)\right]^2, \tag{5}$$

and:

$$g(\varphi,\psi) = \frac{\pi^2}{12}\left(\frac{21}{30}d_q+\frac{16}{15}d_g\right)+\frac{\pi^2}{12}\frac{d_q}{2}\sum_{q=r,b,g}\left[\left(\left(\frac{\theta_q}{\pi^2}\right)^2-1\right)^2-1\right]-\frac{\pi^2}{12}\frac{d_g}{2}\sum_{g=1}^{4}\left[\left(\frac{\theta_g-\pi}{\pi}\right)^2-1\right]^2, \tag{6}$$

here, $d_q = 2N_f$ (N_f: is the number of flavors) and $d_g = 2$ being the degeneracy factors of quarks and gluons respectively. The angles $\theta_q (q = r, b, g)$ are given by:

$$\theta_r = \frac{\varphi}{2} + \frac{\psi}{3}, \ \theta_b = -\frac{2\psi}{3}, \ \theta_g = -\frac{\varphi}{2} + \frac{\psi}{3}, \tag{7}$$

and $\theta_g (g = 1, ..., 4)$ are expressed as follows:

$$\theta_1 = \theta_r - \theta_g, \ \theta_2 = \theta_g - \theta_b, \ \theta_3 = \theta_b - \theta_r, \ \theta_4 = 0. \tag{8}$$

We note that our study is performed at a vanishing chemical potential ($\mu = 0$), taking into account massless u and d quarks, i.e., $N_f = 2$ in the QGP phase, and the pionic degrees of freedom in the HG phase. The value of the bag constant $B^{1/4} = 200\ MeV$ is adopted in all our calculations, with the units chosen as: $k_B = \hbar = c = 1$.

Based on the definition (3), the mean value of any physical quantity within the phase coexistence model may be calculated using the exact total partition function defined in (4). The first quantity of interest for the thermal DPT is the order parameter, which is the mean value of the volume fraction occupied by the HG phase, $\langle h(T, V) \rangle$, given by:

$$\langle h(T, V) \rangle = \frac{\int_0^1 hZ(h)dh}{\int_0^1 Z(h)dh}. \tag{9}$$

The second quantity is the first derivative of the order parameter with respect to temperature T, representing the susceptibility, χ, defined as:

$$\chi(T, V) = \left. \frac{\partial \langle h(T, V) \rangle}{\partial T} \right|_V. \tag{10}$$

On the other hand, we are interested in the cumulants of the probability density, $p(h)$, where the first cumulant is the order parameter $\langle h \rangle$ itself, and the second cumulant is the variance, σ^2, given by [7, 8]:

$$\sigma^2(T, V) = \langle h^2 \rangle - \langle h \rangle^2, \tag{11}$$

where $\langle h^2 \rangle$ is, using the definition (3), expressed as follows:

$$\langle h^2 \rangle = \frac{\int_0^1 h^2 Z(h)dh}{\int_0^1 Z(h)dh}. \tag{12}$$

The involved integrals in the considered response functions may be calculated numerically at each temperature, T, and volume, V, and this allows us to illustrate the variations of the order parameter, the susceptibility and the variance, on the whole adopted range of temperature, for various volumes.

3. Thermal behavior of the studied response functions

In the following, the thermal behavior of the considered response functions, namely, the order parameter, its first derivative with respect to temperature, as well as the second cumulant of the probability density is investigated on a range of temperature around the transition region at different finite-volumes. The considered volumes, i.e., 550, 700, 850 and 1000 fm^3 are included in the RHIC (Relativistic Heavy Ion Collider) realized sizes; (268-2144 fm^3) [9], where the QGP is eventually formed during ultra-relativistic heavy ion collisions.

The order parameter behavior in the range of temperature $143 - 146\ MeV$ at the considered finite volumes is illustrated in figure 1. The curves show a sharp transition in the infinite volume limit, characterized by a step-like jump at a transition temperature noted $T_c(\infty)$. This latter has been found to depend on the number of quark flavors, N_f, as well as on the bag constant, B, and it is obtained for $N_f = 2$ according to the following relationship (see for example [10]):

$$T_c(\infty) = \left(\frac{45B}{17\pi^2}\right)^{1/4}, \tag{13}$$

for $B^{1/4} = 200\ MeV$, we find $T_c(\infty) = 143.93\ MeV$, and this value of the transition temperature agrees very well with that obtained from recent lattice simulations [10, 11]. In small volumes, the transition is smeared out over a transition region that is larger, smaller is the volume. Another characteristic of the obtained curves is also the shift of the transition temperature to higher values for small volumes ($T_c(V) > T_c(\infty)$). This last finding is due to the color-singletness requirement in the QGP partition function [5, 6].

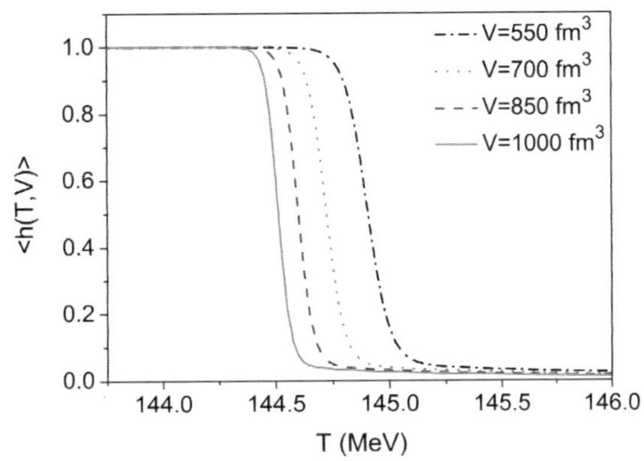

Figure 1. The order parameter variations with temperature for different volumes.

The variations of the susceptibility, $\chi(T,V)$, and its homologue cumulant, i.e., the variance, $\sigma^2(T,V)$, are displayed in figures 2 and 3, respectively. Both quantities behave like bell function, with the peaks broadened, smaller is the volume. The recorded peaks are located at the effective transition temperature $T_c(V)$, seemingly tending to $T_c(\infty)$ with increasing volume.

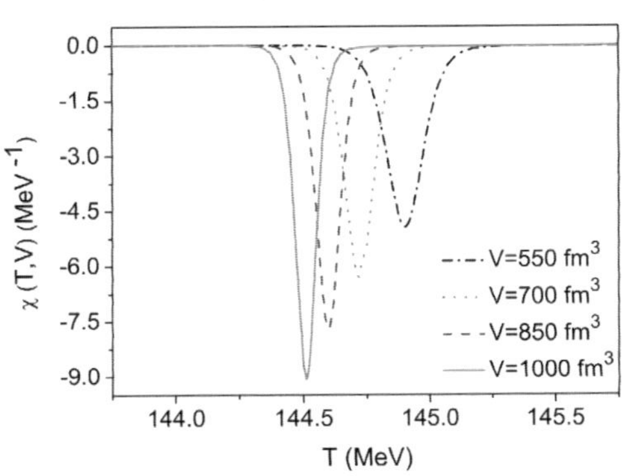

Figure 2. Thermal behavior of the susceptibility χ, for various system volumes.

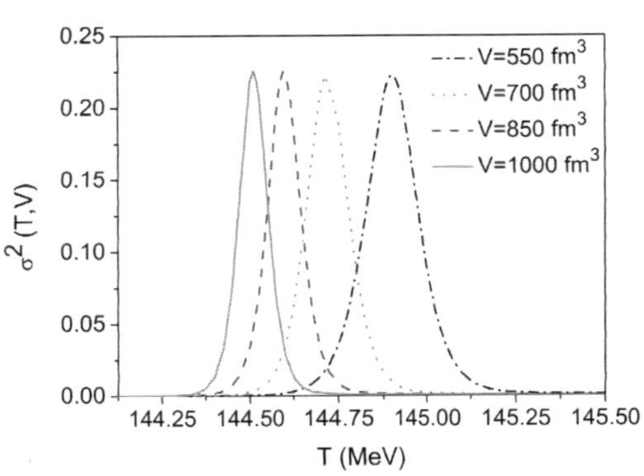

Figure 3. The variance σ^2 vs. temperature for different system volumes.

4. Correlation study and parametric investigation

4.1. Correlation between susceptibility and variance

In the following, let us investigate the similarity between the thermal susceptibility, χ, and the variance, σ^2, for the studied DPT. In a first step, we plot both χ and σ^2 with varying temperature, at a fixed volume ($V = 550\ fm^3$), on figure 4, considering a logarithmic scale, since their order of magnitude are different and do not allow to plot them clearly on the same graph. As expected, a remarkable similarity is observed with the peaks occurring at the same effective transition temperature $T_c(V)$.

In a second step, let us plot σ^2 vs. χ; the variations of the variance with the susceptibility for the considered values of the system volume are illustrated graphically in figure 5. The obtained plots reflect a linear proportionality, where the slope is inversely proportional to the system size. Thus, the variance is proportional to the susceptibility with a proportionality factor depending on the volume V.

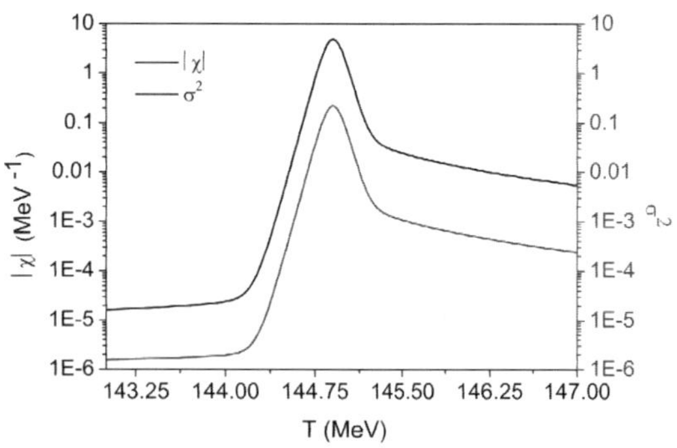

Figure 4. Evolution of the susceptibility and the variance at $V = 550\ fm^3$.

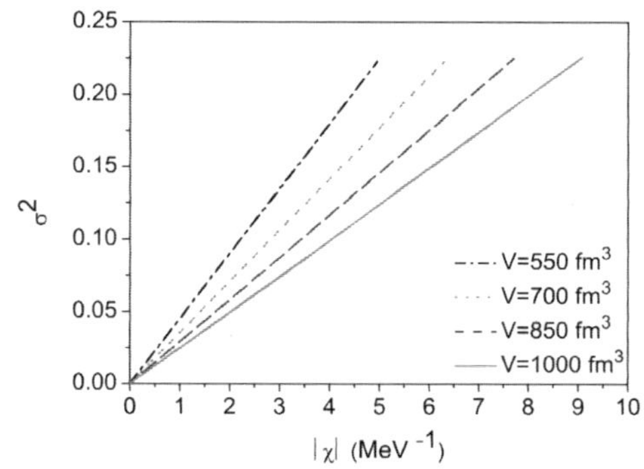

Figure 5. Variation of the variance, σ^2, with the system susceptibility, χ.

4.2. Susceptibility and variance parametrization

In the case of coexisting HM and QGP phases, the order parameter in a finite volume, $\langle h(T,V)\rangle$ may be expressed using one of the known mathematical representations of the Heaviside step-function as follows [3, 12]:

$$\langle h(T,V)\rangle = \frac{1}{2}\left(1 - \tanh\left[\frac{T - T_c(V)}{\Gamma_T(V)}\right]\right), \tag{14}$$

where, $T_c(V)$ is the effective transition temperature and $\Gamma_T(V)$ is the half-width of the rounded transition region. This leads directly to the susceptibility expression:

$$\chi(T,V) = \left.\frac{\partial\langle h(T,V)\rangle}{\partial T}\right|_V = -\frac{1}{2\Gamma_T(V)\cosh^2\left[\frac{T-T_c(V)}{\Gamma_T(V)}\right]}. \tag{15}$$

Based on the parametrization of the susceptibility with equation (15), and since a linear proportionality has been found between variance σ^2 and susceptibility χ, we propose a parametrization for the variance, similarly to the susceptibility, on the form:

$$\sigma^2(T,V) = \frac{1}{\Gamma_{\sigma^2}(V)\cosh^2\left[\frac{T-T_c(V)}{\Gamma_T(V)}\right]}, \tag{16}$$

where, $\Gamma_{\sigma^2}(V)$ is a dimensionless coefficient. An illustration of such parametrizations of the susceptibility and the variance at the volume $V = 1000\ fm^3$, are presented in figure 6 and figure 7, respectively. The regression coefficients are close to one for both performed fits. The obtained fit parameters T_c and Γ_T for $V = 1000\ fm^3$ from the proposed parametrization of the variance are in very good agreement with those obtained with the parametrization of the susceptibility, as it can be noted from the results displayed on figures 6 and 7.

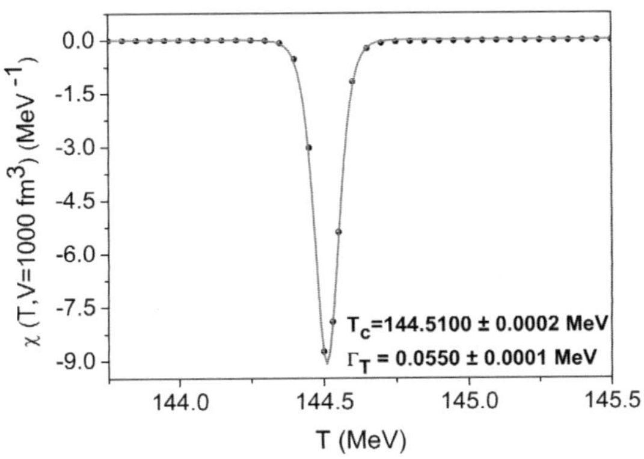

Figure 6. Parameterization of the susceptibility to the equation (15).

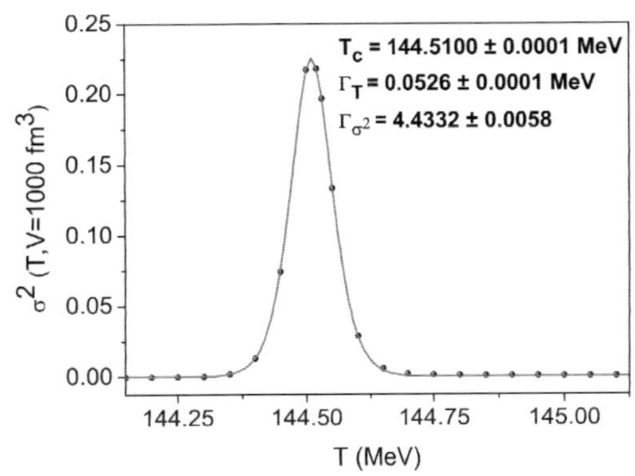

Figure 7. Parameterization of the variance to the equation (16).

5. Conclusion

In this communication, a thermally driven deconfinement phase transition from a Hadronic Gas consisting of massless pions, to a color-singlet QGP, containing gluons and massless up and down quarks, has been studied numerically. Based on a model considering the coexistence of both phases in a finite volume, a statistical approach was used in the framework of the QCD formalism to describe the thermal phase transition occurring between the two phases at vanishing chemical potential ($\mu = 0$). The finite size effects on the DPT have been investigated through the study of the thermal behavior of the order parameter, the susceptibility and the second cumulant of the probability density representing the variance, at different volumes, for $B^{1/4} = 200\ MeV$. The correlation between the susceptibility, χ, and the variance, σ^2, has been probed for the studied DPT, and the linearity between the two quantities is obtained. Furthermore, we proposed a parametrization of the variance, and this is done for the first time, to the best of our knowledge. The obtained results show that the proposed parametrization may be considered as an authentic method to estimate the effective transition temperature.

References

[1] Sarkar S, Satz H and Sinha B 2010 *The Physics of the Quark-Gluon Plasma* (Berlin: Springer)
[2] Cabibbo N and Parisi G 1975 *Phys. Lett.* B **59** 67
[3] Ladrem M and Ait El Djoudi A 2005 *Eur. Phys. J.* C **44** 257
[4] Ait El Djoudi A 2013 *Can. J. Phys.* **91** 793
[5] Spieles C, Stöcker H and Greiner C 1998 *Phys. Rev.* C **57** 908
[6] Elze H-T and Greiner W 1986 *Phys. Lett.* B **179** 385
[7] Tsai S-H and Salinas S R A 1998 *Braz. J. Phys.* **28** 58
[8] Ma S-K 1985 *Statistical Mechanics* (Philadelphia: Wspc)
[9] Zhang Q H 2001 Constraints on the size of the quark gluon plasma *Preprint* hep-ph/0106242
[10] Fraga E S 2013 QCD under extreme conditions: an informal discussion *Preprint* 1310.6656 [hep-ph]
[11] Philipsen O 2013 *Prog. Part. Nucl. Phys.* **70** 55
[12] Blaizot J P and Ollitrault J Y 1987 *Phys. Rev.* D **36** 916

Journal of Physics: Conference Series **766** (2016) 012010 doi:10.1088/1742-6596/766/1/012010

Evaluation of Quantum Corrections to Second Virial Coefficient with Lennard-Jones (12-6) Potential

B. A. Mamedov[1], E. Somuncu[2]

[1]Department of Physics, Faculty of Arts and Sciences, Gaziosmanpaşa University, Tokat, Turkey

[2]Department of Physics, Faculty of Arts and Sciences, Giresun University, Giresun, Turkey

E-mail: bamamedov@yahoo.com, elf_smnc@hotmail.com

Abstract. The efficient method is developed for analytical evalualtion of quantum corrections to second virial coefficient with Lennard-Jones (12-6) potential. The quantum corrections to second virial coefficient played a significant role in the investigation of real gases in the low temperatures range. In this paper, we obtained a series of formula in terms of gamma and exponential integral functions which is one of the useful approaches for analytical evaluation of quantum corrections to second virial coefficient. The calculation results excellently agreement with of the literature data.

1. Introduction

A comprehensive study of quantum corrections to the second virial coefficient is very important for understanding of the thermodynamic and other properties of fluids [1-2]. The accurate investigation of the temperature dependence of the intermolecular integration potentials is of central importance for studying behavior of fluids at low temperature [3-7].Various analytical methods on the evaluation of quantum corrections to the second virial coefficient have been proposed in literatures [8-12]. Gropper et al. [13-15] have done detailed work on the of quantum corrections to the second virial coefficient, which has been successfully applied to the light molecules at low temperature. Also, some researchers proposed analytical and numerical approaches for quantum corrections to the second virial coefficient [16-19]. Although important progress has been made in calculation of quantum corrections to the second virial coefficient, a number of new analytical methods are very important in understanding of various properties of fluids.

In this work, a general method is proposed for evaluating of quantum corrections to the second virial coefficient with the some applications. Compared to previous analytical methods, our obtained expression is more appropriate for studying at low temperature. The proposed algorithm gives a good rate of convergence and numerical stability under a wide range of values of parameters.

2. Expressions for the quantum corrections to second virial coefficient of the Lennard-Jones (12- 6) potential

For evaluation of the quantum corrections to second virial coefficient, we use the Lennard-Jones (12-6) potential and the corresponding intermolecular force in the following forms [20]

$$u(r)=4\varepsilon\left[\left(\frac{\sigma}{r}\right)^{12}-\left(\frac{\sigma}{r}\right)^{6}\right], \tag{1}$$

$$F(r)=\frac{4\varepsilon}{\sigma}\left[12\left(\frac{\sigma}{r}\right)^{13}-6\left(\frac{\sigma}{r}\right)^{7}\right], \tag{2}$$

where ε is the depth of the potential well, σ is the finite distance at which the inter-particle potential is zero, and r is the distance between the particles. Note that the locations where the intermolecular potential $u(r)$ is a minimum and intermolecular force $F(r)$ is exactly zero are given by $r_{\min}$ and σ, respectively and the two parameters are related by $r_{\min} = 2^{1/6}\sigma$ [21]. In this study the quantum corrections to second virial coefficient $B_q(T)$ defined as [22]

$$B_q(T) = \frac{h^2 N_A}{24\pi m (k_B T)^3} \int_0^\infty F^2(r)\left[f(r,T)+1\right] r^2 dr, \tag{3}$$

where the symbols N_A, h, m and k_B denote the number of molecules Avogadro's number, Planck's constant, the molecular mass and Boltzmann constant, respectively. In Eq. (3) the function $f(r,T)$ is the Mayer function and defined as [23]

$$f(r,T) = \exp\left(\frac{u(r)}{k_B T}\right) - 1. \tag{4}$$

The intermolecular force $F(r)$ is correlated to intermolecular potential energy $u(r)$ and written as

$$F(r) = -\frac{d u(r)}{dr} . \tag{5}$$

Eq. (3) is defined in non-dimensional and given by

$$B^*_q(T^*) = \Lambda^* B_q(T^*), \tag{6}$$

for convenience, where

$$B^*_q(T^*) = \Lambda^* B_q(T^*)/b_0, \quad b_0 = 2\pi r_{\min}^3/3, \tag{7}$$

and b_0 is known as the co-volume. The quantum correction to second virial coefficient

$$B_q(T^*) = \frac{3}{T^{*3}} \int_0^\infty F^2(r^*)\left[f(r^*,T^*)+1\right] r^{*2} dr^*, \tag{8}$$

where $T^* = k_B T/\varepsilon$, $r^* = r/r_{\min}$. The quantum mechanical parameter Λ^* given by

$$\Lambda^* = \frac{\hbar^2 R}{12 M k_B \varepsilon r_{\min}^2} , \tag{9}$$

in which $M = mR/k_B$ molar mass, and $\hbar = h/2\pi$.
Substituting Eqs. (1)- (2) into Eq.(8) we obtain

$$B_q\left(T^*\right)=\frac{36}{T^*}\sum_{m=0}^{N}(-1)^m F_m(2)\sum_{k=0}^{N'}\frac{\left(2/T^*\right)^k}{k!}$$
$$\left[\left(\frac{1}{T^*}\right)^{-\frac{23}{12}+\frac{m}{2}-\frac{k}{2}}\left(\Gamma\left[\frac{23}{12}-\frac{m}{2}+\frac{k}{2}\right]-\Gamma\left[\frac{23}{12}-\frac{m}{2}+\frac{k}{2},\frac{1}{T^*}\right]\right)+E_{\left(-\frac{11}{12}+\frac{m}{2}-\frac{k}{2}\right)}\left(\frac{1}{T^*}\right)\right]. \quad (10)$$

In Eq. (10) the indices N and N' are the upper limits of summations respectively. The $\Gamma(\alpha),\Gamma(\alpha,u)$ and the $E_n(z)$ in the Eq.(10) are well known gamma and incomplete gamma functions and exponential integral functions defined by [24]

$$\Gamma(\alpha)=\int_0^\infty t^{\alpha-1}e^{-t}\mathrm{d}t \ , \quad (11)$$

$$E_n(z)=\int_1^\infty t^{-n}e^{-zt}\mathrm{d}t, \quad (12)$$

$$\Gamma(\alpha,u)=\int_u^\infty t^{\alpha-1}e^{-t}\mathrm{d}t. \quad (13)$$

3. Numerical Results and Discussion

Closed analytical formula for quantum corrections to the second virial coefficient with Lennard-Jones (12-6) potential is developed. The obtained expression is formulated in terms of infinite summations involving special functions. Our calculation results, generated using Turbo Pascal 7.0 compiler, are demonstrated in Table 1.

Table 1. Quantum correction to second virial coefficient [2, 25-26]

T^*	$B_q\left(T^*\right)$	T^*	$B_q\left(T^*\right)$
9.9343	3.9654	3.4324x10^2	0.0651322
1.3124x10^1	2.79024	4.6581x10^2	0.0464939
1.7429 x10^1	1.96814	6.3235x10^2	0.0332084
2.3243x10^1	1.39147	8.5844x10^2	0.0237382
3.1105x10^1	0.985627	1.1664x10^3	0.0216798
4.1731x10^1	0.699579	15852.x10^1	0.0121265
5.6123x10^1	0.497188	2.1543x10^1	0.00867326
7.5617x10^1	0.353803	2.9270x10^1	0.0062076
1.0212x10^2	0.251808	3.9784x10^1	0.00444256
1.3802x10^2	0.179476	5.4104x10^1	0.00317847
1.8682x10^2	0.127955	7.3544x10^3	0.00227581
2.5313x10^2	0.0912692	1.0000x10^4	0.0016293

The temperature dependence of the quantum correction second virial coefficients are shown in Fig.1.

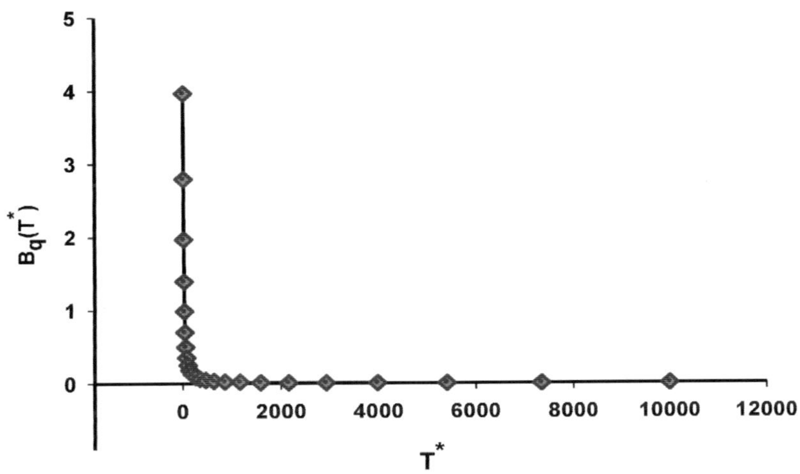

Figure 1. The quantum correction to second virial coefficient

As seen from Table 1 and Fig.1, the method of the present calculation is the useful approximation for analytical evaluation of quantum corrections to the second virial coefficient. The comparison between our results with the literature approximation calculations by Ref. [2, 25-26] is satisfactory. The approach is based on the efficient analytical formula and can be easily implemented to the atom and molecules.

3. References

[1] Gray C G, Gubbins K E, Joslin C G 2011 *Theory of Molecular Fluids* (Oxford University Press)
[2] Hirschfelder J O, Curtiss C F and Bird R B 1954 *Molecular Theory of Gases and Liquids* (Jonh Wiley &Sons)
[3] Lucas K 1991 *Applied Statistical Thermodynamics* (Springer-Verlag)
[4] Uhlenbeck G E and Gropper L 1932 *Phys. Rev.* **41** 79
[5] Gropper L 1936 *Phys. Rev.* **50**, 963
[6] Buckingham R A and Corner J 1947 *Proc. R. Soc. Lond. A* **189**, 118
[7] Kihara T 1955 Rev. Mod. Phys. **27** 412
[8] Corbin N, Meath W J and Allnatt A R 1984 *Mole. Phys.* **53** 225
[9] Nilsen T S 1969 Phys. Rev. **186** 262
[10] D'Arruda J J and Hill R N 1970 *Phys. Rev. A* **1** 1791
[11] McCarty M and Babu S V K 1970 *J. Phys. Chem.* **74** 1113
[12] Ghassib H B, Sandouqa A S, Joudeh B R and Mosameh S M 2014 *Can. J. Phys.* **92** 997
[13] Gropper L 1937 *Theory, Phys. Rev.* **51** 1108
[14] Beth E and Uhlenbeck G E 1937 *Physica* **4** 915
[15] Uhlenbeck G E and Beth E 1936 *Physica* **3** 729
[16] Bruch L W and Kariotis R 2006 *Phys. Rev. B* **73** 193404

[17] Akin-Ojo O, Harvey A H and Szalewicz K 2006 *J. Chem. Phys.* **125** 014314
[18] Marun R A B, Coronada E A and Ferrero J C 2005 *Chem. Phys. Lett.* **405**, 203
[19] Wang W F 2003 *J. Quant. Spec. and Rad. Transfer* **76** 23
[20] Kaplan I G 2006 *Intermolecular Interactions: Physical Picture, Computational Methods and Model Potentials* (Jonh Wiley & Sons)
[21] Hutem A and Boonchui S 2012 *J. Math. Chem.* **50** 1262
[22] Lucas K 1991 *Applied Statistical Thermodynamics* (Springer-Verlag)
[23] McQuarine D A 1973 *Statistical Mechanics* (Harper & Row)
[24] Gradshteyn I S and Ryzhik I M 1965 *Tables of Integrals, Series and Products* (Acedemic, London)
[25] Fowler R H 1936 *Statistical Mechanic: The Theory of the Properties of Matter in Equilibrium* (Cambridge University Press)
[26] Kim S and Henderson D 1966 *Proc. Natl. Acad. Sci. U.S.A.* **55**

Journal of Physics: Conference Series **766** (2016) 012011 doi:10.1088/1742-6596/766/1/012011

Quantum Fisher Information of an Open and Noisy System in the Steady State

Azmi Ali Altintas
Faculty of Engineering and Architecture, Okan University, Istanbul, Turkey
E-mail: altintas.azmiali@gmail.com

Abstract. In this work, we study the quantum Fisher information (QFI) per particle of an open (particles can enter and leave the system) and dissipative (far from thermodynamical equilibrium) steady state system of two qubits in noisy channels. We concentrate on two noisy channels these are dephasing and non-dephasing channels. We will show that under certain conditions QFI per particle is slightly greater than 1 for both systems. This means that both systems can be slightly entangled.

1. Introduction

In quantum mechanics, it is hard to measure observables, instead we try to estimate them. Estimating a parameter is subject of information theory. Because of that quantum mechanics gets some tools from estimation theory to guess the parameters. Also we can design some experiments to estimate parameters, for example Mach-Zender interferometer. It is used to determine the relative phase shift between two collimated beams. It is well known that entanglement can increase the sensitivity of interferometer. Quantum Fisher information (QFI) characterizes the sensitivity of a quantum system with respect to the changes of a parameter of the system. It can be taken as a multipartite entanglement criteria[1, 2]: If the mean quantum Fisher information per particle of a state exceeds the so called *shot-noise limit* i.e. the ultimate limit that separable states can provide, then the state is multipartite entangled. *Shot-noise limit* is $\Delta\theta \equiv \frac{1}{\sqrt{N}}$, where N is the number of particles[3]. Only for N=2 case, any entangled state can be made useful by local operations [4]. It is also shown that GHZ states provide the largest sensitivity, achieving the fundamental, so called Heisenberg limit [5]. Mean QFI determines the phase sensitivity of state with respect to SU(2) rotations. Recently the quantum Fisher information has been further studied both theoretically and experimentally [6, 7, 8, 9, 10, 11, 12, 13, 14, 15, 16, 17, 18, 19, 20, 21, 22, 23, 24, 25, 26].

Usually the natural systems are open and noisy. If a quantum system interacts with environment it is thought as the quantum system is in a noisy channel. It is obvious that noise decreases the entanglement of quantum system. If the system is open the decrease can be balanced by determining a reset mechanism. With the help of reset mechanism an entangled steady state can be established. The reset mechanism replaces randomly system particles with particles from the environment in some standard, sufficiently pure, single-particle state [27]. Reset mechanism, itself, can not produce entanglement. To create entanglement, the fresh particles must interact with the system. Reset mechanism requires particle exchange from the environment, this brings that the system must be open. Hartmann et.al shown that for both gas type and strongly coupled quantum systems the effect of decoherence can be vanished with the help of reset mechanism [28].

In this work, we study the quantum Fisher information per particle of open and dissipative noisy system of two qubits with reset mechanism. We will concentrate on two different types of noise. Firstly we take dephasing channel[29], then secondly we look at non-dephasing channel and we try to find states whose mean QFI is greater than 1. We examine the effect of reset mechanism for dephasing case, and for non-dephasing case we will determine a temperature dependent parameter "s" then we will examine effect of "s" on QFI.

2. QFI of open noisy system in a steady state

Quantum Fisher information of a given quantum system can be written from [4] as;

$$F(\rho, J_{\vec{n}}) = \sum_{i\neq j} \frac{2(p_i - p_j)^2}{p_i + p_j} |\langle i|J_{\vec{n}}|j\rangle|^2 = \vec{n}\mathbf{C}\vec{n}^T. \tag{1}$$

Here p_i and $|j\rangle$ are the eigenvalue and eigenvector of state ρ respectively. Also $\vec{n}$ is a normalized three dimensional vector and $J_{\vec{n}} = \sum_{\alpha=x,y,z} \frac{1}{2} n_\alpha \sigma_\alpha$, the angular momentum operator in $\vec{n}$ direction. σ_α are Pauli matrices. Also, $p_i + p_j = 0$ terms are not included to summation. After some calculations the matrix elements of the symmetric matrix $\mathbf{C}$ can be found as;

$$C_{kl} = \sum_{i\neq j} \frac{(p_i - p_j)^2}{p_i + p_j} [\langle i|J_k|j\rangle\langle j|J_l|i\rangle + \langle i|J_l|j\rangle\langle j|J_k|i\rangle] \tag{2}$$

If ρ is a pure state the equation 2 is written as

$$C_{kl} = 2\langle J_k J_l + J_l J_k\rangle - 4\langle J_k\rangle\langle J_l\rangle, \tag{3}$$

and the QFI is also expressed as $F(\rho, J_{\vec{n}}) = 4(\Delta J_{\vec{n}})^2$. The mean QFI is found as in [6]

$$\overline{F}_{max} = \frac{F_{max}}{N} = \frac{\lambda_{max}}{N} \tag{4}$$

here λ_{max} is maximum eigenvalue of matrix C and N is the number of particles. Also λ_{max} is the maximum value of QFI. It has recently been shown that, the QFI for separable states is [1]

$$\overline{F}_{max} \leq 1 \tag{5}$$

and for general states the mean QFI of the system is

$$\overline{F}_{max} \leq N \tag{6}$$

where the bound $\overline{F}_{max} = N$ can be saturated by maximally entangled states.

Now, we define an open quantum system with reset mechanism in a noisy channel. The total master equation which defines the quantum system is given by [27]

$$\dot{\rho} = -i[H, \rho] + L_{noise}\rho + r\sum_{i=1}^{N}(|\chi_i\rangle_i\langle\chi_i|tr_i\rho - \rho) \tag{7}$$

The master equation is in form of Lindbald equation. The first term in the right hand side of eq. (7) is just about total Hamiltonian of the quantum system, the second term describes the noisy channel and the third term describes the reset mechanism and N is the number of particle.

$$\begin{aligned} L_{noise}\rho &= \sum_{i=1}^{N} -\frac{B}{2}(1-s)[\sigma_+^i\sigma_-^i\rho + \rho\sigma_+^i\sigma_-^i - 2\sigma_-^i\rho\sigma_+^i] \\ &- \frac{B}{2}s[\sigma_-^i\sigma_+^i\rho + \rho\sigma_-^i\sigma_+^i - 2\sigma_+^i\rho\sigma_-^i] - \frac{2C-B}{4}[\rho - \sigma_z^i\rho\sigma_z^i] \end{aligned} \tag{8}$$

Here B is inversion and C is polarization parameter. Also s is a temperature dependent parameter which is

$$s = (e^{\omega\beta} + 1)^{-1} \in [0, 1] \tag{9}$$

Since $\beta = \frac{1}{T}$, when $T \to \infty$ s will be 0.5.

2.1. Dephasing Case

In this case we choose dephasing channel as a noisy channel. The Hamiltonian of two qubit steady state is

$$H = g\vec{\sigma}_z^{(1)}\vec{\sigma}_z^{(2)}, \tag{10}$$

here $g \geq 0$ is the coupling strength. Since the noise is a dephasing channel the second term in equation (7) is

$$L_{noise}\rho = \frac{\gamma}{2}\sum_{i=1,2}(\sigma_z^{(i)}\rho\sigma_z^{(i)} - \rho) \tag{11}$$

here γ is strength of decoherence which is a positive real number. One can get this expression by choosing $B = 0$ and $C = \gamma$ in equation (8). The expression of reset mechanism is written by taking $N = 2$ as

$$L_{reset}\rho = r\sum_{i=1,2}(|\chi_i\rangle_i\langle\chi_i|tr_i\rho - \rho). \tag{12}$$

Since the reset state should be able to produce entanglement from the resulting product state, the reset state must be depend on the Hamiltonian. For example for our Hamiltonian we can not choose the reset state as $|\chi_i\rangle_i = |0\rangle$, since the state does not create any entanglement. Then our two qubit master equation becomes

$$\dot{\rho} = -i[H, \rho] + \frac{\gamma}{2}\sum_{i=1,2}(\sigma_z^{(i)}\rho\sigma_z^{(i)} - \rho) + r\sum_{i=1,2}(|+\rangle_i\langle+|tr_i\rho - \rho). \tag{13}$$

When $r = 0$, the noise destroys the entanglement. For $r \to \infty$ case Hamiltonian and noise parts are neglected and the reset part injects fresh particles to the system. Thus the entanglement between two qubits will be zero eventually. ρ is the density state and it can be expressed as matrix form. In our case it is 4x4 matrix. The matrix is written from [27],

$$\rho_{11} = \rho_{22} = \rho_{33} = \rho_{44} = \frac{1}{4}, \tag{14}$$

$$\rho_{14} = \rho_{23} = \rho_{32} = \rho_{41} = \frac{r^2(r + \gamma/2)}{4(r+\gamma)[2g^2 + (r+\gamma/2)(r+\gamma)])}, \tag{15}$$

$$\rho_{12} = \rho_{13} = \rho_{42} = \rho_{43} = \frac{r(-ig + r + \gamma/2)}{4[2g^2 + (r+\gamma/2)(r+\gamma)]}, \tag{16}$$

$$\rho_{21} = \rho_{24} = \rho_{31} = \rho_{34} = \frac{r(ig + r + \gamma/2)}{4[2g^2 + (r+\gamma/2)(r+\gamma)]}. \tag{17}$$

Here r, γ and g are reset, decoherence and coupling strength parameters respectively and they are real parameters.

Now by using ρ matrix in equation (2) we find QFI per particle of the system depending on parameters r, γ and g.

To understand the behavior of the QFI per particle first we fixed the noise parameter to $\gamma = 0.5$.

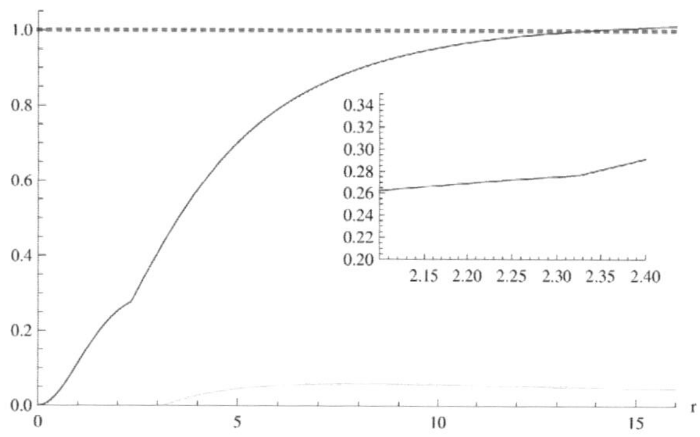

Figure 1. (QFI per particle (blue) and negativity (green) vs reset with $\gamma = 0.5$. Red dotted line represents the shot noise limit. Inset shows the critical point where the optimal direction changes.)

As one can see from the figure at $r = 0$ the QFI of the system is 0 as expected. When r is at 14 QFI per particle has a value as 1.00226. Well known entanglement criteria is negativity and it can take values between 0 and 1. For our entangled state negativity is 0.0496243. It means that the chosen state is weakly entangled. When negativity is 0 the state is separable, when the negativity equals to 1 the state is maximally entangled.

For the figure (1) ($g = 2.5$ and $\gamma = 0.5$ case) the optimal direction $\vec{n}^o = \vec{n}_1$ when $r \leq 2.3$. For $r > 2.3$ the optimal direction $\vec{n}^o = \vec{n}_2 sin(\frac{\pi}{2}) + \vec{n}_3 cos(\frac{\pi}{2})$. Here $\vec{n}_1$ is unit vector in x axis. $\vec{n}_2$ and $\vec{n}_3$ are unit vectors in y and z directions respectively.

2.2. Non-Dephasing Case

In this case, the noisy channel is a non-dephasing channel. In equation 7 the hamiltonian contains free and interaction terms.

$$H_{free} = \frac{\omega}{2}\sum_{i}^{N}\sigma_z^i \text{ and } H_{int} = g\sigma_x^1\sigma_x^2. \tag{18}$$

Also in noise part we take $C = \frac{B}{2}$ and $\omega = B$. In the light of above, one can solve the master equation and the elements of density matrix will be

$$\rho_{11} = \frac{B^2s^2w^2 + (B+2r)((B+r)g^2 + B^2(B+2r)s^2)}{(B+r)((B+r)\omega^2 + (B+2r)(4g^2 + (B+r)(B+2r)))}, \tag{19}$$

$$\rho_{14} = \rho_{41}^* \frac{g(2sB - B - r)(-i(B+2r)+\omega)}{((B+r)\omega^2 + (B+2r)(4g^2 + (B+r)(B+2r)))}, \tag{20}$$

$$\rho_{22} = \rho_{33} = \frac{(B+2r)((B+r)g^2 - B^2(B+2r)s^2 + B(B+r)(B+2r)s) - Bs(sB - B - r)\omega^2}{(B+r)((B+r)\omega^2 + (B+2r)(4g^2 + (B+r)(B+2r)))}, \tag{21}$$

$$\rho_{44} = \frac{(-Bs + B + r)\omega^2 + (B+2r)(B^2(B+2r)s^2 - 2B(B+r)(B+2r)s + (B+r)(g^2 + (B+r)(B+2r))}{(B+r)((B+r)\omega^2 + (B+2r)(4g^2 + (B+r)(B+2r)))}, \tag{22}$$

The other terms of density matrix is are 0.

Now by using the density matrix in equation (2) we find the mean QFI of the system. To understand the effect of temperature we take s as a free parameter, it can have values between 0 and 0.5. We fix the reset parameter as $r = 10$ and we choose as $\omega = 1$

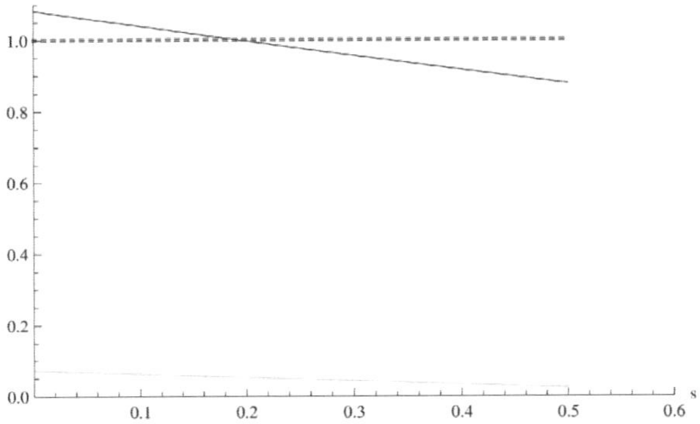

Figure 2. (QFI per particle (blue) and negativity (green) vs s with r = 10. Red dotted line represents the shot noise limit.)

It can be easily seen that when s is greater than 0.193 mean QFI is smaller than 1 although negativity is still greater than 0. This means that QFI does not recognize all the entangled states.

3. Cocnlusion

We have studied the quantum Fisher information of a noisy open quantum system of two qubits which is in a steady state. In the system we use an interaction Hamiltonian in a noisy channel and a reset mechanism. By solving master equation we have defined density state in matrix form and with the help of equation 2 we describe mean QFI of the system depending on reset and noise parameters.

We have shown the change of QFI per particle depending on reset parameter. In figure 1 we have chosen γ as 0.5 and have observed that at $r = 14$ QFI per particle is greater than 1. The negativity of such state is greater than 0. It means that the state is entangled[29].

Also we take non-dephasing noisy channel as second example. At that case we take reset parameter constant as $r = 10$ and we define a temperature dependent parameter s. The change of QFI per particle depending on parameter s is investigated and figure 2 shows that until s be 0.193 QFI per particle is greater than 1. After that value of s QFI per particle is smaller than 1 although negativity is greater than 0.

These two example show us that QFI does not recognize all entangled states. Despite the QFI is entanglement witness it can not be taken as entanglement measure.

There are some recent works on q-deformation and quantum information theory[30, 31], in the light of these works as a further work one can calculate the QFI of q deformed version of the states that we introduce this study.

References

[1] Pezze L, Smerzi A 2009 Entanglement, Non-linear Dynamics, and the Heisenberg Limit *Phys. Rev. Lett.* **102** 100401

[2] Hyllus P, et al. 2012 Fisher Information and Multiparticle Entanglement. *Phys. Rev. A* **85** 022321

[3] Xiong H N, Ma J, Liu W F and Wang X 2010 Quantum Fisher Information for Superpositions of Spin States. *Quant. Inf. Comp.* **10** 498

[4] Hyllus P, Ghne O, and Smerzi A 2010 Not All Pure Entangled States are Useful for Sub Shot-Noise Interferometry. *Phys. Rev. A* **82** 012337

[5] Giovannetti V, Lloyd S, and Maccone L 2004 Quantum-Enhanced Measurements:Beating the Standard Quantum Limit. *Science* **306** 1330

[6] Ma J, Huang Y, Wang X and Sun C P 2011 Quantum Fisher Information of the Greenberger-Horne-Zeilinger State in Decoherence Channels. *Phys. Rev. A* **84** 022302
[7] Ji Z, et. al 2008 Parameter Estimation of Quantum Channels *IEEE Trans. Info. Theory* **54** 5172
[8] Escher B M, Filho M and Davidovich L 2011 General Framework for Estimating the Ultimate Precision Limit in Noisy Quantum-Enhanced Metrology *Nat. Phys.* **7** 406
[9] Yi X, Huang G and Wang J 2012 Quantum Fisher Information of a 3-Qubit State *Int. J. Theor. Phys.* **51** 3458
[10] Spagnalo N, et al. 2010 Quantum Interferometry With Three-Dimensional Geometry *Scientific Reports* **2** 862
[11] Liu Z 2013 Spin Squeezing in Superposition of Four-Qubit Symmetric State and W States *Int. J. Theor. Phys.* **52** 820
[12] Ozaydin F, Altintas A A, Bugu S and Yesilyurt C 2013 Quantum Fisher Information of N Particles in the Superposition of W and GHZ States *Int. J. Theor. Phys.* **52** 2977
[13] Ozaydin F, Altintas A A, Bugu S and Yesilyurt C 2014 Quantum Fisher Information of Several Qubits in the Superposition of a GHZ and two W States with Arbitrary Relative Phase *Int. J. Theor. Phys.* **53** 3219
[14] Ozaydin F, Altintas A A, Bugu S and Yesilyurt C 2014 Behavior of Quantum Fisher Information of Bell Pairs Under Decoherence Channels *Acta Physica Polonica A* **125** 606
[15] Ozaydin F 2014 Phase Damping Destroys Quantum Fisher Information of W states *Phys.Lett. A* **378** 43
[16] Gibilisco P, Imparato D, and Isola T 2007 Uncertainty Principle and Quantum Fisher Information II *J. Math. Phys.* **48** 072109
[17] Andai A 2008 Uncertainty Principle with Quantum Fisher Information *J. Math. Phys.* **49** 012106
[18] Berrada K, Khalek S B and Obada A S F 2012 Quantum Fisher Information for a Qubit System Placed Inside a Dissipative Cavity *Phys. Lett. A* **376** 1412
[19] Luo S 2003 Wigner-Yanase Skew Information and Uncertainty Relations *Phys. Rev. Lett.* **91** 180403
[20] Kacprowicz M. et al. 2010 Experimental Quantum-Enhanced Estimation of a Lossy Phase Shift *Nat. Photon.* **4** 357
[21] Krischek R et al. 2011 Useful Multiparticle Entanglement and Sub-Shot-Noise Senitivity in Experimental Phase Estimation *Phys. Rev. Lett.* **107** 080504
[22] Strobel H et al. 2014 Fisher Information and Entanglement of non-Gaussian Spin States *Science* **345** 424
[23] Erol V, Ozaydin F and Altintas A A 2014 Analysis of Entanglement Measures and LOCC Maximized Quantum Fisher Information of General Two Qubit Systems *Scientific Reports* **4** 5422
[24] Jing X X, Liu J, Xiong H and Wang X 2015 Maximal Quantum Fisher Information for General SU(2) Parametrization Process *Phys. Rev. A* **92** 012312
[25] Ozaydin F, Altintas A A, Yesilyurt C, Bugu S and Erol V 2015 Quantum Fisher Information of Bipartitions of W States *Acta Physica Polonica A* **127** 1233
[26] Ozaydin F, Altintas A A 2015 Quantum Metrology: Surpassing the shot-noise limit with Dzyaloshinskii-Moriya interaction *Scientific Reports* **5** 16360
[27] Hartmann L, Dür W and Briegel H J 2006 Steady-State Entanglement in Open and Noisy Quantum Systems *Phys. Rev. A* **74** 052304
[28] Hartmann L, Dür W and Briegel H J 2007 Entanglement an its Dynamics in Open Dissipative Systems *NJP* **9** 230
[29] Altintas A A 2016 Quantum Fisher information of an open and noisy system in the steady state *Annals of Physics* **367** 192
[30] Gavrilik A M, Mishchenko Yu A 2012 Entanglement in composite bosons realized by deformed oscillators *Phys. Lett. A* **376** 1596
[31] Altintas A A, Ozaydin F, Yesilyurt C, Bugu S and Arik 2014 M Constructing quantum logic gates using q-deformed harmonc oscillator algebras *Quantum Inf. Process* **13** 1035

Journal of Physics: Conference Series **766** (2016) 012012 doi:10.1088/1742-6596/766/1/012012

Comparing concurrence and negativity in time-dependent ionic-phononic system with fifteen dimensional density matrix

Rasim Dermez[1,2,3]

[1]Department of Physics, Afyon Kocatepe University, Afyonkarahisar 03200, Turkey
[2]Department of Electrical and Computer Engineering, Michigan Technological University, Houghton, Michigan 49931, USA
[3]Afyon Vocational School, Afyon Kocatepe University, Afyonkarahisar 03200, Turkey

E-mail: `dermez@aku.edu.tr`

Abstract. It is solved a time-dependent Hamiltonian using a unitary transformation method which $\Lambda(t)$ type is used to engineer a cascade Ξ scheme interaction between the vibrational phonons and trapped three-level ion. Quantum entanglement is characterized by comparing concurrence and negativity of the time-dependent ionic-phononic system. In this quantum system, we obtain that the amount of concurrence can be tuned between 0 and 0.99 while the amount of negativity changes between 0 and 0.49.

1. Introduction

The importance of quantum entanglement has particularly increased during the last thirty years. Doubtless, quantum entanglement is a distinct quantum mechanical correlation between subsystem. Measurements on an entangled state cannot be determined by classical correlations. Entangled state has been shown by different physical setups such as photon [1]. Nonclassical correlations have been explained in the EPR paradox [2], which are related to the Bell theorem [3]. Abdel-Aty focuses on the quantum entanglement of the electronic and vibrational modes of trapped ion [4]. Thus, the quantum entanglement may be modelled physically by the states of a single and three-level trapped ions [5]. It plays in quantum optics, the physics of multilevel systems can be helpful in developing quantum information theory [6]. Furthermore, entangled pure states can be produced by lasers and trapped three-level ions [7, 8]. Qudit entangled states in second order corrections can help to answer some important questions about quantum linearity [9]. Dermez et. al. have worked out the quantum correlations of a full trapped ion interacting with two stationary laser beams [10] and the three-different pure-state entanglement in the time-dependent ionic-phononic system [11]. It considered quantum dynamics of a three-level trapped ion using the quantum entropy of the system with time-dependent Rabi frequency [12].

We propose a scheme to generate quantum entanglement among time-dependent ionic-phononic system. In addition to, we determine the degree of entanglement using the concurrence and negativity which is extended with respect to weak excitation regime [13].

2. Dynamics of time-dependent ionic-phononic system

We assume the lower levels of the three-level ion, r and g are degenerate and the beams are of different polarizations so that each beam can drive only one transition. There are three crucial time-dependent parameters: Rabi frequency $\Omega(t)$, harmonic trap frequency $\nu(t)$, and Lamb-Dicke (LD) parameter $\eta(t)$. The time-dependent Rabi frequencies of the corresponding transitions $e-g$ and $e-r$, where e represents the excited level of the ion, are also considered to be the same and denoted by $\omega_1 = \omega_2$ (see Fig. 1). Two time-dependent laser beams are characterized as their frequencies ω_1 and ω_2. The parameters obtained for Figure 1 are $\omega = \omega_1 = \omega_2$ and $\Omega(t) = \Omega_1(t) = \Omega_2(t)$. The annihilation ($a$) and creation ($a^\dagger$) operators of the vibrational phonons such that $a = (m\nu(t)/2\hbar)^{1/2}(x + ip/m\nu(t))$, where $\nu(t)$ is the harmonic trap frequency, x is the direction along the trap axis, m is the mass of the ion, and p is the momentum of the center of mass of the ion.

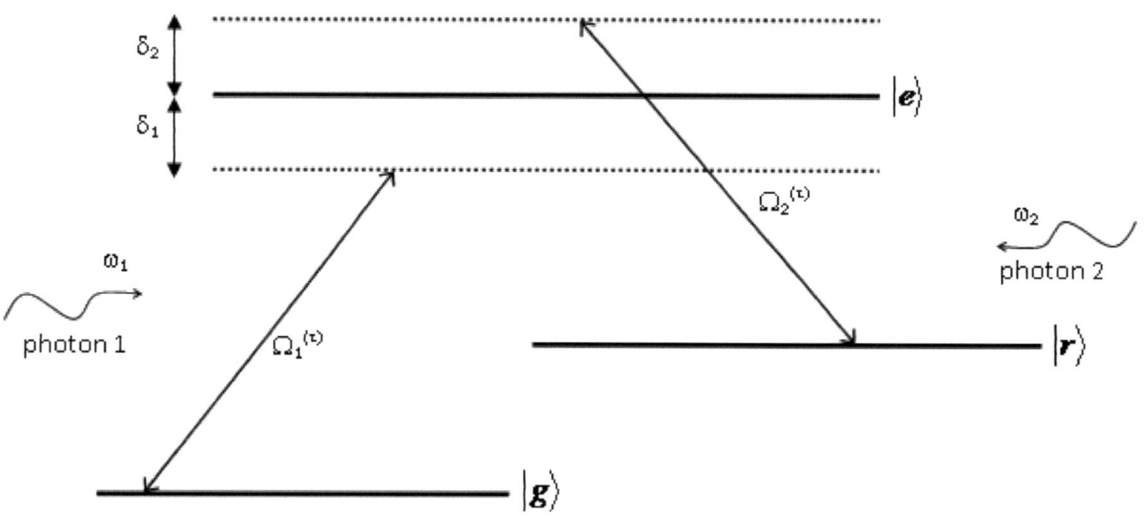

Figure 1. The graphic of time dependent ionic-phononic system. It is shown that excited level (e), ground lever (g) and the Raman level (r) become the intermediate and upper levels of the cascade, respectively. The other parameters are taken to be $\Omega(t) = \Omega_1(t) = \Omega_2(t)$, $\omega = \omega_1 = \omega_2$, $\delta = \delta_1 = \delta_2$, $\omega_1 = \omega_{eg} - \delta_1$,and $\omega_2 = \omega_{eg} - \delta_2$.

Our system has been developed for studying macroscopic quantum superposition states, or so called Schrodinger cat states, of the vibrational phonons [13]. Λ scheme of optical transitions is described by a total Hamiltonian. The total Hamiltonian for the time-dependent ionic-phononic quantum system is $H(t) = H_{\text{ion}}(t) + H_1(t) + H_2(t)$. The trapped ion Hamiltonian and the interacting Hamiltonians are given as ($\hbar = 1$)

$$H_{\text{ion}}(t) = \frac{p^2}{2m} + \frac{1}{2}m\nu^2(t)x_{\text{ion}}^2 + \omega_{eg}|e\rangle\langle e| \tag{1}$$

$$H_1(t) = \frac{\Omega_1(t)}{2}\mathrm{e}^{i(k_1 x_{\text{ion}} - \omega_1 t)}|e\rangle\langle g| + h.c. \tag{2}$$

$$H_2(t) = \frac{\Omega_2(t)}{2}\mathrm{e}^{i(-k_2 x_{\text{ion}} - \omega_2 t)}|e\rangle\langle r| + h.c. \tag{3}$$

where both wavelength of laser beams $\lambda = \lambda_1 = \lambda_2$, are slightly red detuned from the $e - g$ transition angular frequency of the ion ω_{eg} by the same amount $\delta = \nu(t)\eta^2(t)$, so that $\omega = \omega_{eg} - \delta$. $\eta(t) = \frac{k}{\sqrt{2m\nu(t)}}$ is the time-dependent LD parameter and $k = \frac{2\pi}{\lambda}$. Additionally, the weak excitation regime is received to satisfy $\nu(t) = \frac{\Omega(t)}{2}$. It used growing time-dependent Rabi frequency $\Omega(t) = \sinh(\omega_{eg}t)$ [4]. In the weak excitation regime, above optical Λ scheme has been shown to be equivalent to a cascade Ξ scheme for the vibrational phonon transitions, under a unitary transformation, characterized by the transformation matrix U [13]. The time evolution of an initial state $\psi(0)$ is written as:

$$|\psi(t)\rangle = U_0^{\dagger}Ue^{-\int i\tilde{H}_0(t)dt}K(t)U^{\dagger}|\psi(0)\rangle, \tag{4}$$

where $K(t)$ is the well-known the propagator for the cascade Hamiltonian, $\hbar = 1$, and $\exp[-\int i\tilde{H}_0(t)dt]$ are the interaction picture transformation. A given state of the vibrational phonons and the ion evolves in the Ξ configuration according to the propagator $K(t)$ in the interaction picture [13]. Initial state for the ionic-phononic system becomes:

$$|\psi(0)\rangle = \frac{1}{\sqrt{2}}[(|g\rangle - |r\rangle) \otimes (|0\rangle + \alpha|1\rangle)]. \tag{5}$$

With acting the propagator $K(t)$ to the initial state of the cascade [i.e., $U^{\dagger}|\psi(0)\rangle$], we obtain the state $|\psi_K(t)\rangle = \sum_{\text{ip}}(A_{\text{ip}}(t)\,|\text{i},\text{p}\rangle)$ where i stands for the electronic state (e, r, g) and p stands for the vibrational quantum number $(0, 1, 2, 3, 4)$. The Hilbert space $\mathbf{H} = \mathbf{H}_{\text{i}} \otimes \mathbf{H}_{\text{p}}$ of dimension is 15. Thus, we need the fifteen probability amplitudes. Two version of the new final state can be written as

$$|\psi(t)\rangle = \sum_{\text{ip}} A_{\text{ip}}(t)\,|\text{i},\text{p}\rangle, \tag{6}$$

$$\begin{aligned}|\psi(t)\rangle = {} & A_{e0}(t) + A_{e1}(t) + A_{e2}(t) + A_{e3}(t) + A_{e4}(t) + \\ & A_{r0}(t) + A_{r1}(t) + A_{r2}(t) + A_{r3}(t) + A_{r4}(t) + \\ & A_{g0}(t) + A_{g1}(t) + A_{g2}(t) + A_{g3}(t) + A_{g4}(t).\end{aligned} \tag{7}$$

With the fifteen probability amplitudes, we have calculated for Eq. (8) of concurrence and Eq. (11) of negativity in the next section.

3. Generalized concurrence and negativity

Quantum entanglement can be described by comparing concurrence and negativity of time-dependent ionic-phononic system. The three-level ion subsystem are associated with a three-dimensional Hilbert space $\mathbf{H}_{\mathrm{i}}$ of a qutrit spanned by the basis states $\mathbf{H}_{\mathrm{i}}$, while the vibrational phonon subsystem is described by a five-dimensional Hilbert space $\mathbf{H}_{\mathrm{p}}$. Therefore, the initial state in the previous section evolves in the Hilbert space $(C^3 \otimes C^5 = C^{15})$. Concurrence for the ionic-phononic system $(\mathrm{i}-\mathrm{p})$ is written as [14]

$$C = \sqrt{2(Tr(\rho_{\mathrm{i-p}}) - Tr(\rho^2_{\mathrm{ion}}))}. \tag{8}$$

The full density matrix $\rho_{\mathrm{i-p}}$ can be represented by the basis of $|\mathrm{i},\mathrm{p}\rangle$. By tracing over the phononic variable, we get a 3×3 square reduced density matrix ρ_{ion} such that

$$\rho_{\mathrm{ion}} = \mathrm{Tr}_{phonon}(\rho_{\mathrm{i-p}}) = \begin{pmatrix} Tr|e\rangle\langle e| & Tr|e\rangle\langle r| & Tr|e\rangle\langle g| \\ Tr|r\rangle\langle e| & Tr|r\rangle\langle r| & Tr|r\rangle\langle g| \\ Tr|g\rangle\langle e| & Tr|g\rangle\langle r| & Tr|g\rangle\langle g| \end{pmatrix}, \tag{9}$$

where $|e\rangle\langle e|$ is a 5×5 square matrix. The full density matrix is $\rho_{\mathrm{i}-p} = |a\rangle\langle a|$ that $|a\rangle\langle a|$ is a 15×15 square matrix. Also, other measurement of entanglement is negativity. According to the Schmidt decomposition theorem, the pure state $|\psi\rangle$ can be written as

$$|\psi\rangle = \sum_j \sqrt{\mu_j}|a_j\rangle|b_j\rangle \tag{10}$$

where $\sqrt{\mu_j}$ $(j=1,\ldots,d)$ are the Schmidt coefficients (SCs) obtained from the three eigenvalues of the reduced density matrix ρ_{ion}. $|a_j\rangle$ and $|b_j\rangle$ are the orthogonal basis in $\mathbf{H}_{\mathrm{i}}$ and $\mathbf{H}_p$ respectively. The negativity of any arbitrary bipartite pure state $|\psi\rangle$ is given by [15]

$$N(|\psi\rangle) = \frac{2}{d-1}\sum_{i<j}\sqrt{\mu_i}\sqrt{\mu_j} = \frac{2}{3-1}(\sqrt{\mu_1}\sqrt{\mu_2} + \sqrt{\mu_1}\sqrt{\mu_3} + \sqrt{\mu_2}\sqrt{\mu_3}), \tag{11}$$

with $d = min(d_1 = 3, d_2 = 5)$, and μ_1, μ_2, μ_3 are three eigenvalues.

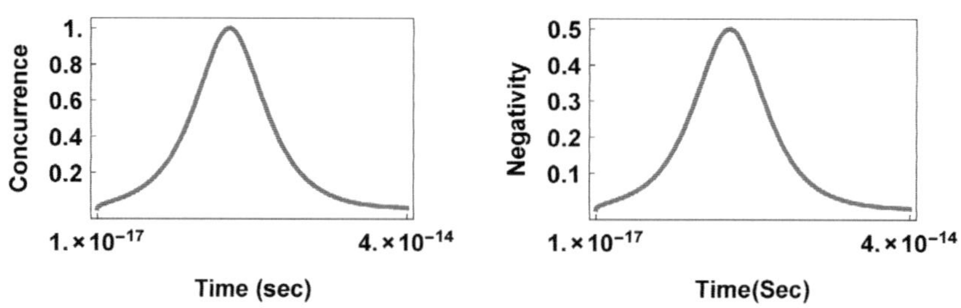

Figure 2. The time evolution of the concurrence and negativity of time-dependent ionic-phononic system in the fifteen dimensional Hilbert space. The time-dependent harmonic trap frequency $\nu(t) = \frac{\sinh(\omega_{eg}t)}{2}$ and Rabi frequency $\Omega(t) = \sinh(\omega_{eg}t)$ are used with $\omega_{eg} = 5\times 10^{14}Hz$. The system is initially prepared in the state $\psi(0) = (1/\sqrt{2})(|g\rangle - |r\rangle)\otimes(|0\rangle + \alpha|1\rangle)$ with $\alpha = 0.01$.

Concurrence and negativity are given in Figure 2 with the help of Equations (8, 11). The time-dependent harmonic trap frequency $\nu(t) = \frac{\sinh(\omega_{eg}t)}{2}$ and Rabi frequency

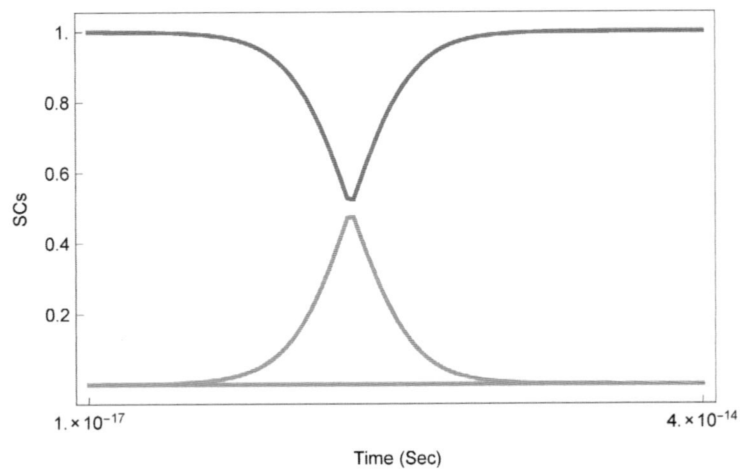

Figure 3. The time dependence of SCs in time-dependent ionic-phononic system using $\hbar = 1$. The other parameters are same as Fig. 2

$\Omega(t) = \sinh(\omega_{eg} t)$ are used with $\omega_{eg} = 5 \times 10^{14} Hz$. Our calculations are presented in Figures 2 − 3, for the ionic-phononic system. Concurrence values (in Fig. 2, left panel) determined to be

$$\begin{aligned} &C(t = 1.0 \times 10^{-17}s = 0.01fs) = 0.002, \\ &C(t = 1.7 \times 10^{-14}s = 17fs) = 0.999, \\ &C(t = 4.0 \times 10^{-14}s = 40fs) = 0.007. \end{aligned} \tag{12}$$

Negativity values (in Fig. 2, right panel) determined to be

$$\begin{aligned} &N(t = 1.0 \times 10^{-17}s = 0.01fs) = 0.001, \\ &N(t = 1.7 \times 10^{-14}s = 17fs) = 0.49, \\ &N(t = 4.0 \times 10^{-14}s = 40fs) = 0.003. \end{aligned} \tag{13}$$

These curves in Fig. 2 both resemble the Gaussian profile . This state (C=0.999, N=0.49) is maximally entangled state at a specific time of $17fs = 1.7 \times 10^{-14}s$, in the middle of these curves (see left and right panels in Fig. 2). For instance, the system is completely disentangled at $t = 0.01fs$ and $t = 40fs$, in the begin and in the end of these curves, where the system reaches the separable state. Figure 3 shows the time dependence of Schmidt coefficients (SCs) which are the three eigenvalues of the reduced density matrix in Eq. (9). Therefore, normalization of the quantum mechanical is equal to $\mu_1 + \mu_2 + \mu_3 = 1$.

4. Conclusion

We have theoretically studied the quantum entanglement process as a result of a time-dependent interaction of a three-level trapped ion with two laser beams in the Λ scheme using a unitary transformation. Measurements of entanglement, concurrence and negativity are too difficult to understand because of complicated structure of the ionic-phononic system and infinite-dimensional Hilbert space. In this circumstances, the system of exact solution seem foggy. Afterwards, we restrict the Hilbert space that the new final state in the Eq. (6) can be easily understood. In this way, the fog has lifted around ionic-phononic system.

Our suggestions exhibit an explicit expression in time-dependent fifteen-dimensional Hilbert space. We have seen the hidden entanglement into structure of the ionic-phononic system. They are clearly calculated concurrence and negativity in order to the ionic-phononic system. Therefore, it is obtained that the new final state $|\psi(t)\rangle$ is superposition of fifteen function. So the new final function can be a robust entangled state.

In our approach, time evolution of concurrence and negativity obtained high amount of entanglement. In the Equations (12, 13), we have found the degree of entanglement created in the system are rather big values, $C(17fs) = 0.999$ and $N(17fs) = 0.49$. Entangled states as demonstrated through first order corrections by time-dependent coupling parameters and optimum entanglement can be beneficial for researchers in quantum computation and quantum optics. We hope such a practical availability of entangled states can be helpful for the ongoing efforts in ion-phonon based quantum information where a high degree of entanglement can be useful for practical quantum computation applications.

5. Acknowledgments

This work has been supported by Afyon Kocatepe University 07-FENED.09 project. I thank to T. Dermez and C. Bulutay for their patience and also I would like to thank Prof. Dr. O. E. Mustecaplioglu and D. Güney for their inspired communications.

6. References

[1] Tittel W and Weihs G 2001, Quantum Inf. Comput. **2,3**
[2] Einstein A , Podolsky B and Rosen N 1935, *Phys. Rev.* **47**, 777.
[3] Bell J S 1964, *Physics* **1**.
[4] Abdel-Aty M 2006, Optics Communications **266** 225-230.
[5] Wang Z-J and Chen F 2007, *Chin. Phys. Lett.* **24** 1570.
[6] Can M A, Cakir O, Klyachko A, Shumousky A 2004, J. Op. B:Quantum Semiclass. Opt. **6** 513.
[7] Dermez R and Müstecaplıoğlu Ö E 2009, *Phys. Scr.* **71** 015304.
[8] Dermez R and Özen S 2010, *Eur. Phys. J. D.* **57** 431.
[9] Dermez R and Özen S 2012, *Phys. Scr.* **85** 055009.
[10] Dermez R and Khalek S A 2011, *J. Russ. Laser Res.* **32** 287.
[11] Dermez R and Khalek S A, Kara K, Deveci B and Gunaydin G N 2012, *J. Russ. Laser Res.* **33** 42.
[12] Dermez R, Deveci B and Güney D Ö 2013, *Eur. Phys. J. D.* **67** 120.
[13] Müstecaplıoğlu Ö E 2003, *Phys. Rev. A* **68** 023811.
[14] Hill S and Wooteers W K 1997, *Phys. Rev. Lett.* **78** 5022.
[15] Lee S, Chi D P, Oh S D and Kim J 2003, *Phys. Rev. A* **68**, 062304.

An investigation for quantum qutrit entanglements through colour wheel compass

Volkan Duran and Azmi Gençten

Department of Physics, Faculty of Arts and Sciences, Ondokuz Mayıs University, Samsun, Turkey

E-mail: volkan.duran8@gmail.com

Abstract: In this study qutrits which have 3-level quantum systems will be used. The investigation of quantum qutrit states was done in a mathematical thought experiment in which most of the quantum phenomena (such as decoherence as well as charge and mass) are neglected. Hence, the model based on the investigation of qutrit states by using colour wheel was given in terms of set theory. Then the results of quantum qutrit entanglements in a hypothetical entangled universe will be analysed through the colour wheel compass.

1. Introduction

Quantum entanglement plays a fundamental role in many applications of quantum information theory, such as quantum teleportation [1]. As the effects of quantum characteristics are taken into account, those classifications and analyses of those quantum entangled states becomes much more complicated [2]. Local quantum operations and classical communication cannot fundamentally change quantum entanglements without destroying it. Therefore, entanglement can be classified by dividing quantum states into equivalence classes [3]. Those studies are also done with different quantum systems. The most common and the simple one is qubits that refer to the state of a 2-level quantum systems. A qutrit is a 3-level quantum system which are labelled as $|0\rangle$, $|1\rangle$, $|2\rangle$ in Hilbert space [4]. A colour wheel or colour circle is an abstract illustrative organization of colour groups around a circle that shows relationships between primary colours, secondary colours, tertiary colours etc. RYB (red–yellow–blue) is the most common example of those kinds of illustrative maps [5].

In this study, first the qutrit states are introduced by using colour wheel in section 2. In section 3, a hypothetical device as a colour wheel compass is explained. Then, the quantum qutrit entanglements are also analysed by using colour wheel compass in section 3.

2. Theory

Set theory can be used to classify many mathematical patterns and mathematical calculations [6]. In this study, set theory is used for the classifications of qutrits. Therefore, set points and corresponding qutrit states are given in Table 1. These classifications can be represented as given in Figure 1. In this Figure, the set points and corresponding qutrit states are placed on the Fano plane. Those classifications can also be transformed into colour wheel in which $|0\rangle$, $|1\rangle$ and $|2\rangle$ can be represented by red, yellow and blue colours, respectively. Superposition states can be represented as the combination of those colours as shown in Figure 2. Negative qutrit states are represented in the outer part of the colour wheel as well as their superposition with positive quantum states. The superpositions of $|0\rangle + |1\rangle + |2\rangle$ and $-|0\rangle - |1\rangle - |2\rangle$ are represented by z coordinate passing through the centre of colour wheel.

Journal of Physics: Conference Series **766** (2016) 012013 doi:10.1088/1742-6596/766/1/012013

Table 1. Set points and corresponding qutrits.

Set points	Corresponding qutrits
100	$\|0\rangle$
010	$\|1\rangle$
001	$\|2\rangle$
011	$\|1\rangle+\|2\rangle$
101	$\|0\rangle+\|2\rangle$
110	$\|0\rangle+\|1\rangle$
111	$\|0\rangle+\|1\rangle+\|2\rangle$

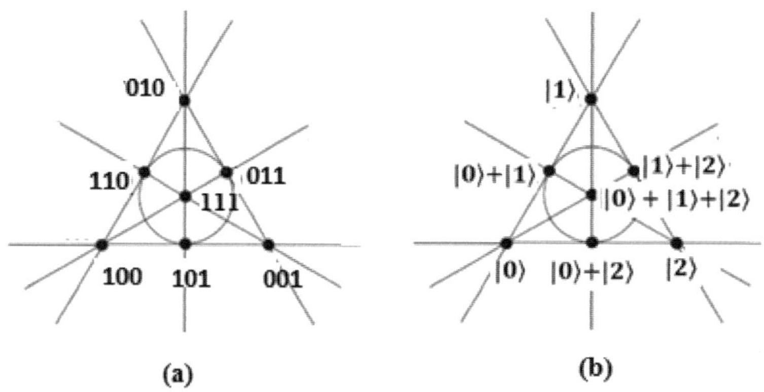

Figure 1. (a) Set points and (b) corresponding qutrit states in the Fano plane.

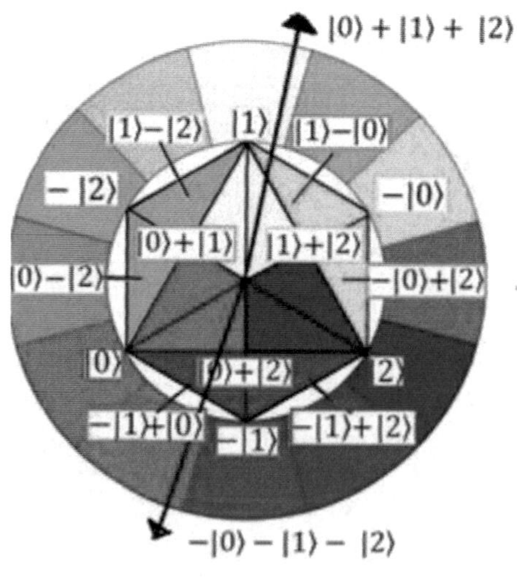

Figure 2. Colour Wheel Compass and corresponding qutrit states.

There exist many quantum qutrit gates. In this study, following ternary selective Hadamard gates are used [7]:

Journal of Physics: Conference Series **766** (2016) 012013 doi:10.1088/1742-6596/766/1/012013

$$H^{01} = \frac{1}{\sqrt{2}}\begin{bmatrix}1 & 1 & 0\\ 1 & -1 & 0\\ 0 & 0 & \sqrt{2}\end{bmatrix},\ H^{02} = \frac{1}{\sqrt{2}}\begin{bmatrix}1 & 0 & 1\\ 0 & \sqrt{2} & 0\\ 1 & 0 & -1\end{bmatrix} \text{ and } H^{12} = \frac{1}{\sqrt{2}}\begin{bmatrix}\sqrt{2} & 0 & 0\\ 0 & 1 & 1\\ 0 & 1 & -1\end{bmatrix} \quad (1)$$

There are many mathematical methods for the analysis of entangled states. One simple method for the analysis is the scalar product of tensors in this respect. Gluing product can be defined as a scalar product in which two vector multiplied based on their main axis or regions. Let i and j =0,1,2 are the indices of quantum qutrit states, the quantum states $|\Psi_i\rangle$ and $|\Psi_j\rangle$ can be defined as following [8]:

$$|\Psi_i\rangle\Omega\,|\Psi_j\rangle = |\Psi_i\rangle\,|\Psi_j\rangle\delta_{ij}\hat{\iota} \quad (2)$$

For example let u and v are two vectors below:

$$|\mathrm{u}\rangle = \alpha_{00}\,|0\rangle\,\hat{0} + \alpha_{11}\,|1\rangle\hat{1} + \alpha_{22}\,|2\rangle\hat{2}\ \ |\mathrm{v}\rangle = \alpha_{00}\,|0\rangle\,\hat{0} + \alpha_{11}\,|1\rangle\hat{1} + \alpha_{22}\,|2\rangle\hat{2} \quad (3)$$

They can be entangled by gluing product as below:

$$|\mathrm{u}\rangle\Omega\,|\mathrm{v}\rangle = \alpha_{00}\,|0\rangle\,|0\rangle + \alpha_{11}\,|1\rangle\,|1\rangle + \alpha_{22}\,|2\rangle\,|2\rangle \quad (4)$$

3. Results and Discussion

The route beginning with the initial qutrit $|0\rangle$ was created by the acting operators H^{01} from the left H^{02} from the right for each sub-cells in the same order represented as in Figure 3. The route restricted by ternary selective Hadamard gates finished with four quantum states A= $|0\rangle$, B= $|0\rangle$+ $|1\rangle$, C= $|0\rangle$+$|2\rangle$ and D= $|0\rangle$+ $|1\rangle$+ $|2\rangle$ in the relation given by A→BC, B→AD, C→DA, D→CB. The route is finished when it reaches the loop points in which no new quantum states emerges and the points in route repeat itself. If the fixed states are represented by the coordinates of colour wheel, the symmetry axis passing through the line combines $|0\rangle$, and $|0\rangle$+ $|1\rangle$+ $|2\rangle$. It can be seen in Figure 4 that the centre of these coordinates is in the direction pointing red colour because there are more quantum states than the opposite symmetrical colour area direction.

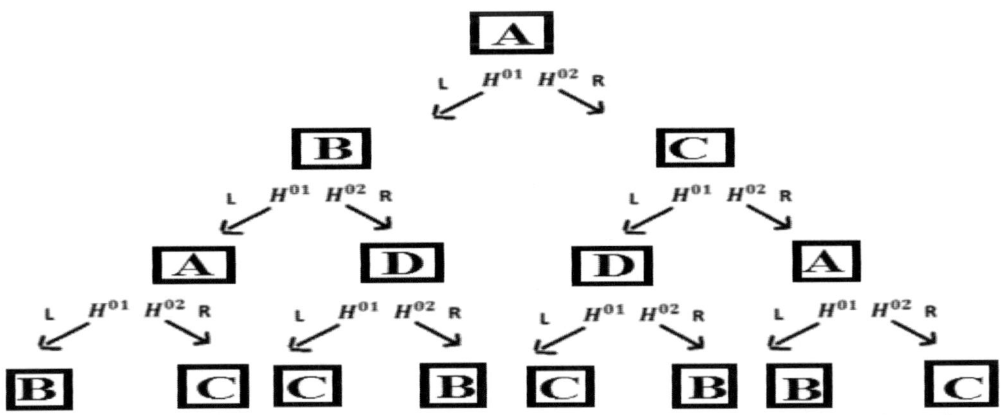

Figure 3.The route beginning with the initial qutrit $|0\rangle$ restricted by H^{01} from the left H^{02} from the right.

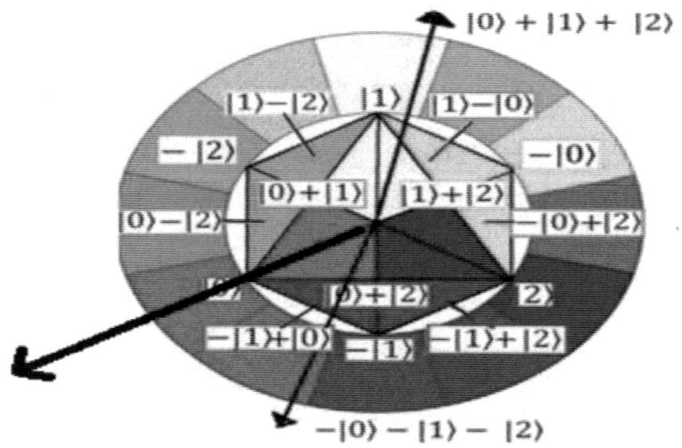

Figure 4. The symmetry axis of the route beginning with |0⟩ and its direction.

The results of the routes beginning with |0⟩, |1⟩, |2⟩ restricted by H^{01} from the left H^{02} from the right can be given in Table 2 and the relationships among variables for each route is given in Table 3.

Table 2. The results of the routes beginning with |0⟩, |1⟩, |2⟩ restricted by H^{01} from the left H^{02}.

The beginning of route	The main points of route
|0⟩	A= |0⟩, B= |0⟩+ |1⟩, C= |0⟩+|2⟩ and D= |0⟩+ |1⟩+ |2⟩
|1⟩	E= |1⟩ F= |0⟩ − |1⟩ G= |0⟩ − |1⟩+|2⟩ H= |1⟩ + |2⟩, K= |0⟩ − |2⟩, L= |0⟩ + |1⟩ − |2⟩, M= |2⟩
|2⟩	M= |2⟩ K= |0⟩ − |2⟩ L= |0⟩ + |1⟩−|2⟩ H= |1⟩ + |2⟩, G= |0⟩ − |1⟩+|2⟩, F= |0⟩ − |1⟩, E= |1⟩

Table 3. The relationships among variables of the routes beginning with |0⟩, |1⟩, |2⟩ restricted by H^{01} from the left H^{02}.

The beginning of route	The relationships among the main points of route
|0⟩	A→BC, B→AD, D→CB,C→DA,
|1⟩	E→FE, F→EG, G→HF, K→LM, H→GL, L→KH, M→MK
|2⟩	M→MK, K→LM, L→KH, G→HF, H→GL, F→EG, E→FE

The results of the routes can be glued in many ways. For example if the quantum states for all routes are glued based on their colour area, no entangled state can be observed. If the results of the qutrits states for the results of the routes beginning with |1⟩, |2⟩ can be glued based on their colour region, the results are given in Table 4. The entangled states can be also achieved by the rotation or changing the colour region in various ways. Then, they might be used for the classification of entangled states. Those states might be formulated based on the colour region area as well as indices and the angle between the colour wheels. If we modify the scalar product based on our colour wheel, that is vectors can be glued based on the colour areas rather their vector positions on the space, the result can be given as in Table 4. This can be regarded as the skeleton of our glued results which might be interpreted as the net result of entanglement of different vectors. Hence entangled systems can be

Journal of Physics: Conference Series **766** (2016) 012013 doi:10.1088/1742-6596/766/1/012013

classified and formulized based the locations of vectors in each colour wheel which might physically the indicator of the extension of a multidimensional hologram.

Table 4. The gluing products of the fixed qutrit points for the routes beginning with $|1\rangle$, $|2\rangle$.

Gluing region	Results
Red Area	$|0\rangle\ \Omega|0\rangle=|0\rangle\ |0\rangle$
Upper Purple Area	$(|0\rangle - |1\rangle+|2\rangle)\Omega(|0\rangle - |1\rangle+|2\rangle) = |0\rangle\ |0\rangle+|1\rangle\ |1\rangle+|2\rangle\ |2\rangle$
Yellow Area	$|1\rangle\ \Omega|1\rangle=|1\rangle\ |1\rangle$
Green Area	$(|1\rangle + |2\rangle)\Omega\ (|1\rangle + |2\rangle) = |1\rangle\ |1\rangle + |2\rangle\ |2\rangle$
Upper Orange Area	$(|0\rangle + |1\rangle-|2\rangle)\Omega\ (|0\rangle + |1\rangle-|2\rangle)=|0\rangle\ |0\rangle+|1\rangle\ |1\rangle+|2\rangle\ |2\rangle$
Pink Area	$(|0\rangle-|1\rangle)\Omega(|0\rangle-|1\rangle) = |0\rangle\ |0\rangle+|1\rangle\ |1\rangle$

4. Conclusion

In this study, first colour wheel compass was defined for qutrits based on set theory. Then the patterns beginning with qutrits $|0\rangle$, $|1\rangle$, $|2\rangle$ restricted by ternary Hadamard gates were analysed. The stable points were determined and located into colour wheel compass. Results for the route beginning with $|0\rangle$, $|1\rangle$, $|2\rangle$ indicates that the direction of the symmetry axis of each routes is the same which is in the direction in red area. And also the fixed points of routes beginning with qutrits $|1\rangle$, $|2\rangle$ are glued with each other and some entangled states for qutrits are obtained. As a suggestion multilayer colour wheel compass can be used as a map for the analysis of quantum systems. Further analysis can be done including the mass, charge and other classical variables as well as the quantum effects.

References

[1] Horodecki R, Horodecki P, Horodecki M and Horodecki K 2009 *Rev. Mod. Phys.* **81** 865
[2] Caves C M and Milburn G J 2000 *Optics Communications* **179** 439
[3] Honda K 2012 preprint arXiv:1210.0617
[4] Das R, Mitra A, Kumar S V and Kumar A 2003 *Int. J. Quantum Inf.* **1** 387
[5] Gaines K S and Curry Z D 2011 *Journal of Family & Consumer Sciences Education* **29,** 46
[6] Moktefi A and Pietatinen A V 2015 J Log. Lang. Inf. **24** 361
[7] Di Y M and Wei H M 2012 preprint arXiv:1105.5485
[8] Duran V and Gençten A 2016 *AIP Conference Proceedings* **1722**, 100001

Construction of two qutrit entanglement by using magnetic resonance selective pulse sequences

Sevcan Çorbacı [1], Mikail Doğuş Karakaş [2] and Azmi Gençten [1]

[1] Department of Physics, Faculty of Arts and Sciences, Ondokuz Mayıs University, Samsun, Turkey
[2]Department of Computer Technology, Vocational School of Design, Amasya University, Amasya, Turkey

E-mail: gencten@omu.edu.tr

Abstract: Quantum entanglement is essential for some applications of quantum information processing such as quantum cryptography, quantum teleportation and superdence coding. A qubit is a two level quantum system and four two−qubit entangled states called Bell states can be easily obtained for two−qubit states. A qutrit is a three level quantum system and Zeeman levels of spin−1 electron or nucleus can be referred as qutrit. For SI (S=1, I=1) spin system there exist nine two−qutrit states. So nine two−qutrit entangled states can be obtained by using the Hadamard and CNOT logic gates. In this study by considering $N^{+}@C_{60}$ molecule as SI (S=1, I=1) spin system, two−qutrit entangled states are also obtained by using the magnetic resonance selective pulse sequences of Hadamard and CNOT logic gates. Then it is shown that these entangled states can be transformed into each other by the suggested transformation operators.

1. Introduction

Quantum information has a great interest in different areas such as physics, mathematics and computer science and engineering as the new technologies are expected in computing, cryptography and communication [1−3]. In these technologies qubits or qutrits (in general qudits) will be used instead of bits [3,4]. Quantum entanglement is essential for some applications of Quantum Information Processing (QIP) such as superdence coding, quantum cryptography and quantum teleportation [5]. In quantum entanglement there is a correlation between the quantum states of entangled particles. For SI (S=1/2, I=1/2) spin system two−qubit entangled states are obtained and they are called Bell states [6].

In this study, first two−qutrit entangled states are introduced in section 2. Then, in section 3, two qutrit entangled states are also obtained by using magnetic resonance selective pulse sequences for SI (S=1, I=1) spin system. In section 3, transformations of two qutrit entangled states into each other are also achieved by using suggested transformation operators.

2. Theory

A qutrit is a three level quantum system. As shown in Table 1, Zeeman levels of spin−1 electron or nucleus are referred as qutrit [4]. The matrix representation of one qutrit Hadamard gate can be written as [7]

$$\mathrm{H} = \frac{1}{\sqrt{3}}\begin{pmatrix} 1 & 1 & 1 \\ 1 & c & c^2 \\ 1 & c^2 & c \end{pmatrix}. \tag{1}$$

Where $c = e^{i\frac{2\pi}{3}}$, $c^2 = e^{-i\frac{2\pi}{3}} = c^*$. By applying this Hadamard gate, superpositions of single qutrit states are found as given in Table 2.

Table 1. Single qutrit states for spin−1.

m_I	qutrit
1	$\|0\rangle=\begin{pmatrix}1\\0\\0\end{pmatrix}$
0	$\|1\rangle=\begin{pmatrix}0\\1\\0\end{pmatrix}$
−1	$\|2\rangle=\begin{pmatrix}0\\0\\1\end{pmatrix}$

Table 2. The superpositions of single qutrit states.

qutrit, $\|a\rangle$	$H\|a\rangle$
$\|0\rangle$	$(\|0\rangle+\|1\rangle+\|2\rangle)/\sqrt{3}$
$\|1\rangle$	$(\|0\rangle+c\|1\rangle+c^2\|2\rangle)/\sqrt{3}$
$\|2\rangle$	$(\|0\rangle+c^2\|1\rangle+c\|2\rangle)/\sqrt{3}$

For two spin−1 system such as SI (S=1, I=1) spin system, nine two−qutrit states of $|00\rangle,|01\rangle,|02\rangle,|10\rangle,|11\rangle,|12\rangle,|20\rangle,|21\rangle$ and $|22\rangle$ are obtained by direct products of single qutrit states. Two qutrit CNOT gates can be found by using the ternary addition of qutrit states:

$$CNOT_a(T)|a,b\rangle=|a,b\oplus a\rangle, \tag{2a}$$

$$CNOT_b(T)|a,b\rangle=|a\oplus b,b\rangle. \tag{2b}$$

These two−qutrit CNOT gates are 9x9 matrices and they can be written in Dirac notation as following:

$$\begin{aligned}CNOT_a(T)=&|00\rangle\langle 00|+|01\rangle\langle 01|+|02\rangle\langle 02|+|10\rangle\langle 12|+|11\rangle\langle 10|\\&+|12\rangle\langle 11|+|20\rangle\langle 21|+|21\rangle\langle 22|+|22\rangle\langle 20|,\end{aligned} \tag{3a}$$

$$\begin{aligned}CNOT_b(T)=&|00\rangle\langle 00|+|01\rangle\langle 21|+|02\rangle\langle 12|+|10\rangle\langle 10|+|11\rangle\langle 01|\\&+|12\rangle\langle 22|+|20\rangle\langle 20|+|21\rangle\langle 11|+|22\rangle\langle 02|.\end{aligned} \tag{3b}$$

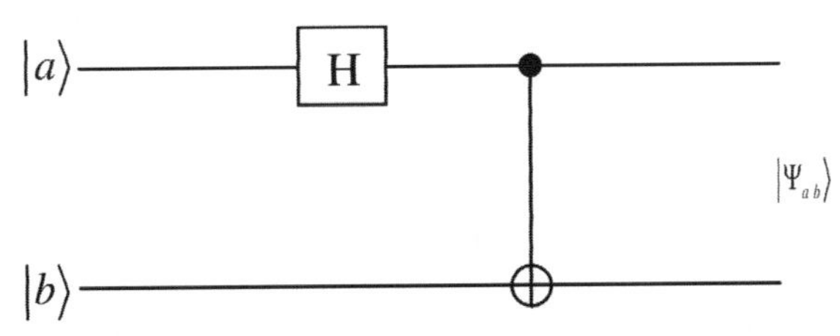

Figure 1. Quantum circuit for generating entangled states.

By using the quantum circuit given in Figure 1, two−qutrit entangled states can be generated. Obtained two−qutrit entangled states are presented in Table 3.

Table 3. Two–qutrit entangled states.

Input qutrit, $\lvert ab\rangle$	Output, $\lvert\Psi_{ab}\rangle$
$\lvert 00\rangle$	$(\lvert 00\rangle+\lvert 11\rangle+\lvert 22\rangle)/\sqrt{3}$
$\lvert 01\rangle$	$(\lvert 01\rangle+\lvert 12\rangle+\lvert 20\rangle)/\sqrt{3}$
$\lvert 02\rangle$	$(\lvert 02\rangle+\lvert 10\rangle+\lvert 21\rangle)/\sqrt{3}$
$\lvert 10\rangle$	$(\lvert 00\rangle+c\lvert 11\rangle+c^2\lvert 22\rangle)/\sqrt{3}$
$\lvert 11\rangle$	$(\lvert 01\rangle+c\lvert 12\rangle+c^2\lvert 20\rangle)/\sqrt{3}$
$\lvert 12\rangle$	$(\lvert 02\rangle+c\lvert 10\rangle+c^2\lvert 21\rangle)/\sqrt{3}$
$\lvert 20\rangle$	$(\lvert 00\rangle+c^2\lvert 11\rangle+c\lvert 22\rangle)/\sqrt{3}$
$\lvert 21\rangle$	$(\lvert 01\rangle+c^2\lvert 12\rangle+c\lvert 20\rangle)/\sqrt{3}$
$\lvert 22\rangle$	$(\lvert 02\rangle+c^2\lvert 10\rangle+c\lvert 21\rangle)/\sqrt{3}$

3. Results and Discussion

For spin−1 electron and nucleus, magnetic quantum numbers and corresponding one qutrit states are given in Figure 2. Selective transitions are also shown in this figure. Selective magnetic resonance pulses can be used to prepare some quantum logic gates. Modified matrix representation of some selective pulse operators for spin−1 is as follows [8]:

$$R_x^{12}(\theta)=\begin{pmatrix}\cos\left(\frac{\theta}{2}\right) & \sin\left(\frac{\theta}{2}\right) & 0\\ \sin\left(\frac{\theta}{2}\right) & \cos\left(\frac{\theta}{2}\right) & 0\\ 0 & 0 & 1\end{pmatrix},\quad R_x^{23}(\theta)=\begin{pmatrix}1 & 0 & 0\\ 0 & e^{i\frac{2\pi}{3}}\cos\left(\frac{\theta}{2}\right) & e^{-i\frac{2\pi}{3}}\sin\left(\frac{\theta}{2}\right)\\ 0 & e^{-i\frac{2\pi}{3}}\sin\left(\frac{\theta}{2}\right) & e^{i\frac{2\pi}{3}}\cos\left(\frac{\theta}{2}\right)\end{pmatrix}. \tag{4}$$

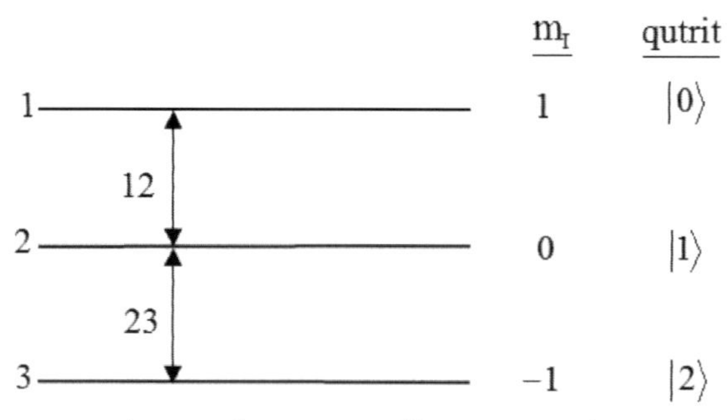

Figure 2. Magnetic quantum numbers and corresponding one qutrit states with selective transition for spin-1.

$N^+@C_{60}$ molecule can be considered as SI (S=1, I=1) spin system [9−10]. In this molecule total electron spin of N^+ in the ground state is 1 and nuclear spin of ^{14}N is also 1 with the abundance of 99,63%. For this spin system total Hamiltonian in an external magnetic field is

$$\hat{H}=\Omega_S\hat{S}_Z-\Omega_I\hat{I}_Z+A\hat{S}_Z\hat{I}_Z. \tag{5}$$

Where $\Omega_S = g\mu_B B$, $\Omega_I = \gamma_I B$. Energies of the spin Hamiltonian and corresponding two-qutrit states for SI (S=1, I=1) spin system are presented in Table 4.

Table 4. Energies of the spin Hamiltonian for SI (S=1, I=1) spin system and their two-qutrit states.

Level Number	$\lvert M_S\rangle$	$\lvert M_I\rangle$	Energy	Two−qutrit state, $\lvert ab\rangle$
$\boxed{1}$	$\lvert 1\rangle$	$\lvert 1\rangle$	$\Omega_S - \Omega_I + A$	$\lvert 00\rangle$
$\boxed{2}$	$\lvert 1\rangle$	$\lvert 0\rangle$	Ω_S	$\lvert 01\rangle$
$\boxed{3}$	$\lvert 1\rangle$	$\lvert -1\rangle$	$\Omega_S + \Omega_I - A$	$\lvert 02\rangle$
$\boxed{4}$	$\lvert 0\rangle$	$\lvert 1\rangle$	$-\Omega_I$	$\lvert 10\rangle$
$\boxed{5}$	$\lvert 0\rangle$	$\lvert 0\rangle$	0	$\lvert 11\rangle$
$\boxed{6}$	$\lvert 0\rangle$	$\lvert -1\rangle$	Ω_I	$\lvert 12\rangle$
$\boxed{7}$	$\lvert -1\rangle$	$\lvert 1\rangle$	$-\Omega_S - \Omega_I - A$	$\lvert 20\rangle$
$\boxed{8}$	$\lvert -1\rangle$	$\lvert 0\rangle$	$-\Omega_S$	$\lvert 21\rangle$
$\boxed{9}$	$\lvert -1\rangle$	$\lvert -1\rangle$	$-\Omega_S + \Omega_I + A$	$\lvert 22\rangle$

Two−qutrit entangled states can be also obtained by using magnetic resonance pulse sequences of Hadamard and CNOT logic gates. For the construction of entanglement for two−qutrit state of $\lvert 00\rangle$, microwave pulse sequence of the Hadamard gate

$$\mathrm{H} = \left(\frac{\pi}{2}\right)_x^{\boxed{1}-\boxed{7}} - \left(70.52^{\circ}\right)_x^{\boxed{1}-\boxed{4}} \tag{6}$$

should be applied. Where the numbers such as $\boxed{1}$, $\boxed{4}$ and $\boxed{7}$ represent the level number as given in Table 4. Matrix representation of selective microwave pulses are as following:

$$\left(70.52^{\circ}\right)_x^{\boxed{1}-\boxed{4}} = R_x^{12}(70.52)\otimes I^{1} + S_E\otimes\left(I_E - I^{1}\right), \tag{7a}$$

$$\left(\frac{\pi}{2}\right)_x^{\boxed{1}-\boxed{7}} = R_x^{13}\left(\frac{\pi}{2}\right)\otimes I^{1} + S_E\otimes\left(I_E - I^{1}\right). \tag{7b}$$

Matrices used in the microwave and radio frequency pulses for spin-1 are given as following:

$$S_E = I_E = \begin{pmatrix} 1 & 0 & 0 \\ 0 & 1 & 0 \\ 0 & 0 & 1 \end{pmatrix}, \quad S^{1} = I^{1} = \begin{pmatrix} 1 & 0 & 0 \\ 0 & 0 & 0 \\ 0 & 0 & 0 \end{pmatrix},$$

$$S^{0} = I^{0} = \begin{pmatrix} 0 & 0 & 0 \\ 0 & 1 & 0 \\ 0 & 0 & 0 \end{pmatrix}, \quad S^{-1} = I^{-1} = \begin{pmatrix} 0 & 0 & 0 \\ 0 & 0 & 0 \\ 0 & 0 & 1 \end{pmatrix}. \tag{8}$$

When the Hadamard pulse sequence is applied to $\lvert 00\rangle$ state,

$$\mathrm{H}\lvert 00\rangle = \left(\lvert 00\rangle + \lvert 10\rangle + \lvert 20\rangle\right)/\sqrt{3} \tag{9}$$

is obtained. Then as a $\mathrm{CNOT_a(T)}$ logic gate, radio frequency pulse sequence

$$CNOT_a(T) = (\pi)_x^{\boxed{7}-\boxed{9}} - (\pi)_x^{\boxed{4}-\boxed{5}} \tag{10}$$

can be applied. Where the matrix representation of selective radio frequency pulses are

$$(\pi)_x^{\boxed{4}-\boxed{5}} = S^0 \otimes R_x^{12}(\pi) + \left(S_E - S^0\right) \otimes I_E, \tag{11a}$$

$$(\pi)_x^{\boxed{7}-\boxed{9}} = S^{-1} \otimes R_x^{13}(\pi) + \left(S_E - S^{-1}\right) \otimes I_E. \tag{11b}$$

After the application of radio frequency pulse sequence of $CNOT_a(T)$ to equation (9), entanglement of $|00\rangle$ state is found:

$$CNOT_a(T)\left(\mathrm{H}|00\rangle\right) = \left(|00\rangle + |11\rangle + |22\rangle\right)/\sqrt{3}. \tag{12}$$

The entanglements of the other two-qutrit states can be also constructed by using suitable Hadamard and $CNOT_a(T)$ pulse sequences. For the construction of two–qutrit entangled states, all suitable Hadamard microwave selective pulse sequences and $CNOT_a(T)$ radio frequency pulse sequences are presented in Table 5. It can be shown that after the application of these pulse sequences to two qutrit states, the same entangled states (given in Table 3) are found.

Table 5. Suitable Hadamard and $CNOT_a(T)$ pulse sequences for the entanglement of two–qutrit states.

Two–qutrit state, $\|ab\rangle$	Suitable H microwave pulse sequence	Suitable $CNOT_a(T)$ radio frequency pulse sequence
$\|00\rangle$	$\left(\frac{\pi}{2}\right)_x^{\boxed{1}-\boxed{7}} - (70.52^\circ)_x^{\boxed{1}-\boxed{4}}$	$(\pi)_x^{\boxed{7}-\boxed{9}} - (\pi)_x^{\boxed{4}-\boxed{5}}$
$\|01\rangle$	$\left(\frac{\pi}{2}\right)_x^{\boxed{2}-\boxed{8}} - (70.52^\circ)_x^{\boxed{2}-\boxed{5}}$	$(\pi)_x^{\boxed{7}-\boxed{8}} - (\pi)_x^{\boxed{5}-\boxed{6}}$
$\|02\rangle$	$\left(\frac{\pi}{2}\right)_x^{\boxed{3}-\boxed{9}} - (70.52^\circ)_x^{\boxed{3}-\boxed{6}}$	$(\pi)_x^{\boxed{8}-\boxed{9}} - (\pi)_x^{\boxed{4}-\boxed{6}}$
$\|10\rangle$	$\left(\frac{\pi}{2}\right)_x^{\boxed{4}-\boxed{7}} - (70.52^\circ)_x^{\boxed{1}-\boxed{4}}$	$(\pi)_x^{\boxed{7}-\boxed{9}} - (\pi)_x^{\boxed{4}-\boxed{5}}$
$\|11\rangle$	$\left(\frac{\pi}{2}\right)_x^{\boxed{5}-\boxed{8}} - (70.52^\circ)_x^{\boxed{2}-\boxed{5}}$	$(\pi)_x^{\boxed{8}-\boxed{9}} - (\pi)_x^{\boxed{5}-\boxed{6}}$
$\|12\rangle$	$\left(\frac{\pi}{2}\right)_x^{\boxed{6}-\boxed{9}} - (70.52^\circ)_x^{\boxed{3}-\boxed{6}}$	$(\pi)_x^{\boxed{8}-\boxed{9}} - (\pi)_x^{\boxed{4}-\boxed{6}}$
$\|20\rangle$	$\left(\frac{\pi}{2}\right)_x^{\boxed{4}-\boxed{7}} - (70.52^\circ)_x^{\boxed{4}-\boxed{7}}$	$(\pi)_x^{\boxed{7}-\boxed{9}} - (\pi)_x^{\boxed{4}-\boxed{5}}$
$\|21\rangle$	$\left(\frac{\pi}{2}\right)_x^{\boxed{5}-\boxed{8}} - (70.52^\circ)_x^{\boxed{2}-\boxed{8}}$	$(\pi)_x^{\boxed{7}-\boxed{8}} - (\pi)_x^{\boxed{5}-\boxed{6}}$
$\|22\rangle$	$\left(\frac{\pi}{2}\right)_x^{\boxed{6}-\boxed{9}} - (70.52^\circ)_x^{\boxed{3}-\boxed{9}}$	$(\pi)_x^{\boxed{8}-\boxed{9}} - (\pi)_x^{\boxed{4}-\boxed{6}}$

Weyl operators (also called d^2 operators) can be obtained for dimension d as following [11]:

$$U_{nm} = \sum_{k=0}^{d-1} e^{i\frac{2\pi}{d}km} |k\rangle\langle (k+m) \bmod d|. \tag{13}$$

Where n, m=0, 1,…, d−1. For qutrits (d=3) these operators can be generated as given in Table 6. These operators can be called as Pauli bases for qutrits. In order to transform the two–qutrit entangled states between each other, transformation operators can be defined as $U_{nm} \otimes I_E$. For example $U_{01} \otimes I_E$

transformation operator can be used to transform the $|\psi_{01}\rangle$ entangled state to $|\psi_{02}\rangle$ entangled state. Applications of all the transformation operators to all two–qutrit entangled states can be also achieved.

Table 6. Weyl operators for qutrits [11]. Where $c = e^{i\frac{2\pi}{3}}$, $c^2 = e^{-i\frac{2\pi}{3}} = c^*$.

$U_{00} = \begin{pmatrix} 1 & 0 & 0 \\ 0 & 1 & 0 \\ 0 & 0 & 1 \end{pmatrix}$	$U_{01} = \begin{pmatrix} 0 & 1 & 0 \\ 0 & 0 & 1 \\ 1 & 0 & 0 \end{pmatrix}$	$U_{02} = \begin{pmatrix} 0 & 0 & 1 \\ 1 & 0 & 0 \\ 0 & 1 & 0 \end{pmatrix}$
$U_{10} = \begin{pmatrix} 1 & 0 & 0 \\ 0 & c & 0 \\ 0 & 0 & c^2 \end{pmatrix}$	$U_{11} = \begin{pmatrix} 0 & 1 & 0 \\ 0 & 0 & c \\ c^2 & 0 & 0 \end{pmatrix}$	$U_{12} = \begin{pmatrix} 0 & 0 & 1 \\ c & 0 & 0 \\ 0 & c^2 & 0 \end{pmatrix}$
$U_{20} = \begin{pmatrix} 1 & 0 & 0 \\ 0 & c^2 & 0 \\ 0 & 0 & c \end{pmatrix}$	$U_{21} = \begin{pmatrix} 0 & 1 & 0 \\ 0 & 0 & c^2 \\ c & 0 & 0 \end{pmatrix}$	$U_{22} = \begin{pmatrix} 0 & 0 & 1 \\ c^2 & 0 & 0 \\ 0 & c & 0 \end{pmatrix}$

4. Conclusion

For SI (S=1, I=1) spin system, there exist 9 two–qutrit states. By using the matrix representation of one qutrit Hadamard and two–qutrit CNOT(T) gates, two–qutrit entangled states are acquired for this spin system. Two–qutrit entangled states are also obtained by using the magnetic resonance selective pulse sequences of Hadamard and CNOT(T) logic gates. In order to transform two–qutrit entangled states between each other, transformation operators are suggested by using the Weyl operators. Then, the applications of these operators are presented.

References

[1] Nielsen M A and Chuang I L 2001 *Quantum Computation and Quantum Information* (Cambridge: Cambridge University Press)
[2] DiVincenzo D P 1995 Quantum computation *Science* **270** 255
[3] Lloyd S 1995 Quantum-mechanical computers *Scientific American* **273** 140
[4] Das R, Mitra A, Kumar S V and Kumar A 2003 Quantum information processing by NMR: preparation of pseudopure states and implementation of unitary operations in a single–qutrit system *Int. J. Quantum Inf.* **1** 387
[5] Horodecki R, Horodecki P, Horodecki M and Horodecki K 2009 Quantum entanglement *Rev. Mod. Phys.* **81** 865
[6] Bell J S 1964 On the Einstein–Podolsky–Rosen Paradox *Physics* **1** 195
[7] Anwar H, Campbell E T and Browne D E 2012 Qutrit magic state distillation *New Journal of Physics* **14** 063006
[8] Shauro V 2015 Exact solutions for time–optimal control of spin I=1 by NMR *Quantum Inf. Process.* **14** 2345
[9] Harneit W 2002 Fullerene-based electron-spin quantum computer *Phys. Rev.* A **65** 032322.
[10] Oliveira I S, Bonagamba T J, Sarthour R S, Freitas J C C and deAzevede E R 2007 *NMR Quantum Information Processing* (Amsterdam:Elsevier)
[11] Bertlmann R A and Krammer P 2008 Bloch Vectors for qudits *J. Phys. A: Math. Theor.* **41** 235303

Quantum calculus of classical vortex images, integrable models and quantum states

Oktay K Pashaev

Department of Mathematics, Izmir Institute of Technology, Gulbahce Campus, Urla-Izmir, 35430, Turkey

E-mail: oktaypashaev@iyte.edu.tr

Abstract. From two circle theorem described in terms of q-periodic functions, in the limit $q \to 1$ we have derived the strip theorem and the stream function for N vortex problem. For regular N-vortex polygon we find compact expression for the velocity of uniform rotation and show that it represents a nonlinear oscillator. We describe q-dispersive extensions of the linear and nonlinear Schrödinger equations, as well as the q-semiclassical expansions in terms of Bernoulli and Euler polynomials. Different kind of q-analytic functions are introduced, including the pq-analytic and the golden analytic functions.

1. Introduction

The quantum or the so called q-calculus rises from classical works of Euler, Gauss, Fermat etc., but only recently, after discovering quantum integrable models and quantum groups, the subject has attracted much attention. In present paper we are going to study several applications of this calculus to problems of classical hydrodynamics, quantum theory and integrable systems.

1.1. Fermat partition

The Fermat partition plays the central role in q-calculus. Introduced first for calculation of integrals, it divides an interval in geometric progression and has many applications. For guitar frets it gives the scales $L, Lq, Lq^2, ..., Lq^{12}$, where $q = e^{-\ln 2/12} = 0.94387645$. Another example is given by intervals for vertical distance traveled by a bouncing ball if the height of each rebound is reduced by the factor $q < 1$. The total distance is $h+2hq+2hq^2+... = -h+2h(1+q+q^2+...) = -h + 2h[\infty]_q = h(1+q)/(1-q)$.

In general, any real analytic function $f(q) = a_0 + a_1 q + a_2 q^2 + ...$, convergent for $0 < q < 1$, has geometrical meaning as an area on Fermat partition $(1, q, q^2, ...)$,

$$A = a_0(1-q) + a_1(q-q^2) + a_2(q^2-q^3) + ... = (1-q)(a_0 + a_1 q + a_2 q^2 + ...) = (1-q)f(q).$$

In particular case, when $a_n = F(q^n)$, this area becomes just the Jackson integral

$$A = (1-q)\sum_{n=0}^{\infty} F(q^n)q^n = \int_0^1 F(x)d_q x.$$

One more example is given by problem of point vortex in concentric annular domain, where the set of vortex images is distributed according to the Fermat partition.

2. Hydrodynamic images

As a first application we consider the hydrodynamic flow in annular domain and the two circle theorem [1]. For incompressible and irrotational planar flow in annular domain $r_1 < |z| < r_2$, between two concentric circles $C_1 : |z| = r_1^2$ and $C_2 : |z| = r_2^2$, the complex potential $F(z)$ is analytic function, q-periodically extended to the whole complex plane:$F(qz) = F(z)$ or $D_q^z F(z) = 0$. Explicit form of this function is given in terms of $f(z)$, as a complex potential of flow in the plain without boundaries:

$$F(z) = f_q(z) + \overline{f_q(z^*)} = \sum_{n=-\infty}^{\infty} f(q^n z) + \sum_{n=-\infty}^{\infty} \bar{f}\left(q^n \frac{r_1^2}{z}\right), \tag{1}$$

where $z^* = r_1^2/\bar{z}$ is inversion of the point z, $f_q(z) \equiv \sum_{n=-\infty}^{\infty} f(q^n z)$ - q-periodic extension of the flow to even annular, and $\overline{f_q(z^*)} = \bar{f}_q(r_1^2/z) = \sum_{n=-\infty}^{\infty} \bar{f}\left(q^n \frac{r_1^2}{z}\right)$ - q-periodic extension to the odd ones. This expression shows that restriction of the flow from whole plane to annular domain is equivalent to the q-periodic flow in the plane. It happens due to the boundary conditions $\Im F(z)|_{C_{1,2}} = 0$, identifying imaginary parts of the analytic function on the circles and as follows, identifying values of this function on the boundaries (up to irrelevant constant). In addition to q-periodicity, function $F(z)$ must be Q- periodic, with $Q = e^{2\pi i}$, under rotations on 2π around the origin. The real and imaginary parts of this function in polar coordinates $F(z) = \Phi(r,\theta) + i\Psi(r,\theta)$ are the velocity potential and the stream function respectively. These functions are q-periodic in r, and periodic in θ: $\Phi(qr,\theta) = \Phi(r,\theta)$, $\Phi(r,\theta+2\pi) = \Phi(r,\theta)$; $\Psi(qr,\theta) = \Psi(r,\theta)$, $\Psi(r,\theta+2\pi) = \Psi(r,\theta)$. In explicit form

$$\Phi(r,\theta) = \sum_{n=-\infty}^{\infty} \left(\varphi(q^n r,\theta) + \varphi(q^n \frac{r_1^2}{r},\theta)\right), \quad \Psi(r,\theta) = \sum_{n=-\infty}^{\infty} \left(\psi(q^n r,\theta) - \psi(q^n \frac{r_1^2}{r},\theta)\right), \tag{2}$$

where $f(z) = \varphi(r,\theta) + i\psi(r,\theta)$ is the flow in plane without boundaries, periodic in θ: $\varphi(r,\theta+2\pi) = \varphi(r,\theta)$, $\psi(r,\theta+2\pi) = \psi(r,\theta)$. From the above formulas, the stream function vanishes at boundary circles: $\Psi(r_1,\theta) = 0$ and $\Psi(r_2,\theta) = 0$. Moreover, if $\Gamma_k : |z| = r_1 q^k$, and $\gamma_k : |z| = r_2 q^k$, $k = 0, \pm1, \pm2, ...$ are the boundary circles images, then $\Psi|_{\Gamma_k} = 0$ and $\Psi|_{\gamma_k} = 0$ as well.

Several extensions of the two circle theorem, requiring different type of q-calculus were considered in [2] . The wedge theorem for the domain with angle $\alpha = 2\pi/N$ was given with the base $Q^N = 1$. Geometry of the double circular wedge theorem requires calculus with two bases, $q = r_2^2/r_1^2$ and $Q^N = 1$. By using these theorems, the Kummer kaleidoscope of vortices was described as well. In the limit $q \to \infty$ our two circles theorem reduces to the Milne-Thomson one circle theorem. However, more common in q-calculus limit $q \to 1$ was not studied. It happens due to the fact that for finite r_1 and r_2, at this limit the annular domain vanishes and problem has no meaning. But, as we describe below, for infinite radiuses $r_1 \to \infty$, $r_2 \to \infty$, the finite limit $q = r_2^2/r_1^2 \to 1$ of the domain exists and it is the strip domain.

2.1. *The strip theorem*

For incompressible and irrotational flow in the strip domain S: $\{z = x + iy; -h/2 < y < h/2\}$, the complex potential is

$$F(z) = \sum_{n=-\infty}^{\infty} f(z + (2n)ih) + \sum_{n=-\infty}^{\infty} \bar{f}(z + (2n-1)ih), \tag{3}$$

where $f(z)$ is flow in the whole plane. The proof is straightforward by checking the boundary conditions: $\Im F(z)|_{z=x+i\frac{h}{2}} = 0$ and $\Im F(z)|_{z=x-i\frac{h}{2}} = 0$. The flow (3) satisfies periodicity and the combined periodicity conditions, respectively

$$F(z+2ih)=F(z),\quad \bar{F}(z+ih)=F(z). \tag{4}$$

2.1.1. Vortex in strip For the point vortex $f(z)=\frac{\Gamma}{2\pi i}\ln(z-z_0)$ in the strip, we have

$$F(z)=\frac{-i\Gamma}{2\pi}\left(\ln(z-z_0)+\sum_{n=1}^{\infty}\ln\frac{(z-z_0)^2+(2n)^2h^2}{(z-\bar{z}_0)^2+(2n-1)^2h^2}\right). \tag{5}$$

By using infinite product representation

$$\sinh z = z\prod_{n=1}^{\infty}\left(1+\frac{z^2}{\pi^2 n^2}\right)$$

it can be rewritten as

$$F(z)=\frac{-i\Gamma}{2\pi}\left(\ln\prod_{n=1}^{\infty}\frac{4h}{\pi}\left(\frac{2n}{2n-1}\right)^2+\ln\frac{\sinh\frac{\pi}{2h}(z-z_0)\sinh\frac{\pi}{2h}(z-\bar{z}_0)}{\sinh\frac{\pi}{h}(z-\bar{z}_0)}\right)$$

and due to $\sinh\frac{\pi}{h}(z-\bar{z}_0)=2\sinh\frac{\pi}{2h}(z-\bar{z}_0)\cosh\frac{\pi}{2h}(z-\bar{z}_0)$, up to irrelevant constant we get

$$F(z)=\frac{-i\Gamma}{2\pi}\ln\frac{\sinh\frac{\pi}{2h}(z-z_0)}{\cosh\frac{\pi}{2h}(z-\bar{z}_0)}. \tag{6}$$

For N vortices $\Gamma_1,...,\Gamma_N$ at positions $z_1,...,z_N$ this gives the complex potential

$$F(z)=\sum_{k=1}^{N}\frac{-i\Gamma_k}{2\pi}\ln\frac{\sinh\frac{\pi}{2h}(z-z_k)}{\cosh\frac{\pi}{2h}(z-\bar{z}_k)}, \tag{7}$$

and the stream function $\Psi=\Im F(z)$,

$$\Psi(x,y)=\sum_{k=1}^{N}\frac{-\Gamma_k}{4\pi}\ln\frac{\sinh^2\frac{\pi}{2h}(x-x_k)+\sin^2\frac{\pi}{2h}(y-y_k)}{\sinh^2\frac{\pi}{2h}(x-x_k)+\cos^2\frac{\pi}{2h}(y+y_k)}. \tag{8}$$

For $N=2$ this function was obtained in [3], though formula in the paper has some typos.

2.1.2. Gamma function To express (5) in terms of the Gamma function we split

$$F(z)=\frac{-i\Gamma}{2\pi}\left(\ln(z-z_0)+\ln\prod_{n=1}^{\infty}\frac{(z-z_0)+2nih}{(z-\bar{z}_0)+(2n-1)ih}\frac{(z-z_0)-2nih}{(z-\bar{z}_0)-(2n-1)ih}\right).$$

Rewriting the second term as

$$C_\infty+\ln\prod_{n=1}^{\infty}\frac{(1+\frac{z-z_0}{2ihn})(1+\frac{z-\bar{z}_0}{2ihn})}{(1+\frac{z-\bar{z}_0}{ihn})}\frac{(1+\frac{z_0-z}{2ihn})(1+\frac{\bar{z}_0-z}{2ihn})}{(1+\frac{\bar{z}_0-z}{ihn})} \tag{9}$$

we find

$$F(z)=\frac{-i\Gamma}{2\pi}\left(\ln(z-z_0)+\ln\frac{\Gamma(1+\frac{z-\bar{z}_0}{ih})\Gamma(1-\frac{z-\bar{z}_0}{ih})}{\Gamma(1+\frac{z-z_0}{2ih})\Gamma(1-\frac{z-z_0}{2ih})\Gamma(1+\frac{z-\bar{z}_0}{2ih})\Gamma(1-\frac{z-\bar{z}_0}{2ih})}\right),$$

where we have used definition of the Gamma function as an infinite product

$$\frac{1}{\Gamma(z)}=ze^{\gamma z}\prod_{n=1}^{\infty}\left(1+\frac{z}{n}\right)e^{-\frac{z}{n}}.$$

2.2. Strip theorem as the limit $q \to 1$

Now we are going to show that our two circle theorem (1) in the limit $q \to 1$ reduces to the strip theorem (3). Let $r_1 \to \infty$, $r_2 \to \infty$, so that $r_2 = r_1 + h$. Then $q = \frac{r_2^2}{r_1^2} = (1 + \frac{h}{r_1})^2$ and $q^n = (1 + \frac{h}{r_1})^{2n} = 1 + 2n\frac{h}{r_1} + O(\frac{h^2}{r_1^2})$. To get the strip domain we need to shift our coordinate system z with origin z=0, to the new one z' with origin at $i(r_1 + \frac{h}{2})$. These coordinates are related by equation $z = z' + i(r_1 + \frac{h}{2})$. Then $q^n z = (1 + 2n\frac{h}{r_1} + O(\frac{h^2}{r_1^2}))(i(r_1 + \frac{h}{2}) + z') = i(r_1 + \frac{h}{2}) + z' + 2nhi + O(\frac{h}{r_1})$. For analytic in plane function $f(z)$ we have Taylor expansion $f(z) = f(i(r_1 + \frac{h}{2}) + z') = \sum_{n=0}^{\infty} \frac{f^{(n)}(i(r_1+\frac{h}{2}))}{n!}(z')^n \equiv g(z')$, so that

$$f(q^n z) = f((1 + \frac{h}{r_1})^n(i(r_1 + \frac{h}{2}) + z')) = f(i(r_1 + \frac{h}{2}) + z' + 2nih + O(\frac{h}{r_1})) = g(z' + 2nih).$$

This gives the first part of formula (1): $f_q(z) = \sum_{-\infty}^{\infty} g(z' + 2nih)$. To get the second part we expand $r_1^2/z = r_1^2/(z' + i(r_1 + \frac{h}{2})) = r_1/i(1 + z'/ir_1 + h/2r_1) = -i(r_1 + \frac{h}{2}) + ih + z' + O(h/r_1)$, and $q^n r_1^2/z = -i(r_1 + \frac{h}{2}) + ih + z' + O(h/r_1)$. Then $\bar{f}(r_1^2/z) = \bar{g}(z' + ih)$ and $\bar{f}(q^n r_1^2/z) = \bar{g}(z' - (2n-1)ih)$. For the second part of (1) it gives: $\bar{f}_q(r_1^2/z) = \sum_{n=-\infty}^{\infty} \bar{f}(q^n r_1^2/z) = \sum_{n=-\infty}^{\infty} \bar{g}(z' - (2n-1)ih)$. Combining both terms together we find that in the limit $q \to 1$ the two circle theorem (1) reduces to the strip theorem (3), $(n \to -n,\ n \to n-1)$,

$$G(z') = \sum_{-\infty}^{\infty} g(z' + 2nih) + \sum_{n=-\infty}^{\infty} \bar{g}(z' + (2n-1)ih).$$

3. N vortex polygon as nonlinear oscillator

From two circle theorem it is easy to get equations of motion for N point vortices [4]. When vortices with equal strength Γ are located at vertices of the regular polygon in annular domain, the problem admits exact solution $z_k(t) = r\exp(i\omega t + i2\pi k/N)$, $k = 1, ..., N$, with uniform rotation frequency

$$\omega(r) = \frac{\Gamma}{2\pi r^2}\frac{N-1}{2} + \frac{\Gamma}{2\pi r^2}\frac{1}{q-1}\sum_{j=1}^{N}\left[Ln_q\left(1 - \frac{r_2^2}{r^2}e^{i\frac{2\pi}{N}j}\right) - Ln_q\left(1 - \frac{r^2}{r_1^2}e^{i\frac{2\pi}{N}j}\right)\right]. \tag{10}$$

Summation in this formula can be performed explicitly if we notice that $\zeta = e^{i\frac{2\pi}{N}}$ is the primitive root of unity and $\zeta^N = 1$ implies $\zeta^N - 1 = (\zeta - 1)(1 + \zeta + \zeta^2 + ... + \zeta^{N-1}) = 0$. Then, $1 + \zeta + \zeta^2 + ... + \zeta^{N-1} = 0$ and for the sum of q-logarithm functions we have

$$\sum_{n=1}^{N} Ln_q(1 - x\zeta^n) = -\sum_{k=1}^{\infty}\frac{x^k}{[k]_q}(1 + \zeta^k + \zeta^{2k} + ... + \zeta^{k(N-1)}). \tag{11}$$

The sum $1 + \zeta^k + \zeta^{2k} + ... + \zeta^{k(N-1)} = N$, if $k = Nl$ and vanishes for $k \neq Nl$, where $l = 1, 2, ...$ It can be easily undersood if we notice that ζ^2, ζ^3,...ζ^{N-1} are primitive roots and $(\zeta^k)^N - 1 = (\zeta^k - 1)(1 + \zeta^k + \zeta^{2k} + ... + \zeta^{(N-1)k}) = 0$ for $k \neq Nl$. For $k = Nl$, $1 + \zeta^k + \zeta^{2k} + ... + \zeta^{k(N-1)} = 1 + 1 + ... + 1 = N$. Thus, in (11) nonvanishing terms are

$$\sum_{n=1}^{N} Ln_q(1 - x\zeta^n) = -\sum_{l=1}^{\infty}\frac{x^{Nl}}{[Nl]_q}N = -N(q-1)\sum_{l=1}^{\infty}\frac{x^{Nl}}{q^{Nl}-1} = -N(q-1)\sum_{l=1}^{\infty}\frac{(q^N-1)x^{Nl}}{(q^{Nl}-1)(q^N-1)}$$

and

$$\sum_{n=1}^{N} Ln_q(1 - x\zeta^n) = -\frac{N}{[N]_q}\sum_{l=1}^{\infty}\frac{(x^N)^l}{[l]_{q^N}} = \frac{N}{[N]_q}Ln_{q^N}(1 - x^N). \tag{12}$$

By applying formula (12) finally we rewrite frequency (10) in a simple form

$$\omega(r) = \frac{\Gamma(N-1)}{4\pi r^2} + \frac{\Gamma N}{2\pi r^2 (q^N - 1)} \left[Ln_{q^N}\left(1 - \frac{r_2^{2N}}{r^{2N}}\right) - Ln_{q^N}\left(1 - \frac{r^{2N}}{r_1^{2N}}\right) \right]. \quad (13)$$

The result shows that frequency $\omega(r)$ is function of r, and uniformly rotating polygon of vortices represents nonlinear oscillator. Quantization of this polygon as a nonlinear oscillator can be done in a similar way as for the one vortex case [2].

4. q-semiclassical expansion

Quantization of nonlinear oscillators as q-dispersive equations has been studied in [2]. Here we develop the q-semiclassical expansion for the q-deformed theories. Expansion formulas in powers of $\lambda = \ln q$ are based on the Bernoulli polynomials and the Euler polynomials with generating functions correspondingly

$$\frac{te^{xt}}{e^t - 1} = \sum_{n=0}^{\infty} B_n(x) \frac{t^n}{n!}, \quad |t| < 2\pi; \qquad \frac{2e^{xt}}{e^t + 1} = \sum_{n=0}^{\infty} E_n(x) \frac{t^n}{n!}, \quad |t| < \pi. \quad (14)$$

4.1. Non-symmetric q-calculus

We have expansion of q-number

$$[n]_q = \frac{q^n - 1}{q - 1} = \sum_{m=0}^{\infty} (B_m(n) - B_m) \frac{(\ln q)^{m-1}}{m!}, \quad (15)$$

for $e^{-2\pi} < q < e^{2\pi}$, where $B_m = B_m(0)$ are the Bernoulli numbers. By using $B_0(x) = 1, B_1(x) = x - 1/2$ for any real or complex number, or even an arbitrary operator A we get

$$[A]_q = A + \sum_{k=1}^{\infty} (B_{k+1}(A) - B_{k+1}) \frac{\lambda^k}{(k+1)!}. \quad (16)$$

4.1.1. Nonsymmetric q-dispersive Schrödinger equation The q-dispersive linear Schrödinger equation is derived as

$$i\hbar \frac{\partial \Psi}{\partial t} = \left[-\frac{\hbar}{2m} \frac{\partial^2}{\partial x^2} \right]_q \Psi,$$

and it admits the q-deformed Galilean boost operator

$$K = x + t \frac{\ln q}{q - 1} \frac{i\hbar}{m} \frac{d}{dx} q^{-\frac{\hbar^2}{2m}\frac{d^2}{dx^2}},$$

generating dynamical symmetry and producing solutions of the equation. By expansion (16) we get higher derivative corrections to the Schrödinger equation

$$i\hbar \frac{\partial \Psi}{\partial t} + \frac{\hbar}{2m} \frac{\partial^2}{\partial x^2} \Psi = \sum_{k=1}^{\infty} \frac{(\ln q)^k}{(k+1)!} \left(B_{k+1}\left(-\frac{\hbar}{2m} \frac{\partial^2}{\partial x^2} \right) - B_{k+1} \right)$$

and to the Galilean boost

$$K = \left(x + \frac{t}{m} i\hbar \frac{d}{dx} \right) + \frac{t}{m} i\hbar \frac{d}{dx} \sum_{k=1}^{\infty} \frac{(\ln q)^k}{k!} B_k \left(-\frac{\hbar}{2m} \frac{d^2}{dx^2} \right).$$

4.2. Symmetric q-calculus

For symmetric q-numbers, in a similar way we get

$$[n]_{\tilde{q}} = \frac{q^n - q^{-n}}{q - q^{-1}} = n + \sum_{m=1}^{\infty} \left(B_{m+1}\left(\frac{1+n}{2}\right) - B_{m+1}\left(\frac{1-n}{2}\right)\right) \frac{2^m (\ln q)^m}{(m+1)!}. \tag{17}$$

Due to symmetry $B_m(1-x) = (-1)^m B_m(x), m = 0, 1, 2...$ this formula can be simplified and extended to any operator

$$[A]_{\tilde{q}} = \frac{\sinh(A \ln q)}{\sinh(\ln q)} = A + \sum_{k=1}^{\infty} B_{2k+1}\left(\frac{I+A}{2}\right) \frac{2^{2k+1}(\ln q)^{2k}}{(2k+1)!}. \tag{18}$$

4.2.1. q-symmetric quantum oscillator For the spectrum of q-oscillator [5], [6], [7], ($\hbar = 1, \omega = 1$), we have q-semiclassical expansion

$$E_n = \frac{1}{2}([n]_{\tilde{q}} + [n+1]_{\tilde{q}}) = \frac{1}{2}[2n+1]_{\widetilde{\sqrt{q}}} = (n + \frac{1}{2}) + \sum_{k=1}^{\infty} B_{2k+1}(n+1)\frac{(\ln q)^{2k}}{(2k+1)!}.$$

4.2.2. Symmetric q-dispersive Schrödinger equation For the linear Schrödinger equation with symmetric q-dispersion [2] we find following expansion with higher order derivatives

$$i\hbar \frac{\partial \Psi}{\partial t} + \frac{-\hbar}{2m}\frac{\partial^2 \Psi}{\partial x^2} = \sum_{k=1}^{\infty} \frac{2^{2k+1}(\ln q)^{2k}}{(2k+1)!} B_{2k+1}\left(\frac{1}{2} - \frac{-\hbar}{4m}\frac{\partial^2}{\partial x^2}\right)\Psi. \tag{19}$$

4.2.3. q-symmetric dispersive NLS Nonlinearization of model (19) consists in replacement of momentum operator by the recursion operator of NLS hierarchy [2], so that

$$i\sigma_3 \begin{pmatrix} \psi \\ \bar{\psi} \end{pmatrix}_t = \frac{\sinh \lambda \frac{-\hbar}{2m}\mathcal{R}^2}{\sinh \lambda} \begin{pmatrix} \psi \\ \bar{\psi} \end{pmatrix} = \left[\frac{-\hbar}{2m}\mathcal{R}^2\right]_{\tilde{q}} \begin{pmatrix} \psi \\ \bar{\psi} \end{pmatrix}.$$

Then, we have expansion

$$i\sigma_3 \begin{pmatrix} \psi \\ \bar{\psi} \end{pmatrix}_t - \frac{-\hbar}{2m}\mathcal{R}^2 \begin{pmatrix} \psi \\ \bar{\psi} \end{pmatrix} = \sum_{k=1}^{\infty} \frac{2^{2k+1}(\ln q)^{2k}}{(2k+1)!} B_{2k+1}\left(\frac{I}{2} + \frac{-\hbar}{4m}\mathcal{R}^2\right)\begin{pmatrix} \psi \\ \bar{\psi} \end{pmatrix},$$

which gives q-corrections to the NLS equation ($\hbar = 1$, $m = 1/2$),

$$\begin{pmatrix} i\psi_t + \psi_{xx} + 2\kappa^2|\psi|^2\psi \\ -i\bar{\psi}_t + \bar{\psi}_{xx} + 2\kappa^2|\psi|^2\bar{\psi} \end{pmatrix} = \sum_{k=1}^{\infty} \frac{2^{2k+1}(\ln q)^{2k}}{(2k+1)!} B_{2k+1}\left(\frac{I}{2} + \frac{1}{2}\mathcal{R}^2\right)\begin{pmatrix} \psi \\ \bar{\psi} \end{pmatrix},$$

preserving integrability at any order of $\ln q$. The Lax pair can be expanded in a similar way.

4.3. Fibonacci polynomials for symmetric q-fermions

For $p = -1/q$ in pq calculus we have

$$[n]_{q,-q^{-1}} = \frac{q^n - (-q)^{-n}}{q - (-q)^{-1}} = \frac{q^n - (-1)^n q^{-n}}{q + q^{-1}} = \frac{e^{(1+n)\ln q} - (-1)^n e^{(1-n)\ln q}}{e^{2\ln q} + 1}$$

and expanding in terms of Euler polynomials

$$[n]_{q,-q^{-1}} = \frac{1+(-1)^{n+1}}{2} + \frac{1}{2}\sum_{m=1}^{\infty}\left(E_m\left(\frac{1+n}{2}\right) - (-1)^n E_m\left(\frac{1-n}{2}\right)\right)\frac{2^m (\ln q)^m}{m!}. \tag{20}$$

By the relation $E_m(1-x) = (-1)^m E_m(x)$ for even and odd numbers we get

$$[2k]_{q,-q^{-1}} = \sum_{s=0}^{\infty} E_{2s+1}\left(k+\frac{1}{2}\right)\frac{2^{2s+1}(\ln q)^{2s+1}}{(2s+1)!},$$

$$[2k+1]_{q,-q^{-1}} = 1 + \sum_{s=1}^{\infty} E_{2s}\left(k+1\right)\frac{2^{2s}(\ln q)^{2s}}{(2s)!}.$$

From (20) for $q=1$ we find that the "classical" fermion number $[n]_{1,-1} = \frac{1+(-1)^{n+1}}{2}$ is the parity number, vanishing if $n=2s$ is an even number, and equal to 1, if $n=2s+1$ is an odd number. In terms of it we find following compact expression

$$[n]_{q,-q^{-1}} = [n]_{1,-1} + \sum_{m=1}^{\infty} E_m\left(\frac{n+1}{2}\right)[n+m]_{1,-1}\frac{2^m(\ln q)^m}{(m)!}. \tag{21}$$

4.4. Fibonacci numbers and Euler polynomials
From Binet formula for Fibonacci numbers

$$F_n = \frac{\varphi^n - \varphi'^n}{\varphi - \varphi'} = [n]_{\varphi,\varphi'} \equiv [n]_F \tag{22}$$

we find

$$F_n = [n]_{1,-1} + \sum_{m=1}^{\infty} E_m\left(\frac{n+1}{2}\right)[n+m]_{1,-1}\frac{2^m(\ln\varphi)^m}{(m)!}. \tag{23}$$

This provides an expansion of Fibonacci numbers in powers of $\ln\varphi$ and Euler polynomials

$$F_{2k} = \sum_{s=0}^{\infty} E_{2s+1}\left(k+\frac{1}{2}\right)\frac{2^{2s+1}(\ln\varphi)^{2s+1}}{(2s+1)!},$$

$$F_{2k+1} = 1 + \sum_{s=1}^{\infty} E_{2s}\left(k+1\right)\frac{2^{2s}(\ln\varphi)^{2s}}{(2s)!}.$$

4.4.1. Golden oscillator By these formulas spectrum of the golden oscillator [8], $E_n = \frac{\bar{h}\omega}{2}(F_{n+1}+F_n) = \frac{\bar{h}\omega}{2}F_{n+2}$, can be expanded in golden ratio as

$$E_{2k} = \frac{\bar{h}\omega}{2}\sum_{s=0}^{\infty} E_{2s+1}\left(k+\frac{3}{2}\right)\frac{2^{2s+1}}{(2s+1)!}\left(\ln\frac{1+\sqrt{5}}{2}\right)^{2s+1},$$

$$E_{2k+1} = \frac{\bar{h}\omega}{2} + \frac{\bar{h}\omega}{2}\sum_{s=1}^{\infty} E_{2s}\left(k+2\right)\frac{2^{2s}}{(2s)!}\left(\ln\frac{1+\sqrt{5}}{2}\right)^{2s}.$$

5. Diversity of q-analytic functions

Here we introduce several types of q-analytic functions. These functions are determined by the q-binomials and represent quantum states in the q-analytic Fock-Bargman representation. The q-binomials itself correspond to states $|n>$ and are the generalized analytic functions.

Journal of Physics: Conference Series **766** (2016) 012015 doi:10.1088/1742-6596/766/1/012015

5.1. Nonsymmetric q-analytic functions

By q-translation

$$e_{1/q}^{iyD_q^x} x^n = (x+iy)_q^n$$

from any real analytic function $f(x)$ we get the q-analytic function [9],

$$f(x+iy)_q = e_{1/q}^{iyD_q^x} f(x) = \sum_{n=0}^{\infty} a_n (x+iy)_q^n,$$

satisfying $\bar{\partial}_q$ equation, $(D_q^x + iD_{1/q}^y) f(x+iy)_q = 0$. The real part $u(x,y) = \cos_{1/q}(yD_q^x) f(x)$ and imaginary part $v(x,y) = \sin_{1/q}(yD_q^x) f(x)$ of this function are q-harmonic and satisfy the q-Cauchy-Riemann equations

$$D_q^x u(x,y) = D_{1/q}^y v(x,y), \quad D_{1/q}^y u(x,y) = -D_q^x v(x,y).$$

5.2. Symmetric q-analytic functions

For symmetric q-binomials we have q-translation

$$e_{\tilde{q}}^{iyD_{\tilde{q}}^x} x^n = (x+iy)_{\tilde{q}}^n,$$

which determines symmetric q-analytic function

$$f(x+iy)_{\tilde{q}} = e_{\tilde{q}}^{iyD_{\tilde{q}}^x} f(x) = \sum_{n=0}^{\infty} a_n (x+iy)_{\tilde{q}}^n,$$

satisfying $\bar{\partial}_{\tilde{q}}$ equation

$$\frac{1}{2}(D_{\tilde{q}}^x + iD_{\tilde{q}}^y) f(x+iy)_{\tilde{q}} = 0.$$

As an example,

$$e(z;\tilde{q}) = \sum_{n=0}^{\infty} \frac{(x+iy)_{\tilde{q}}^n}{[n]_{\tilde{q}}!}$$

is an entire symmetric q-analytic function. The real part $u(x,y) = \cos_{\tilde{q}}(yD_{\tilde{q}}^x) f(x)$ and imaginary part $v(x,y) = \sin_{\tilde{q}}(yD_{\tilde{q}}^x) f(x)$ of the function satisfy the q-Cauchy-Riemann equations

$$D_{\tilde{q}}^x u(x,y) = D_{\tilde{q}}^y v(x,y), \quad D_{\tilde{q}}^y u(x,y) = -D_{\tilde{q}}^x v(x,y),$$

and are q-harmonic: $(D_{\tilde{q}}^x)^2 u(x,y) + (D_{\tilde{q}}^y)^2 u(x,y) = 0$.

5.3. pq-analytic function

For pq-calculus we have binomials

$$e_{\frac{1}{p}\frac{1}{q}}^{iyD_{pq}^x} x^n = (x+iy)_{pq}^n,$$

where

$$(x+iy)_{pq}^n = (x+ip^{n-1}y)(x+ip^{n-2}qy)...(x+ipq^{n-2}y)(x+iq^{n-1}y) = \sum_{k=0}^{n} \begin{bmatrix} n \\ k \end{bmatrix}_{pq} (pq)^{\frac{k(k-1)}{2}} x^{n-k} i^k y^k$$

and

$$e_{pq}(x) = \sum_{n=0}^{\infty} \frac{x^n}{[n]_{pq}!}.$$

Then, the pq-analytic function generated by pq-translation

$$f(x+iy)_{pq} = e_{\frac{1}{p}\frac{1}{q}}^{iyD^x_{pq}} f(x) = \sum_{n=0}^{\infty} a_n (x+iy)^n_{pq},$$

satisfies $\bar{\partial}_{pq}$ equation

$$\frac{1}{2}(D^x_{pq} + iD^y_{\frac{1}{p}\frac{1}{q}}) f(x+iy)_{pq} = 0.$$

For $u(x,y) = \cos_{\frac{1}{p}\frac{1}{q}}(yD^x_{pq})f(x)$ and $v(x,y) = \sin_{\frac{1}{p}\frac{1}{q}}(yD^x_{pq})f(x)$ we have the pq-Cauchy-Riemann equations

$$D^x_{pq}u(x,y) = D^y_{\frac{1}{p}\frac{1}{q}}v(x,y), \quad D^y_{\frac{1}{p}\frac{1}{q}}u(x,y) = -D^x_{pq}v(x,y)$$

and the pq-Laplace equation

$$(D^x_{pq})^2 u(x,y) + (D^y_{\frac{1}{p}\frac{1}{q}})^2 u(x,y) = 0.$$

5.4. Golden analytic function

Complex golden binomials, defined as

$$(x+iy)^n_F = (x + i\varphi^{n-1}y)(x - i\varphi^{n-3}y)...(x + i(-1)^{n-1}\varphi^{1-n}y) = \sum_{k=0}^{n} \left[\begin{array}{c} n \\ k \end{array} \right]_F (-1)^{\frac{k(k-1)}{2}} x^{n-k} i^k y^k,$$

can be generated by the golden translation

$$E_F^{iyD^x_F} x^n = (x+iy)^n_F,$$

where

$$E^x_F = \sum_{n=0}^{\infty} (-1)^{\frac{n(n-1)}{2}} \frac{x^n}{F_n!}.$$

They determine the golden analytic function

$$f(z,F) = E_F^{iyD^x_F} f(x) = \sum_{n=0}^{\infty} a_n \frac{(x+iy)^n_F}{F_n!},$$

satisfying the golden $\bar{\partial}_F$ equation $\frac{1}{2}(D^x_F + iD^y_{-F})f(z;F) = 0$, where $D_{-F} = (-1)^{x\frac{d}{dx}} D_F$. For $u(x,y) = Cos_F(yD^x_F)f(x)$ and $v(x,y) = Sin_F(yD^x_F)f(x)$, the golden Cauchy-Riemann equations are

$$D^x_F u(x,y) = D^y_{-F} v(x,y), \quad D^y_{-F} u(x,y) = -D^x_F v(x,y),$$

and the golden-Laplace equation is $(D^x_F)^2 u(x,y) + (D^y_{-F})^2 u(x,y) = 0$.

References

[1] Pashaev O K 2014 *J. Phys.: Conf. Series* **482** 012033
[2] Pashaev O K 2015 *Phys. Scr.*. **90** 074010
[3] Erdakova N N and Mamaev I S 2014 *Fluid Dyn. Res.*. **46** 031420
[4] Pashaev O K and Yilmaz O 2008 *J. Phys. A: Math. Theor.*. **41** 135207
[5] Biedenharn L C 1989 *J. Phys. A.* **22** L873
[6] Macfarlane A J 1989 *J. Phys. A.* **22** 4581
[7] Sun C P and Fu H C 1989 *J. Phys. A.* **22** L983
[8] Pashaev O K and Nalci S 2012 *J. Phys. A: Math. Theor.*. **45** 015303
[9] Pashaev O K and Nalci S 2014 *J. Phys. A: Math. Theor.*. **47** 045204

Some Results on The Incomplete q-Gamma Function and Its First Derivative

İnci Ege, Emrah Yıldırım

Department of Mathematics, Adnan Menderes University, 09100 Aydın, Turkey

E-mail: iege@adu.edu.tr

E-mail: emrahyildirim@adu.edu.tr

Abstract. In this paper, we use neutrix calculus in order to obtain some results on the q-analogue of the incomplete gamma function and its first derivative for all real numbers.

1. Introduction and Preliminaries

Quantum calculus, the q-analogue of classic calculus, has been used in many research areas such as physics, mathematics, statistic etc. recently [1, 2, 3]. The q-analogue $\mathcal{M}_q$ of mathematical object $\mathcal{M}$ has not been given any specific definition, but common admission is that as q tends to 1, the limit of $\mathcal{M}_q$ approaches to $\mathcal{M}$. Therefore many mathematical objects have got more then one q-analogue such as exponential function. One of the q-analogue of the exponential function e^x is denoted by E_q^x and defined by

$$E_q^x = \sum_{n=0}^{\infty} \frac{q^{\frac{n(n+1)}{2}} x^n}{[n]!} = (1-(1-q)x)_q^{\infty}$$

where $[x] = \frac{1-q^x}{1-q}$; $[n]! = [1][2]\dots[n-1][n]$ $n \in \mathbb{Z}^+$ with $[0]! = 1$ and q-analogue of $(x-a)^n$ is defined by

$$(x-a)_q^n = \begin{cases} 1, & n = 0, \\ (x-a)(x-qa)\dots(x-q^{n-1}a), & n \geq 1. \end{cases}$$

Let us take a function f on any interval I of real numbers such that if $x \in I$ then $qx \in I$. The q-derivative of the function f is defined by

$$D_q f(x) = \frac{f(qx) - f(x)}{(q-1)x} \text{ for } x = 0 \text{ and } (D_q f)(0) = f'(0)$$

provided $f'(0)$ exists. Note that $D_q E_q^x = E_q^{qx}$.
The q-integral of a function f from zero to a is defined by

$$\int_0^a f(x) d_q x = a(1-q) \sum_{n=0}^{\infty} q^n f(q^n a)$$

provided the sum converges absolutely and one of the differences between quantum calculus and classic calculus is that the q-derivative of the product of two functions is not symmetric. Because of that, the q-integration by parts can be given by two ways. First one can be found in [4] and the second one, which will be used in this paper, is given for suitable functions f and g as

$$\int_a^b f(qx)d_q g(x) = f(b)g(b) - f(a)g(a) - \int_a^b g(x)d_q f(x). \tag{1}$$

The reader may find more information about quantum calculus in [4, 5] and references therein. In our study we also need the definitions of neutrix and neutrix limit given by Van der Corput in [6].

Definition 1. (Neutrix) Let N' be a nonempty set and let $\mathcal{N}$ be a commutative, additive group of functions mapping N' into a commutative, additive group N''. The group $\mathcal{N}$ is called neutrix if the function which is identically equal to zero is the only constant function occurring $\mathcal{N}$. The function which belongs to $\mathcal{N}$ is called "negligible function" in $\mathcal{N}$.
Let $N^{'}$ be a domain lying in a topological space with a limit point b not belonging to $N^{'}$ and $\mathcal{N}$ be a commutative additive group of functions defined on N' with the following property:

$$\text{"}f \in \mathcal{N}\ , \lim_{x\to b} f(x) = c\,(\text{constant}) \text{ for } x \in N^{'} \text{ then } c = 0\text{"}.$$

Then this group $\mathcal{N}$ is a neutrix.

Definition 2. (Neutrix Limit) Let f be a real valued function defined on N' and suppose that it is possible to find a constant c such that $f(x) - c$ is negligible in $\mathcal{N}$. Then c is called the neutrix limit of $f(x)$ as x tends to b and denoted by

$$\operatorname*{N-lim}_{x\to b} f(x) = c. \tag{2}$$

In [7], authors defined the incomplete q-gamma function for $\alpha, x > 0$ by q-integral as

$$\gamma_q(\alpha, x) = \int_0^x t^{\alpha-1} E_q^{-qt} d_q t. \tag{3}$$

In this paper, we let $\mathcal{N}$ be the neutrix having domain the open interval $N' = \{\varepsilon : 0 < \varepsilon < \infty\}$ and range N'' as the real numbers with negligible functions being finite linear sums of the functions

$$\varepsilon^\lambda \ln^{r-1}\varepsilon, \quad \ln^r \varepsilon, \quad [\varepsilon]^\lambda \quad (\lambda < 0, \quad r = 1, 2, \ldots)$$

and all being functions $f(\varepsilon)$ which converges to zero in the usual sense as ε tends to zero.
In [8], the author used the neutrix limit in order to define the incomplete q-gamma function with its derivatives as

$$\gamma_q^{(r)}(\alpha, x) = \operatorname*{N-lim}_{\varepsilon\to 0} \int_\varepsilon^x t^{\alpha-1} \ln^r t E_q^{-qt} d_q t \tag{4}$$

for all real values of α and $x > 0$ and showed that

$$\gamma_q(-n, x) = \int_1^x t^{-n-1} E_q^{-qt} d_q t + \int_0^1 t^{-n-1}\Big[E_q^{-qt} - \sum_{i=0}^{n} \frac{(-1)^i q^{\frac{i(i+1)}{2}}}{[i]!} t^i\Big] d_q t + \sum_{i=0}^{n-1} \frac{(-1)^i q^{\frac{i(i+1)}{2}}}{[i-n][i]!} \tag{5}$$

for $n \in \mathbb{Z}^+$, $x > 0$.

2. Main Results

In this section, we aim to give some equalities for the incomplete q-gamma function with its first derivative for all real values of α.
First we show the following result.

Theorem 1. *We have*

$$\gamma_q(0,x)=\frac{q-1}{\ln q}\gamma_q'(1,x)+\frac{(q-1)\ln x}{\ln q}E_q^{-x} \tag{6}$$

and for $n=1,2,\ldots$

$$\gamma_q(-n,x)=\frac{1}{[-n]}\gamma_q(-n+1,x)+\frac{x^{-n}E_q^{-x}}{[-n]}-\frac{(-1)^n q^{\frac{n(n-1)}{2}}}{[-n][n]!}. \tag{7}$$

Proof. The proof of equation (6) is straightforward from equation (1). Now for equation (7), we have

$$\begin{aligned}\gamma_q(-n,x) &= \frac{1}{[-n]}\left\{\int_1^x t^{-n}E_q^{-qt}d_qt+\int_0^1 t^{-n}\left[E_q^{-qt}-\sum_{i=0}^{n-1}\frac{(-1)^i q^{\frac{i(i+1)}{2}}}{[i]!}t^i\right]d_qt\right\}\\ &+ \frac{x^{-n}E_q^{-x}}{[-n]}-\frac{E_q^{-1}}{[-n]}+\frac{E_q^{-1}}{[-n]}-\sum_{i=0}^{n}\frac{(-1)^i q^{\frac{i(i-1)}{2}}}{[-n][i]!}+\sum_{i=0}^{n-1}\frac{(-1)^i q^{\frac{i(i+1)}{2}}}{[i]![i-n]}\end{aligned}$$

by using q-integration by parts (1) in equation (5). By calculating the last two terms, we get desired result. □

The recurrence formula of incomplete q-gamma function for negative integers is also obtained in [8] as

$$\gamma_q(-n,x)+\frac{q^n}{[n]}\gamma_q(-n+1,qx)=\frac{(-1)^n q^{\frac{n(n+1)}{2}}}{[n][n]!}-\frac{q^n}{[n]}x^{-n}E_q^{-qx} \tag{8}$$

by q-integration by parts in [4]. Remark that, the classical version is given by

$$\gamma(-n,x)+\frac{1}{n}\gamma(-n+1,x)=\frac{(-1)^n}{n\,n!}-\frac{1}{n}e^{-x}x^{-n} \tag{9}$$

in [9]. Nevertheless to say, both of the equations (7) and (8) tend to (9) as $q\to 1$ and the equations (6) and (7) tends to the result in [10] as $x\to\infty$.
Also we want to note that by mathematical induction one can find that

$$\gamma_q(-n,x)=\frac{(-1)^n q^{\frac{n(n+1)}{2}}}{[n]!}\left[\varphi_q(n)+\gamma(0,x)\right]+\mu_q(n)E_q^{-x}. \tag{10}$$

for $n=1,2,\ldots$ where the functions are defined by

$$\mu_q(n)=\sum_{i=0}^{n-1}(-1)^{i+1}\eta_q(i)x^{-n+i}\quad \varphi_q(n)=\sum_{i=1}^{n}\frac{1}{[n]}\quad ,\quad \eta_q(i)=\prod_{j=0}^{i}\frac{q^{n-j}}{[n-j]}$$

respectively. The equation (10) approaches to the result in [11] as $x\to\infty$, $q\to 1$.
Before we give the result for the first derivative of incomplete q-gamma function at negative integers we need the following properties.

Lemma 1.

$$\gamma_q(\alpha,x)=\int_1^x t^{\alpha-1}E_q^{-qt}d_qt+\int_0^1 t^{\alpha-1}\left(E_q^{-qt}-\sum_{i=0}^{n-1}\frac{(-1)^i q^{\frac{i(i+1)}{2}}}{[i]!}t^i\right)d_qt+\sum_{i=0}^{n-1}\frac{(-1)^i q^{\frac{i(i+1)}{2}}}{[i+\alpha][i]!} \tag{11}$$

and

$$\gamma_q'(\alpha,x)=\int_1^x t^{\alpha-1}\ln t E_q^{-qt}d_qt+\int_0^1 t^{\alpha-1}\ln t\left[E_q^{-qt}-\sum_{i=0}^{n-1}\frac{(-1)^i q^{\frac{i(i+1)}{2}}}{[i]!}t^i\right]d_qt$$

$$+\sum_{i=0}^{n-1}\frac{(-1)^i q^{\frac{i(i+1)}{2}}}{[i]!}\left[\frac{\ln q^{-1}}{[\alpha+i]}+\frac{\ln q^{-1}}{(q-1)[\alpha+i]^2}\right] \tag{12}$$

for $-n<\alpha<-n+1$, $n=1,2,\ldots$, $x>0$ and

$$\begin{aligned}\gamma'_q(-n,x) &= \int_1^x t^{-n-1}\ln t E_q^{-qt} d_qt + \int_0^1 t^{-n-1}\ln t\left[E_q^{-qt}-\sum_{i=0}^{n}\frac{(-1)^i q^{\frac{i(i+1)}{2}}}{[i]!}t^i\right]d_qt\\ &+\sum_{i=0}^{n-1}\frac{(-1)^i q^{\frac{i(i+1)}{2}}}{[i]!}\left[\frac{\ln q^{-1}}{[i-n]}+\frac{\ln q^{-1}}{(q-1)[-n+i]^2}\right]\end{aligned} \tag{13}$$

for $n=1,2,\ldots$.

Proof. For $\varepsilon>0$, we have

$$\begin{aligned}\int_\varepsilon^x t^{\alpha-1}E_q^{-qt}d_qt &= \int_1^x t^{\alpha-1}E_q^{-qt}d_qt + \int_\varepsilon^1 t^{\alpha-1}\left[E_q^{-qt}-\sum_{i=0}^{n-1}\frac{(-1)^i q^{\frac{i(i+1)}{2}}}{[i]!}t^i\right]d_qt\\ &+\sum_{i=0}^{n-1}\frac{(-1)^i q^{\frac{i(i+1)}{2}}}{[i]!}\left[\frac{1}{[\alpha+i]}-\frac{\varepsilon^{\alpha+i}}{[\alpha+i]}\right].\end{aligned}$$

Since the last term is a negligible function, we get equation (11). The proofs of the equation (12) and (13) can be obtained similarly. □

We can now prove the following theorem.

Theorem 2.

$$\gamma_q^{(r)}(\alpha,x)=\mathrm{N-}\lim_{\varepsilon\to0}\gamma_q^{(r)}(\alpha+\varepsilon,x)$$

for all real values of α *and* $r=0,1$.

Proof. Since $\gamma_q^{(r)}(\alpha,x)$ is continuous function for $\alpha\neq0,-1,-2,\ldots$, the result follows immediately for $\alpha\neq0,-1,-2,\ldots$. If $0<\varepsilon<1$ and $r=0$, we have from the equation (11) that

$$\gamma_q(-n+\varepsilon,x)=\int_1^x t^{-n+\varepsilon-1}E_q^{-qt}d_qt+\int_0^1 t^{-n+\varepsilon-1}\left[E_q^{-qt}-\sum_{i=0}^{n-1}\frac{(-1)^i q^{\frac{i(i+1)}{2}}}{[i]!}t^i\right]d_qt+\sum_{i=0}^{n-1}\frac{(-1)^i q^{\frac{i(i+1)}{2}}}{[i-n+\varepsilon][i]!}$$

$$=\int_1^x t^{-n+\varepsilon-1}E_q^{-qt}d_qt+\int_0^1 t^{-n+\varepsilon-1}\left[E_q^{-qt}-\sum_{i=0}^{n}\frac{(-1)^i q^{\frac{i(i+1)}{2}}}{[i]!}t^i\right]d_qt+\sum_{i=0}^{n-1}\frac{(-1)^i q^{\frac{i(i+1)}{2}}}{[i-n+\varepsilon][i]!}+\frac{(-1)^n q^{\frac{n(n+1)}{2}}}{[n]!}\frac{1}{[\varepsilon]}.$$

The last term on the right side of the equation is a negligible function. Because of that, by taking neutrix limit on the both side and using the equation (5) we get

$$\mathrm{N-}\lim_{\varepsilon\to0}\gamma_q(-n+\varepsilon,x)=\gamma_q(-n,x).$$

We also have

$$\begin{aligned}\gamma_q(-n-\varepsilon,x) &= \int_1^x t^{-n+\varepsilon-1}E_q^{-qt}d_qt+\int_0^1 t^{-n+\varepsilon-1}\left[E_q^{-qt}-\sum_{i=0}^{n}\frac{(-1)^i q^{\frac{i(i+1)}{2}}}{[i]!}t^i\right]d_qt\\ &+\sum_{i=0}^{n-1}\frac{(-1)^i q^{\frac{i(i+1)}{2}}}{[i-n+\varepsilon][i]!}+\frac{(-1)^n q^{\frac{n(n+1)}{2}}}{[n]![-\varepsilon]}\end{aligned}$$

Now taking neutrix limits of the both sides, using the equation (5) and the fact that $[-\varepsilon]=-q^{\varepsilon}[\varepsilon]$, we get

$$\mathrm{N-}\lim_{\varepsilon\to0}\gamma_q(-n-\varepsilon,x)=\gamma_q(-n,x).$$

The case $r=1$ can be proved similarly. □

Note that, the results tends to the one in [13, 10] as $x \to \infty$ and also to the one [12] as $q \to 1$. In [7] author showed that the incomplete q-gamma function satisfies recurrence formula

$$\gamma_q(\alpha+1,x) = [\alpha]\gamma_q(\alpha,x) - x^{\alpha}E_q^{-x} \tag{14}$$

for $\alpha > 0$ and $x > 0$. Now we generalize equation (14) for all real numbers.

Theorem 3.

$$\gamma_q(\alpha+1,x) = \mathrm{N}\!-\!\lim_{\varepsilon\to 0}\left([\alpha+\varepsilon]\gamma_q(\alpha+\varepsilon,x) - x^{\alpha+\varepsilon}E_q^{-x}\right)$$

for all real values of α.

Proof. The result follows because of the continuity of $\gamma_q(\alpha,x)$ for $\alpha \neq 0,-1,-2,\ldots$. By using equation (14) we get

$$\gamma_q(-n+\varepsilon+1,x) = [-n+\varepsilon]\gamma_q(-n+\varepsilon,x) - x^{-n+\varepsilon}E_q^{-x}.$$

for $0 < |\varepsilon| < 1$ and $n = 1,2,\ldots$.
Then using theorem 2 we obtain

$$\gamma_q(-n+1,x) = \mathrm{N}\!-\!\lim_{\varepsilon\to 0}\gamma_q(-n+\varepsilon+1,x) = \mathrm{N}\!-\!\lim_{\varepsilon\to 0}\left([-n+\varepsilon]\gamma_q(-n+\varepsilon,x) - x^{-n+\varepsilon}E_q^{-x}\right).$$

□

One can find that the result goes to the one in [13] as $x \to \infty$.
In the next theorem we will show that the first derivative of incomplete q-gamma function at negative integers can be presented by itself.

Theorem 4. *For $n \in \mathbb{Z}^+$, we have*

$$\gamma_q'(-n,x) = \frac{x^{-n}\ln(q^{-1}x)E_q^{-x}}{[-n]} + \frac{\ln q^{-1}}{(q-1)[-n]}\gamma_q(-n,x) + \frac{\ln q^{-1}}{[-n]}\gamma_q(-n+1,x) + \frac{1}{[-n]}\gamma_q'(-n+1,x). \tag{15}$$

Proof. By using q-integration by parts on equation (13) we get

$$\begin{aligned}
\gamma_q'(-n,x) &= \frac{x^{-n}\ln q^{-1}xE_q^{-x}}{[-n]} - \frac{\ln q^{-1}}{[-n]}\sum_{i=0}^{n}\frac{(-1)^i q^{\frac{i(i-1)}{2}}}{[i]!} + \frac{(-1)^n q^{\frac{n(n-1)}{2}}}{[n]!}\frac{\ln q^{-1}}{[-n]} \\
&+ \sum_{i=0}^{n-1}\frac{(-1)^i q^{\frac{i(i+1)}{2}}}{[i]!}\left[\frac{\ln q^{-1}}{[i-n]} + \frac{\ln q^{-1}}{(q-1)[-n+i]^2}\right] \\
&+ \frac{\ln q^{-1}}{(q-1)[-n]}\left\{\int_1^x t^{-n-1}E_q^{-qt}d_qt + \int_0^1 t^{-n-1}\left(E_q^{-qt} + \sum_{i=0}^{n}\frac{(-1)^i q^{\frac{i(i+1)}{2}}}{[i]!}t^i\right)d_qt\right\} \\
&+ \frac{\ln q^{-1}}{[-n]}\left\{\int_1^x t^{-n}E_q^{-qt}d_qt + \int_0^1 t^{-n}\left(E_q^{-qt} + \sum_{i=0}^{n-1}\frac{(-1)^i q^{\frac{i(i+1)}{2}}}{[i]!}t^i\right)d_qt\right\} \\
&+ \frac{1}{[-n]}\left\{\int_1^x t^{-n}\ln tE_q^{-qt}d_qt + \int_0^1 t^{-n-1}\ln t\left(E_q^{-qt} + \sum_{i=0}^{n-1}\frac{(-1)^i q^{\frac{i(i+1)}{2}}}{[i]!}t^i\right)d_qt\right\}.
\end{aligned}$$

Then we add missing series parts of the definition of incomplete q-gamma function and its first derivative into the corresponding parentheses, we obtain

$$\begin{aligned}
\gamma_q'(-n,x) &= \frac{x^{-n}\ln q^{-1}xE_q^{-x}}{[-n]} + \frac{\ln q^{-1}}{(q-1)[-n]}\gamma_q(-n,x) + \frac{\ln q^{-1}}{[-n]}\gamma_q(-n+1,x) + \frac{1}{[-n]}\gamma_q'(-n+1,x) \\
&+ \frac{\ln q}{[-n]}\sum_{i=0}^{n-1}\frac{(-1)^i q^{\frac{i(i-1)}{2}}}{[i]!} + \frac{\ln q}{(q-1)[-n]}\sum_{i=0}^{n-1}\frac{(-1)^i q^{\frac{i(i+1)}{2}}}{[i]![i-n]} + \frac{\ln q}{[-n]}\sum_{i=0}^{n-2}\frac{(-1)^i q^{\frac{i(i+1)}{2}}}{[i]![i-n+1]} + \frac{\ln q}{[-n]}\sum_{i=0}^{n-2}\frac{(-1)^i q^{\frac{i(i+1)}{2}}}{[i]![i-n+1]} \\
&+ \frac{\ln q}{(q-1)[-n]}\sum_{i=0}^{n-2}\frac{(-1)^i q^{\frac{i(i+1)}{2}}}{[i]![i-n+1]^2} - \ln q\sum_{i=0}^{n-1}\frac{(-1)^i q^{\frac{i(i+1)}{2}}}{[i]![i-n]} - \frac{\ln q}{q-1}\sum_{i=0}^{n-1}\frac{(-1)^i q^{\frac{i(i+1)}{2}}}{[i]![i-n]^2}.
\end{aligned}$$

The sums of the first, fourth and sixth series and also the remaining series' sums are equal to zero. Hence it completes the proof. □

This result also tends to the one in [10] as $x \to \infty$.
Differentiating equation (14), we obtain

$$\gamma_q'(\alpha+1,x) = \frac{q^\alpha \ln q}{q-1}\gamma_q(\alpha,x) + [\alpha]\gamma_q'(\alpha,x) - x^\alpha \ln x E_q^{-x} \quad (16)$$

for $\alpha \neq 0,-1,-2,\ldots$.
In the following theorem we use neutrix limits in order to generalize equation (16) for all real numbers.

Theorem 5.

$$\gamma_q'(\alpha+1,x) = \mathrm{N}{-}\lim_{\varepsilon\to 0}\left[\frac{q^{\alpha+\varepsilon}\ln q}{q-1}\gamma_q(\alpha+\varepsilon,x) + [\alpha+\varepsilon]\gamma_q'(\alpha+\varepsilon,x) - x^{\alpha+\varepsilon}\ln x E_q^{-x}\right]$$

for all real values of α.

Proof. The result can easily be obtained because of the continuity of $\gamma_q'(\alpha,x)$ for $\alpha \neq 0,-1,-2,\ldots$. Equation (15) also satisfies for all real values of α. By rewriting this equation for $n \in \mathbb{Z}^+$ and $0<\varepsilon<1$ we have

$$\gamma_q'(-n+\varepsilon+1,x) = [-n+\varepsilon]\gamma_q'(-n+\varepsilon,x) + \frac{\ln q}{q-1}\gamma_q(-n+\varepsilon,x) + \ln q\,\gamma_q(-n+\varepsilon+1,x) - x^{-n+\varepsilon}E_q^{-x}\ln(q^{-1}x)$$

$$= [-n+\varepsilon]\gamma_q'(-n+\varepsilon,x) + \frac{\ln q}{q-1}\gamma_q(-n+\varepsilon,x) + \ln q\Big([-n+\varepsilon]\gamma_q(-n+\varepsilon,x) - x^{-n+\varepsilon}E_q^{-x}\Big)$$

$$+x^{-n+\varepsilon}E_q^{-x}\ln q - x^{-n+\varepsilon}E_q^{-x}\ln x$$

$$= [-n+\varepsilon]\gamma_q'(-n+\varepsilon,x) + \frac{\ln q}{q-1}\Big(1+q^{-n+\varepsilon}-1\Big)\gamma_q(-n+\varepsilon,x) - x^{-n+\varepsilon}E_q^{-x}\ln x$$

Then by taking the neutrix limits of the both sides and theorem 2 the proof is completed. □

References

[1] Algın A., Irk D., Topçu G., Anyonic behavior of an intermediate-statistics fermion gas model, Physical Review E., 22,91(6), 062131, (2015)
[2] Haven, E., Potential functions and the characterization of economics-based information, Foundations of Physics, 45(10), 1394-1406, (2015)
[3] Ren, Mei-Ying., Statistical Approximation of the q-Bernstein-Durrmeyer Type Operators, Fuzzy Systems and Operations Research and Management, Springer International Publishing, 117-124, (2016)
[4] Kac V.G., Cheung P., Quantum Calculus, Universitext, Springer-Verlag, New York, (2002)
[5] Ernst T., A comprehensive treatment of q-calculus. Springer Science & Business Media, (2012)
[6] Corput J.G., Introduction to the neutrix calculus, J. Analyse Math., 7, 291-398, (1959)
[7] El-Shahed M., Salem A., On q-analogue of the Incomplete Gamma Function, Int. J. Pure Appl. Math, 44, 773-780, (2008)
[8] Salem A., Existence of the neutrix limit of the q-analogue of the incomplete gamma function and its derivatives. Applied Mathematics Letters, 25, 363-368, (2012)
[9] Özçağ E., et al., Some remarks on the incomplete gamma function, Mathematical Methods in Engineering. Springer Netherlands, 97-108, (2007)
[10] Ege İ, Yıldırım E., Some equalities on q-Gamma and q-Digamma functions. Filomat, (in press)
[11] Fisher B., Kuribayashi Y., Neutrices and the gamma function, The Jour. of the Fac. of Edu. Tottori Uni., 36, 1-22, (1987)
[12] Fisher B., Kuribayashi Y., Some results on the gamma function, The Journal of Faculty of Education Tottori University, 37, 2, (1988)
[13] Ege İ, Yıldırım E., Some generalized equalities for the q-gamma function. Filomat, 26(6), 1227-1232, (2012)

From q-analytic functions to double q-analytic Hermite binomials and q-traveling waves

Sengul Nalci Tumer and Oktay K. Pashaev

Department of Mathematics, Izmir Institute of Technology, Gulbahce Campus 35430 Urla, Izmir, Turkey

E-mail: `sengulnalci@iyte.edu.tr`, `oktaypashaev@iyte.edu.tr`

Abstract. We extend the concept of q-analytic function in two different directions. First we find expansion of q-binomial in terms of q-Hermite polynomials, analytic in two complex arguments. Based on this representation, we introduce a new class of complex functions of two complex arguments, which we call the double q-analytic functions. As another direction, by the hyperbolic version of q-analytic functions we describe the q-analogue of traveling waves, which is not preserving the shape during evolution. The IVP for corresponding q-wave equation we solved in the q-D'Alembert form.

1. Introduction

Recently, we have introduced a new class of complex valued functions $f(z)$ of one complex variable z according to equation [1],

$$D_q^{\bar{z}} f(z) = \frac{1}{2}\left(D_q^x + i D_{\frac{1}{q}}^y\right) f(z) = 0 \tag{1}$$

and we called it as the *q-analytic function.* In the limit $q \to 1$ it reduces to the standard analytic function satisfying the Cauchy-Riemann equations. The basic example of q-analytic function is given by complex q-binomial

$$(x+iy)_q^n \equiv (x+iy)(x+iqy)(x+iq^2y)...(x+iq^{n-1}y) = \sum_{k=0}^{n} \begin{bmatrix} n \\ k \end{bmatrix}_q q^{\frac{k(k-1)}{2}} x^{n-k}(iy)^k, \tag{2}$$

which is not analytic function in the usual sense $\frac{\partial}{\partial \bar{z}}(x+iy)_q^n = 0$. But, as was shown in [1] it is a generalized analytic function [2], which satisfies the $\bar{\partial}$- equation

$$\frac{\partial \Phi}{\partial \bar{z}} = A(z,\bar{z})\Phi. \tag{3}$$

This can be solved in a closed form [2], [3], [4]:

$$\Phi(z,\bar{z}) = \omega(z) e^{\frac{1}{2\pi i}\int\int_D \frac{A(\zeta,\bar{\zeta})}{\zeta - z} d\zeta \wedge d\bar{\zeta}}, \tag{4}$$

Journal of Physics: Conference Series **766** (2016) 012017 doi:10.1088/1742-6596/766/1/012017

providing following representation for q-binomial [1]:

$$(x+iy)_q^n = (x+iy)^n \left(\prod_{k=0}^{n-1} \frac{[2]_{q^k}}{2}\right) \exp\left[\frac{1}{2\pi i} \int\int_D \frac{1-q}{\zeta - z} \sum_{k=1}^{n-1} \frac{[k]_q}{(1+q^k)\zeta + (1-q^k)\bar{\zeta}} d\zeta \wedge d\bar{\zeta}\right]. \quad (5)$$

It was shown that every complex q-analytic function $\psi(z)$ determines a quantum state in the q-analytic Fock-Bargmann representation [1]. So that the orthonormal basis states $|n>$ in this representation are given by complex q-binomials (2),

$$\psi_n(z) = \frac{(x+iy)_q^n}{\sqrt{[n]_q!}}. \quad (6)$$

In the present paper we are going to extend the concept of q-analytic function in two different directions. First we describe the double q-analytic functions and expansion of them in the q-Hermite binomials. Then, as a hyperbolic version of q-analyticity we introduce the q-traveling waves.

2. Double analytic and q-analytic functions

2.1. Double analytic function

A complex valued function $f(z,w)$ of two complex variables z and w, analytic in both variables $\frac{\partial}{\partial \bar{z}} f(z,w) = \frac{\partial}{\partial \bar{w}} f(z,w) = 0$ we call a double analytic if it satisfies the following equation

$$\frac{1}{2}\left(\frac{\partial}{\partial z} + i\frac{\partial}{\partial w}\right) f(z,w) = 0. \quad (7)$$

As an example, $f(z,w) = (z+iw)^n$ is double analytic function. In general, the double analytic function can be written as a power series in complex binomials $f(z,w) = \sum_{n=0}^{\infty} a_n (z+iw)^n$. For such double analytic binomials we can derive the following Hermite binomial formula

$$(z+iw)^n = \frac{1}{2^n} \sum_{k=0}^{n} \binom{n}{k} i^k H_{n-k}(z) H_k(w). \quad (8)$$

This formula is motivated by description of the Dirac type δ - function for quantum states in the Fock-Bargmann representation [6]. By generalization of this approach to two complex variables we find expansion (8). In the present paper we derive the q-analogue of this identity,

$$(z+iw)_q^n = \frac{1}{[2]_q^n} \sum_{k=0}^{n} \begin{bmatrix} n \\ k \end{bmatrix}_q i^k q^{\frac{k(k-1)}{2}} H_{n-k}(z;q) H_k(qw, \frac{1}{q}), \quad (9)$$

from which (8) follows as a particular case at $q = 1$.

2.2. Double q-analytic function

A complex valued function $f(z,w)$ of two complex variables z and w, analytic in both variables and satisfying equation

$$\bar{D}_{z,w} f(z,w) = \frac{1}{2}\left(D_q^z + i D_{\frac{1}{q}}^w\right) f(z,w) = 0 \quad (10)$$

we call as a *double q-analytic* function, where $D_q^z f(z) = \frac{f(qz)-f(z)}{(q-1)z}$, $D_{\frac{1}{q}}^w f(w) = \frac{f(\frac{w}{q})-f(w)}{(\frac{1}{q}-1)w}$.

As an example, complex q-binomial $(z+iw)_q^2 = z^2 + [2]_q iwz - qw^2$, is analytic in z and w, and double q-analytic due to $\bar{D}_{z,w}(z+iw)_q^2 = 0$.

The complex q-binomial $(z+iw)_q^n$, for n-positive integer, is double q-analytic. This is why, any convergent power series

$$f(z,w)_q \equiv \sum_{n=0}^{\infty} a_n (z+iw)_q^n$$

represents a double q-analytic function. For example, analytic function of two complex variables

$$e_q(z+iw)_q = \sum_{n=0}^{\infty} \frac{(z+iw)_q^n}{[n]_q!} \tag{11}$$

is double q-analytic in the disk $|w| < \frac{q}{q-1}$, for $q > 1$, and admits factorization

$$e_q(z+iw)_q = e_q(z) e_{\frac{1}{q}}(iw) = e_q(z)(\cos_{\frac{1}{q}} w + i \sin_{\frac{1}{q}} w) \tag{12}$$

in terms of Jackson's exponential functions

$$e_q(x) = \sum_{n=0}^{\infty} \frac{x^n}{[n]_q!},$$

where q-numbers and q-factorials are defined as follows:

$$[n]_q = \frac{q^n - 1}{q - 1}, \qquad [n]_q! = [1]_q [2]_q ... [n]_q. \tag{13}$$

As a central result of the present paper, we prove an expansion (9) of the q-binomial in terms of q-Hermite polynomials. Here, the q-Hermite polynomials are defined according to generating function [7]

$$e_q(-t^2) e_q([2]_q tx) = \sum_{n=0}^{\infty} H_n(x;q) \frac{t^n}{[n]_q!}. \tag{14}$$

In the limit $q \to 1$, (9) reduces to the Hermite binomial formula (8).

2.2.1. q-binomial and q-translation operator Our proof is based on representation of q-binomial as a q-translation:

$$e_{\frac{1}{q}}^{aD_q^x} x^n = (x+a)_q^n. \tag{15}$$

This formula can be proved by expanding q-exponential function as follows

$$\begin{aligned} e_{\frac{1}{q}}^{aD_q^x} x^n &= \sum_{k=0}^{\infty} \frac{a^k (D_q^x)^k}{[k]_{\frac{1}{q}}!} x^n = \sum_{k=0}^{n} \frac{a^k (D_q^x)^k}{[k]_{\frac{1}{q}}!} x^n = \sum_{k=0}^{n} a^k \frac{[n]_q [n-1]_q ... [n-k+1]_q}{[k]_q!} q^{\frac{k(k-1)}{2}} x^{n-k} \\ &= \sum_{k=0}^{n} \begin{bmatrix} n \\ k \end{bmatrix}_q q^{\frac{k(k-1)}{2}} x^{n-k} a^k = (x+a)_q^n, \end{aligned} \tag{16}$$

where $(D_q^x)^k = [n]_q [n-1]_q ... [n-k+1] x^{n-k}$, $\qquad [k]_{\frac{1}{q}}! = \frac{1}{q^{\frac{k(k-1)}{2}}} [k]_q!$.

3. q-binomial expansion in terms of q-Hermite polynomials

Complex q-binomial expansion in terms of q-Hermite polynomials we find in the next form:

$$e_{\frac{1}{q}}^{iyD_q^x} x^n = (x+iy)_q^n = \frac{1}{[2]_q^n} \sum_{k=0}^{n} \begin{bmatrix} n \\ k \end{bmatrix}_q i^k q^{\frac{k(k-1)}{2}} H_{n-k}(x;q) H_k(qy;\frac{1}{q}). \quad (17)$$

In order to prove this formula, from the generating function for q-Hermite polynomials (14) we get

$$H_n(x;q) = e_q(-\frac{1}{[2]_q^2}(D_q^x)^2)[2]_q^n x^n. \quad (18)$$

By replacing $n \to k, \quad x \to qy, \quad q \to \frac{1}{q}$ we obtain

$$H_k(qy;\frac{1}{q}) = e_{\frac{1}{q}}(-\frac{1}{[2]_{\frac{1}{q}}^2}(D_{\frac{1}{q}}^{qy})^2)[2]_{\frac{1}{q}}^k (qy)^k. \quad (19)$$

As the next step, to both sides of Gauss' binomial formula (2) we apply the operator

$$\frac{1}{[2]_q^n} e_q(-\frac{1}{[2]_q^2}(D_q^x)^2)[2]_q^{n-k} e_{\frac{1}{q}}(-\frac{1}{[2]_{\frac{1}{q}}^2}(D_{\frac{1}{q}}^{qy})^2)[2]_{\frac{1}{q}}^k q^k. \quad (20)$$

Then, by using (18) and (19), from the RHS of (2) we obtain

$$\frac{1}{[2]_q^n} \sum_{k=0}^{n} \begin{bmatrix} n \\ k \end{bmatrix}_q i^k q^{\frac{k(k-1)}{2}} H_{n-k}(x;q) H_k(qy;\frac{1}{q}). \quad (21)$$

From another side the LHS of (2) becomes

$$e_q(-\frac{1}{[2]_q^2}(D_q^x)^2) e_{\frac{1}{q}}(-\frac{1}{[2]_{\frac{1}{q}}^2}(D_{\frac{1}{q}}^{qy})^2) e_{\frac{1}{q}}^{iyD_q^x} x^n,$$

and due to $(D_{\frac{1}{q}}^{qy})^2 e_{\frac{1}{q}}^{iyD_q^x} x^n = (\frac{i}{q})^2 (D_q^x)^2 e_{\frac{1}{q}}^{iyD_q^x} x^n$ and as follows

$$e_{\frac{1}{q}}(-\frac{1}{[2]_{\frac{1}{q}}^2}(D_{\frac{1}{q}}^{qy})^2) e_{\frac{1}{q}}^{iyD_q^x} x^n = e_{\frac{1}{q}}(\frac{1}{[2]_q^2}(D_q^x)^2) e_{\frac{1}{q}}^{iyD_q^x} x^n,$$

finally we find

$$e_{\frac{1}{q}}^{iyD_q^x} e_q(-\frac{1}{[2]_q^2}(D_q^x)^2 + \frac{1}{[2]_q^2}(D_q^x)^2)_q x^n = e_{\frac{1}{q}}^{iyD_q^x} x^n = (x+iy)_q^n. \quad (22)$$

By changing real variables to the complex ones $x \to z$ and $y \to w$, we obtain the expansion of double q-analytic q-binomial in terms of q-Hermite polynomials (9).

4. q-traveling waves

Here as another, hyperbolic extension of q-analytic functions, we consider the q-analogue of traveling waves as a solution of the q-wave equation. Direct extension of traveling waves to q-traveling waves is not possible. This happens due to the lack of the chain rule in q-calculus and as follows, impossibility to use moving frame as an argument of the wave function. Moreover, if in the Fourier harmonics $f(x,t) = e^{i(kx-\omega t)}$, we try naively to replace the exponential function by Jackson's q-exponential, $f_q(x,t) = e_q(i(kx-\omega t))$, then we find that it doesn't work due to the absence of factorization for q-exponential function $e_q(i(kx-\omega t)) \neq e_q(ikx)\, e_q(i\omega t)$.

Journal of Physics: Conference Series **766** (2016) 012017 doi:10.1088/1742-6596/766/1/012017

4.1. q-wave equation

This is why, here we propose another way. First we observe that q-binomials

$$(x \pm ct)_q^n = (x \pm ct)(x \pm qct)...(x \pm q^{n-1}ct) \tag{23}$$

for $n = 0, \pm 1, \pm 2, ...$, satisfy the first order one-directional q-wave equations

$$\left(D^t_{\frac{1}{q}} \mp c D^x_q\right)(x \pm ct)_q^n = 0, \tag{24}$$

which are hyperbolic analogs of q-analyticity (and anti-analyticity). Then, the Laurent series expansion in terms of these q-binomials determines the q-analog of traveling waves

$$f(x \pm ct)_q = \sum_{n=-\infty}^{\infty} a_n (x \pm ct)_q^n.$$

Due to (24) the q-binomials (23) satisfy the q-wave equation

$$\left((D^t_{\frac{1}{q}})^2 - c^2 (D^x_q)^2\right) u(x,t) = 0 \tag{25}$$

and the general solution of this equation is expressed in the form of q-traveling waves

$$u(x,t) = F(x + ct)_q + G(x - ct)_q, \tag{26}$$

where

$$F(x + ct)_q = \sum_{n=-\infty}^{\infty} a_n (x + ct)_q^n, \quad G(x - ct)_q = \sum_{n=-\infty}^{\infty} b_n (x - ct)_q^n$$

are arbitrary functions.

4.2. q-D'Alembert solution

This general solution can be fixed by the IVP for the q-wave equation

$$\left[\left(D^t_{\frac{1}{q}}\right)^2 - c^2 (D^x_q)^2\right] u(x,t) = 0, \tag{27}$$

$$u(x,0) = f(x), \tag{28}$$

$$D^t_{\frac{1}{q}} u(x,0) = g(x), \tag{29}$$

where $-\infty < x < \infty$. In the D'Alembert form the solution is

$$u(x,t) = \frac{f(x+ct)_q + f(x-ct)_q}{2} + \frac{1}{2c} \int_{(x-ct)_q}^{(x+ct)_q} g(x') d_q x', \tag{30}$$

where the Jackson integral is defined as

$$\int_{(x-ct)_q}^{(x+ct)_q} g(x') d_q x' = (1-q)(x+ct) \sum_{j=0}^{\infty} q^j g(q^j (x+ct))_q - (1-q)(x-ct) \sum_{j=0}^{\infty} q^j g(q^j (x-ct))_q.$$

If the initial velocity is zero, $g(x) = 0$, the formula reduces to

$$u(x,t) = \frac{1}{2} \left(f(x+ct)_q + f(x-ct)_q\right). \tag{31}$$

It should be noted here that the q-traveling wave is not traveling wave in the standard sense, since it is not preserving shape during evolution. It can be seen from simple observation. The traveling wave polynomial $(x - ct)_q^n = (x - ct)(x - qct)(x - q^2ct)...(x - q^{n-1}ct)$ includes the set of moving frames (as zeros of this polynomial) with geometrically re-scaled set of speeds $(c, qc, q^2c, ..., q^{n-1}c)$. It means that zeros of this polynomial are moving with different speeds and therefore the shape of polynomial wave is not preserving. Only in the linear case and in the case $q = 1$, when speeds of all frames coincide, we are getting standard traveling wave.

4.3. Example: q-Gaussian traveling wave

For the initial function in the Gaussian form: $u(x,0) = e^{-x^2}$ in the standard case $q = 1$ we have the Gaussian traveling wave with the same shape $u(x,t) = e^{-(x-ct)^2}$ (see Figure 1). For the q-traveling wave with the same initial condition $u(x,0) = e^{-x^2}$, we have solution

$$u(x,t) = \left(e^{-(x-ct)^2}\right)_q \equiv \sum_{n=0}^{\infty} \frac{(-1)^n}{n!}(x-ct)_q^{2n}$$

(see Figure 2). As we can see, during evolution the shape of the wave is changing as an amplitude is growing.

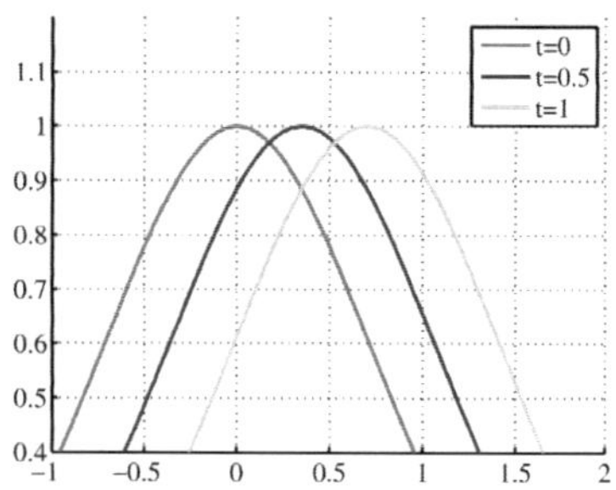

Figure 1. Gaussian traveling wave at time intervals t=0, t=0.5 and t=1.

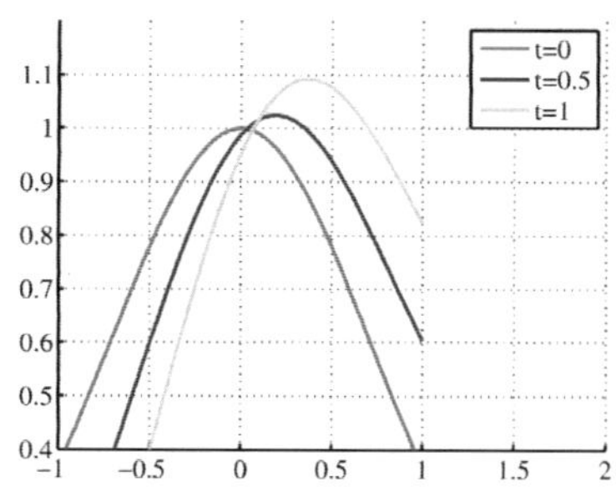

Figure 2. q-Gaussian traveling wave at time intervals t=0, t=0.5 and t=1.

5. Conclusions

As a potential application of our results we should mention that an analytic function of two complex variables can be related to the tensor product of Glauber coherent states. Then, the double analytic functions, as well as the q-double analytic functions correspond to some symmetry restrictions on these states. Expansion of these states in binomial and Hermite binomials form would reflect some entanglement properties of these states. These questions are under the study now. Interesting problem also is to find the symmetry group of the q-wave equation as a q-deformation of the Lorentz group.

5.1. Acknowledgments

This work was supported by Izmir Institute of Technology. One of the authors (S. Nalci Tumer) was partially supported by TUBITAK scholarship for graduate students. We would like to thank Prof. Mourad Ismail for stimulating discussions of q-translation operator and sharing his lecture notes.

6. References

[1] Pashaev O K and Nalci S 2014 q-analytic functions, fractals and generalized analytic functions *J. Phys. A: Math. Theor.* **47** 045204
[2] Vekua I N 1962 *Generalized Analytic Fucntions* (New York: Pergamon)
[3] Theodoresco N 1936 La derivee areolaire *Ann. Roumains Math. Cahier, Bucarest* **3**
[4] Ablowitz M J and Fokas A S 1997 *Complex Variables: Introduction and Applications* (Cambridge: Cambridge University Press)
[5] Perelomov A 1986 *Generalized Coherent states and their applications* (Berlin: Springer)
[6] Floratos E G 1991 The many-body problem for q-oscillators *J. Phys. A: Math. Gen.* **24**
[7] Nalci S and Pashaev O K 2010 q-analog of shock soliton solution *J. Phys. A: Math. Theor.* **43** 445205
[8] Kac V and Cheung P 2002 *Quantum Calculus* (New York: Springer)
[9] Nalci S and Pashaev O K 2016 Double q-analytic q-Hermite binomial formula and q-traveling waves *Preprint* Math.QA/ 1602.00921v1

International Conference on Quantum Science and Applications (ICQSA-2016) IOP Publishing
Journal of Physics: Conference Series **766** (2016) 012018 doi:10.1088/1742-6596/766/1/012018

A polynomial type Jost solution and spectrum of the self-adjoint quantum Dirac system

N Coskun and N Yokus

Department of Mathematics, Karamanoglu Mehmetbey University, Karaman, 70100, Turkey

E-mail: cannimet@kmu.edu.tr

Abstract. This paper is devoted to investigation of a polynomial type Jost solution for the self-adjoint quantum Dirac system. After examination of analytical features and asymptotic behaviour of the Jost solution, continuous and discrete spectrum and some properties of the eigenvalues of the operator L generated by the quantum difference system have been discussed.

1. Introduction

Spectral analysis of differential and difference operators is of great importance for the solutions of certain problems in many areas including engineering, economics, quantum mechanics and mathematical physics [1-5]. In this context, Dirac system of differential and discrete operators have been studied in [6-8].

In the last years, an important effort has been devoted to quantum calculus [9]. As a consequence of developments in quantum theory, quantum difference equations has been subject matter of various studies [10-12]. In particular, spectral analysis of quantum difference equations has been studied in [13-15]. However, the Dirac system of quantum difference equations including a polynomial type Jost solution has not been examined in the known literature yet.

In this paper, we assume $q > 1$ and use the notation

$$q^{\mathbb{N}_0} := \left\{ q^n : n \in \mathbb{N}_0 \right\}$$

where $\mathbb{N}_0$ indicates the set of nonnegative integers. The q-derivative of a function $f : q^{\mathbb{N}} \to \mathbb{C}$ is defined by

$$f^{\Delta}(t) := \frac{f(qt) - f(t)}{\mu(t)}, \ \forall t \in q^{\mathbb{N}},$$

where $\mu(t) = (q-1)t$ is the graininess function [10]. Hereafter, we will denote the Hilbert space $l_2(q^{\mathbb{N}}, \mathbb{C}^2)$ including all sequences $y = \{y(t)\} = \begin{Bmatrix} y^{(1)}(t) \\ y^{(2)}(t) \end{Bmatrix}$ with the inner product,

$$\langle y, f \rangle_{(q^{\mathbb{N}}, \mathbb{C}^2)} := \sum_{t \in q^{\mathbb{N}}} y^{(1)}(t)\overline{f^{(1)}(t)} + y^{(2)}(t)\overline{f^{(2)}(t)}, \quad y, f : q^{\mathbb{N}} \to \mathbb{C}$$

and the norm

$l_2(q^{\mathbb{N}},\mathbb{C}^2)$
$$\|y(t)\|_{(q^{\mathbb{N}},\mathbb{C}^2)} = \left(\sum_{t\in q^{\mathbb{N}}} \left|y^{(1)}(t)\right|^2 + \left|y^{(2)}(t)\right|^2 \right)^{\frac{1}{2}} \quad \text{for} \quad y : q^{\mathbb{N}} \to \mathbb{C}$$

by .

Let us consider the system

$$\begin{cases} (y^{(2)}(t))^{\Delta} + p(t)y^{(1)}(t) = \dfrac{\lambda y^{(1)}(t)}{\mu(t)}, \\ -(y^{(1)}(\frac{t}{q}))^{\Delta} + r(t)y^{(2)}(t) = \dfrac{\lambda y^{(2)}(t)}{\mu(t)}, \end{cases} \quad t\in q^{\mathbb{N}} \tag{1.1}$$

where $\{p(t)\}_{t\in q^{\mathbb{N}}}$, $\{r(t)\}_{t\in q^{\mathbb{N}}}$ and $\{\mu(t)\}_{t\in q^{\mathbb{N}}}$ are real sequences for all $t\in q^{\mathbb{N}}$ and λ is a spectral parameter. Note that, the system of equations (1.1) is quantum analogue of the well-known Dirac system

$$\begin{pmatrix} 0 & 1 \\ -1 & 0 \end{pmatrix}\begin{pmatrix} y_1' \\ y_2' \end{pmatrix} + \begin{pmatrix} p(x) & 0 \\ 0 & q(x) \end{pmatrix}\begin{pmatrix} y_1 \\ y_2 \end{pmatrix} = \lambda \begin{pmatrix} y_1 \\ y_2 \end{pmatrix}$$

([6], Chapter 2). For this reason, the system (1.1) is called a quantum Dirac system. It is worth to point out here that the construction of the quantum Dirac system (1.1) is completely different from other studies [7, 8] because of the quantum derivative of a function. $l_2(q^{\mathbb{N}},\mathbb{C}^2)$

Let us define the operator L generated in by the following system of quantum difference expression:

$$(ly)(t) = \begin{pmatrix} y^{(2)}(qt) - y^{(2)}(t) + p(t)\mu(t)y^{(1)}(t) \\ y^{(1)}(\frac{t}{q}) - y^{(1)}(t) + r(t)\mu(t)y^{(2)}(t) \end{pmatrix}$$

where $\{p(t)\}_{t\in q^{\mathbb{N}}}$, $\{r(t)\}_{t\in q^{\mathbb{N}}}$ and $\{\mu(t)\}_{t\in q^{\mathbb{N}}}$ are real sequences. It is clear that the system of equations (1.1) can be rewritten as

$$(Ly)(t) = \lambda y(t), \quad t\in q^{\mathbb{N}}.$$

The set up of this paper organized as follows: Section 2 is concerned with the investigation of the polynomial type Jost solution of the system (1.1) with the boundary condition

$$y^{(1)}(1) = 0 \tag{1.2}$$

and investigate analytic properties and asymptotic behavior of the Jost solution. Section 3 presents continuous and discrete spectrum and some properties of the eigenvalues of the boundary value problem BVP (1.1)-(1.2).

2. Jost solution and Jost function of (1.1)

Let the real sequences $\{p(t)\}_{t\in q^{\mathbb{N}}}$, $\{r(t)\}_{t\in q^{\mathbb{N}}}$ and $\{\mu(t)\}_{t\in q^{\mathbb{N}}}$ satisfy the condition

$$\sum_{t\in q^{\mathbb{N}}} \frac{\ln t}{\ln q}\left(|p(t)\mu(t)| + |r(t)\mu(t)|\right) < \infty . \tag{2.1}$$

Theorem 2.1. Under the condition (2.1), the system of equations (1.1) has unique solutions

International Conference on Quantum Science and Applications (ICQSA-2016) IOP Publishing
Journal of Physics: Conference Series **766** (2016) 012018 doi:10.1088/1742-6596/766/1/012018

$$f(t,z)=\begin{pmatrix} f^{(1)}(t,z) \\ f^{(2)}(t,z) \end{pmatrix}=\left[\begin{pmatrix} 1 & 0 \\ 0 & 1 \end{pmatrix}+\sum_{r\in q^{\mathbb{N}_0}} A_{tr} z^{2\log_q r}\right]\begin{pmatrix} z \\ -i \end{pmatrix} z^{2\log_q t}, \tag{2.2}$$

$$f^{(1)}(1,z)=z+\sum_{r\in q^{\mathbb{N}_0}} A_{1r}^{\ 11} z^{2\log_q r+1}-i\sum_{r\in q^{\mathbb{N}_0}} A_{1r}^{\ 12} z^{2\log_q r}, \tag{2.3}$$

for $\lambda=-iz-(iz)^{-1}$, $t\in q^{\mathbb{N}}$ and $|z|=1$ where $A_{tr}=\begin{pmatrix} A_{tr}^{\ 11} & A_{tr}^{\ 12} \\ A_{tr}^{\ 21} & A_{tr}^{\ 22} \end{pmatrix}$.

Proof. If we substitute the equation (2.2) in the system of equations (1.1) and take $\lambda=-iz-(iz)^{-1}$ and $|z|=1$, then we obtain

$$A_{tq}^{\ 12}=-\sum_{s\in[qt,\infty)\cap q^{\mathbb{N}}} p(s)\mu(s)+r(s)\mu(s),$$

$$A_{tq}^{\ 11}=\sum_{s\in[qt,\infty)\cap q^{\mathbb{N}}} p(s)\mu(s)A_{sq}^{\ 12},$$

$$A_{tq}^{\ 22}=\sum_{s\in[t,\infty)\cap q^{\mathbb{N}}} p(s)\mu(s)A_{sq}^{\ 12},$$

$$A_{tq}^{\ 21}=A_{tq}^{\ 12}+p(t)\mu(t)A_{tq}^{\ 11}+\sum_{s\in[qt,\infty)\cap q^{\mathbb{N}}}\left[r(s)\mu(s)A_{sq}^{\ 22}+p(s)\mu(s)A_{sq}^{\ 11}\right],$$

$$A_{tq^2}^{\ 12}=-\sum_{s\in[qt,\infty)\cap q^{\mathbb{N}}}\left[p(s)\mu(s)A_{sq}^{\ 11}+r(s)\mu(s)A_{sq}^{\ 22}\right],$$

$$A_{tq^2}^{\ 11}=-A_{qt,q}^{\ 22}+\sum_{s\in[qt,\infty)\cap q^{\mathbb{N}}}\left[p(s)\mu(s)A_{s,q^2}^{\ 12}-r(s)\mu(s)A_{sq}^{\ 21}\right],$$

$$A_{tq^2}^{\ 22}=-A_{tq}^{\ 11}+\sum_{s\in[t,\infty)\cap q^{\mathbb{N}}}\left[p(s)\mu(s)A_{s,q^2}^{\ 12}-r(sq)\mu(sq)A_{sq,q}^{\ 21}\right],$$

$$A_{tq^2}^{\ 21}=A_{t,q^2}^{\ 12}+\sum_{s\in[t,\infty)\cap q^{\mathbb{N}}}\left[p(s)\mu(s)A_{s,q^2}^{\ 11}+r(sq)\mu(sq)A_{sq,q^2}^{\ 22}\right],$$

for $r\geq q^3$

$$A_{tr}^{\ 12}=A_{tq,\frac{r}{q^2}}^{\ 21}-\sum_{s\in[tq,\infty)\cap q^{\mathbb{N}}} r(s)\mu(s)A_{s,\frac{r}{q}}^{\ 22}+p(s)\mu(s)A_{s,\frac{r}{q}}^{\ 11},$$

$$A_{tr}^{\ 11}=-A_{tq,\frac{t}{q^2}}^{\ 22}+\sum_{s\in[tq,\infty)\cap q^{\mathbb{N}}} p(s)\mu(s)A_{sr}^{\ 12}-r(s)\mu(s)A_{s,\frac{r}{q}}^{\ 21},$$

$$A_{tr}^{\ 22}=-A_{t,\frac{t}{q}}^{\ 11}+\sum_{s\in[t,\infty)\cap q^{\mathbb{N}}}\left(p(s)\mu(s)A_{sr}^{\ 12}-r(sq)\mu(sq)A_{sq,\frac{r}{q}}^{\ 21}\right),$$

$$A_{tr}^{\ 21}=A_{tr}^{\ 12}+p(t)\mu(t)A_{tr}^{\ 11}+\sum_{s\in[tq,\infty)\cap q^{\mathbb{N}}}\left(r(s)\mu(s)A_{sr}^{\ 22}+p(s)\mu(s)A_{sr}^{\ 11}\right).$$

Based on the condition (2.1), the series in the definition of $A_{tr}^{\ ij}$ $(i,j=1,2)$ are absolutely convergent. Hence, $A_{tr}^{\ ij}$ $(i,j=1,2)$ can be uniquely determined by $p(t)$, $r(t)$ and $\mu(t)$ $(t\in q^{\mathbb{N}})$, i.e., the system (1.1) has the solution given by the equations (2.2) and (2.3). □

The solution f is called Jost solution of the system of equations (1.1). Using the inequalities for $A_{tr}^{\ ij}$ $(i,j=1,2)$ given in Theorem (2.1), we find

International Conference on Quantum Science and Applications (ICQSA-2016) IOP Publishing
Journal of Physics: Conference Series **766** (2016) 012018 doi:10.1088/1742-6596/766/1/012018

$$\left|A_{tr}^{ij}\right| \le C \sum_{s\in\left[tq^{\left[\frac{\ln r}{2\ln q}\right]},\infty\right)\cap q^{\mathbb{N}}}^{\infty} \left(\left|p(s)\mu(s)\right| + \left|r(s)\mu(s)\right|\right), \quad i,j=1,2, \tag{2.4}$$

by induction, where $\left[\dfrac{\ln r}{2\ln q}\right]$ is the integer part of $\dfrac{\ln r}{2\ln q}$, $C>0$ is a constant and $t,r\in q^{\mathbb{N}}$.

It is clear from the equations (2.2), (2.3) and (2.4) that $\begin{pmatrix} f^{(1)}(t,z) \\ f^{(2)}(t,z)\end{pmatrix}$ has analytic continuation from $\{z:|z|=1\}$ to $D:=\{z:|z|<1\}\setminus\{0\}$.

Theorem 2.2. If the condition (2.1) holds then the Jost solution f satisfies the following asymptotic:

$$\begin{pmatrix} f^{(1)}(t,z) \\ f^{(2)}(t,z)\end{pmatrix} = \left[\begin{pmatrix} 1 & 0 \\ 0 & 1\end{pmatrix} + o(1)\right]\begin{pmatrix} z \\ -i\end{pmatrix} z^{2\log_q t}, \ z\in\overline{D}:=\{z:|z|\le 1\}\setminus\{0\}, t\to\infty. \tag{2.5}$$

Proof. It can be written from the equation (2.2) that

$$f^{(1)}(t,z)z^{-2\log_q t-1} = 1 + \sum_{r\in q^{\mathbb{N}_0}} A_{tr}^{11} z^{2\log_q r} - i\sum_{r\in q^{\mathbb{N}_0}} A_{tr}^{12} z^{2\log_q r-1}. \tag{2.6}$$

From the equations (2.4) and (2.6), it is seen that

$$\left|f^{(1)}(t,z)z^{-2\log_q t-1}\right| \le 1 + \sum_{r\in q^{\mathbb{N}_0}} \left|A_{tr}^{11}\right| + \sum_{r\in q^{\mathbb{N}_0}} \left|A_{tr}^{12}\right|$$

$$\le 1 + 2C\sum_{r\in q^{\mathbb{N}_0}} \sum_{s\in\left[tq^{\left[\frac{\ln r}{2\ln q}\right]},\infty\right)\cap q^{\mathbb{N}}} \left(\left|p(s)\mu(s)\right| + \left|r(s)\mu(s)\right|\right)$$

$$\le 1 + 2C\sum_{r\in q^{\mathbb{N}_0}} \sum_{s\in[tq,\infty)\cap q^{\mathbb{N}}} \left(\frac{\ln s}{\ln q} - \frac{\ln t}{\ln q}\right)\left(\left|p(s)\mu(s)\right| + \left|r(s)\mu(s)\right|\right)$$

$$\le 1 + 2C\sum_{s\in[tq,\infty)\cap q^{\mathbb{N}}} \frac{\ln s}{\ln q}\left(\left|p(s)\mu(s)\right| + \left|r(s)\mu(s)\right|\right) \tag{2.7}$$

where $C>0$ is a constant. Then, we get from the equation (2.7) that

$$f^{(1)}(t,z) = z^{2\log_q t+1}\left[1+o(1)\right], \ z\in\overline{D}, t\to\infty. \tag{2.8}$$

In a similar way to equation (2.8), we have

$$f^{(2)}(t,z) = -iz^{2\log_q t}\left[1+o(1)\right], \ z\in\overline{D}, t\to\infty. \tag{2.9}$$

From equations (2.8) and (2.9), we obtain the equation (2.5).

□

3. Main Results

Now, we shall give some theorems and definitions to prove our main results. We begin with investigating continuous spectrum of the operator L.

Theorem 3.1. Assume the condition (2.1) satisfies. Then $\sigma_c(L)=[-2,2]$, where $\sigma_c(L)$ represents the continuous spectrum of the operator L.

$$l_2(q^{\mathbb{N}},\mathbb{C}^2)$$

Proof. Let L_0 be the operator generated in by the following system of quantum difference expression:

$$\left(l_0 y\right)(t)=\begin{pmatrix} y^{(2)}(qt)-y^{(2)}(t) \\ y^{(1)}(\frac{t}{q})-y^{(1)}(t) \end{pmatrix} \qquad l_2(q^{\mathbb{N}},\mathbb{C}^2)$$

with the boundary condition $y^{(1)}(1)=0$. We also define the operator L_1 in by the following

$$\left(l_1 y\right)(t)=\begin{pmatrix} p(t)\mu(t)y^{(1)}(t) \\ r(t)\mu(t)y^{(2)}(t) \end{pmatrix}.$$

One can easily observe that the operator L_0 is self-adjoint and L_1 is compact operator [16]. Also, $L=L_0+L_1$ holds. From the Weyl Theorem [17] of a compact perturbation, we find

$$\sigma_c(L)=\sigma_c(L_0)=\left[-2,2\right]. \qquad \square$$

Definition 3.1. The Wronskian of two solutions $y=\left\{y(t,z)\right\}_{t\in q^{\mathbb{N}}}$ and $u=\left\{u(t,z)\right\}_{t\in q^{\mathbb{N}}}$ of (1.1) is defined by $W[y,u](t)=y(t,z)u(qt,z)-y(qt,z)u(t,z)$ for $t\in q^{\mathbb{N}}$.

Let $\varphi(z)=\varphi(\lambda)=\left\{\varphi(t,\lambda)\right\}$, $t\in q^{\mathbb{N}}$ be the solution of the system of equations (1.1) subject to the initial conditions

$$\varphi^{(1)}(1,z)=0, \quad \varphi^{(2)}(q,z)=1.$$

Then from definition (3.1)

$$\begin{aligned} W\left[f(z),\varphi(z)\right] &= f^{(1)}(t,z)\varphi^{(2)}(qt,z)-f^{(2)}(qt,z)\varphi^{(1)}(t,z) \\ &= f^{(1)}(1,z). \end{aligned} \qquad (3.1)$$

Since the operator L is self-adjoint, the eigenvalues of L is real valued. By using equation (3.1) and the definition of eigenvalues, we get

$$\sigma_d(L)=\left\{\lambda:\lambda=-iz-(iz)^{-1}, iz\in\left(-1,0\right)\cup\left(0,1\right), f^{(1)}(1,z)=0\right\}, \qquad (3.2)$$

where $\sigma_d(L)$ sembolizes the eigenvalues of the operator L.

Definition 3.2. The multiplicity of a zero of $f^{(1)}(1,z)$ is called the multiplicity of the corresponding eigenvalue or spectral singularity of BVP (1.1), (1.2).

Theorem 3.2. Under the condition (2.1), the operator L has a finite number of real eigenvalues in D.

Proof. In order to prove the theorem, it is necessary and sufficient to show that the function $f^{(1)}(1,z)$ has finite number of real zeros in D. The accumulation points of the zeros of the analytic function $f^{(1)}(1,z)$ can take real values $-i$, 0 and i. It is known that L is self-adjoint bounded operator. So its eigenvalues is different from infinity. Assume $z_0=0$ is a zero of the function $f^{(1)}(1,z)$. But in this case the eigenvalue λ is infinite. Therefore $z_0=0$ is not a zero of the function $f^{(1)}(1,z)$. Now, assume $z=\pm i$. In this case, $\lambda=\pm 2$ and D is bounded. From Theorem (3.1), ± 2 are elements of continuous spectrum of the operator L. From operator theory, discrete spectrum of a self-adjoint operator and continuous spectrum of this operator are distinct. Hence, the set of zeros of the function $f^{(1)}(1,z)$ in D is finite from the Bolzano Weierstrass Theorem. It is obvious from (3.2) that the eigenvalues of the operator L are real. This completes the proof. $\square$

International Conference on Quantum Science and Applications (ICQSA-2016) IOP Publishing
Journal of Physics: Conference Series **766** (2016) 012018 doi:10.1088/1742-6596/766/1/012018

References

[1] Naimark M A 1960 Investigation of the spectrum and expansion of eigenfunctions of a non-selfadjoint operator of second order on a semi-axis *AMS Transl.* **2** pp 103-193

[2] Lyance V E 1967 A differential opeator with spectral singularities I, II *AMS Transl.* **60** pp 185-225

[3] Agarwal R P and Wong P J Y 1997 *Advanced Topics in Difference Euations* (Dordrecht: Kluwer Academic Publishers Group)

[4] Bairamov E, Cakar O and Krall A M 2001 Nonselfadjoint difference operators and Jacobi matrices with spectral singularities *Math. Nachr.* **229** pp 5-14

[5] Kelley W G and Peterson A C 2001 *Difference Equations: An Introduction with Applications* (New York: Academic Press)

[6] Levitan B M and Sargsjan I S 1991 *Sturm-Liouville and Dirac Operators* (Dordrecht: Kluwer Academic Publishers Group)

[7] Bairamov E and Celebi A O 1999 Spectrum and spectral expansion for the non-selfadjoint discrete Dirac operators *Q. J. Math.* **50** pp 371-384

[8] Aygar Y and Olgun M 2014 Investigation of spectrum and the Jost solutions of discrete Dirac system on the whole axis *Journal of Inequalities and Applications* **73**

[9] Kac V and Cheung P 2002 *Quantum Calculus* (New York: Springer)

[10] Bohner M and Peterson A 2003 *Dynamic equations on time scales* (Boston: Birkhauser)

[11] Trijtzinsky W J 1933 Analytic theory of linear q-difference equations *Acta Math.* **61** pp 1-38

[12] Berg C and Ruffing A 2001 Generalized q-Hermite polynomials *Comm. Math. Phys.* **223** pp 29-46

[13] Adıvar M and Bohner M 2006 Spectral analysis of q-difference equations with spectral singularities *Math. Comput. Modelling* **43** pp 695-703

[14] Adıvar M and Bohner M 2006 Spectrum and principal vectors of second order q-difference equations *Indian J. Math.* **48** pp 17-33

[15] Adıvar M 2010 Quadratic Pencil of Difference Equations: Jost Solutions, Spectrum and Principal vectors *Quaestiones Mathematicae* **33** pp 305-323

[16] Lusternik L A and Sobolev V I 1974 *Elements of Functional Analysis* (New York: Halsted)

[17] Glazman I M 1965 *Direct Methods of Qualitative Spectral Analysis of Singular Differential Operators* (Jerusalem: Israel Program for Scientific Translations)

Journal of Physics: Conference Series **766** (2016) 012019 doi:10.1088/1742-6596/766/1/012019

Quantum Codes From Negacyclic Codes over Group Ring $(F_q + vF_q)\,G$ [1]

Mehmet E. Koroglu and Irfan Siap

Yıldız Technical University, Department of Mathematics, Faculty of Art and Sciences, 34220, Esenler, Istanbul-Turkey

E-mail: mkoroglu@yildiz.edu.tr, isiap@yildiz.edu.tr

Abstract. In this paper, we determine self dual and self orthogonal codes arising from negacyclic codes over the group ring $(F_q + vF_q)\,G$. By taking a suitable Gray image of these codes we obtain many good parameter quantum error-correcting codes over F_q.

1. Introduction and Preliminaries

Quantum error-correcting (QEC) codes play a crucial role in protecting quantum information. In recent years many researchers have been working to find quantum codes with good parameters over various fields. The construction of quantum codes via classical codes over F_2 was first introduced by Calderbank and Shor [4] and Steane [13] in 1996. This method, known as CSS construction, has received a lot of attention and it has allowed to find many good quantum stabilizer codes. Later, construction of quantum codes over larger alphabets from classical linear codes over F_q has shown by Ketkar et al. in [10]. One direction of the main research in quantum error correction codes is constructing quantum codes that have large minimum distances [9] for a given size and length. In [14], based on classical quaternary constacyclic linear codes, some parameters for quantum codes are obtained. In [8, 9], respectively based on classical negacyclic and constacyclic linear codes some parameters for quantum MDS codes are presented. In this work, we determine self-dual and self-orthogonal codes arising from constacyclic codes over the group ring $(F_q + vF_q)\,G$. Based on these codes we obtain some quantum codes with promising parameters.

$R = F_q + vF_q$ is a commutative, characteristic 3 ring with $v^2 = v$ or with a ring isomorphism $F_q\,[v] / \langle v^2 - v\rangle$. For a prime p and an integer k take $n = 2p^k$ then the set $G = 2\mathbb{Z}_n^*$ is a cyclic group of order $p^k - p^{k-1}$ and identity element $p^k + 1$. Then, the group ring RG is the set of all linear combinations in the form $u = \sum_{g\in G} \alpha_g g$ such that $\alpha_g \in R$ and only finitely many of the α_g's are non-zero. This set is a commutative ring with respect to the following binary operations

$$u + v = \sum_{g\in G} \alpha_g g + \sum_{g\in G} \beta_g g = \sum_{g\in G} (\alpha_g + \beta_g)\, g$$

[1] This research is supported by Yıldız Technical University Scientific Research Projects Coordination Department. Project Number: 2016-01-03-DOP01.

and

$$uv = \left(\sum_{g\in G} \alpha_g g\right)\left(\sum_{h\in G} \beta_h h\right) = \sum_{g,h\in G} \alpha_g \beta_h gh.$$

A non-zero element $u \in RG$ is a zero-divisor if and only if there exists a non-zero $v \in RG$ such that $uv = 0$. For a fixed listing $\{g_1, g_2, \dots, g_n\}$ of the elements of G the RG matrix of the element $w = \sum_{i=1}^{n} \alpha_{g_i} g_i \in RG$ is defined as

$$W = \begin{pmatrix} \alpha_{g_1^{-1}g_1} & \alpha_{g_1^{-1}g_2} & \cdots & \alpha_{g_1^{-1}g_n} \\ \alpha_{g_2^{-1}g_1} & \alpha_{g_2^{-1}g_2} & \cdots & \alpha_{g_2^{-1}g_n} \\ \vdots & \vdots & \vdots & \vdots \\ \alpha_{g_n^{-1}g_1} & \alpha_{g_n^{-1}g_2} & \cdots & \alpha_{g_n^{-1}g_n} \end{pmatrix}.$$

A group ring RG is isomorphic to a subring of the ring of $n \times n$ matrices over R [6].

The rank of an element $u = \sum_{g\in G} \alpha_g g$ in RG is the rank of the matrix U. The transpose of an element $u = \sum_{g\in G} \alpha_g g$ in RG is $u^T = \sum_{g\in G} \alpha_g g^{-1}$ or equivalently $u^T = \sum_{g\in G} \alpha_{g^{-1}} g$. Given an element $\alpha = \sum_{g\in G} \alpha_g g \in RG$, its support is the set $supp\,(\alpha) = \{g \in G |\, \alpha_g \neq 0\}$. The Hamming weight of an element $\alpha \in RG$ is the number of nonzero coefficient group elements in its support i.e., $w\,(\alpha) = |supp\,(\alpha)|$. The minimum weight of a submodule M in RG is $w\,(M) = \min\{|supp\,(\alpha)||\, 0 \neq \alpha \in M\}$. Let w_L denote the Lee weight and w_H denote the Hamming weight for the codes over $R = F_q + vF_q$. Then, we set

$$w_L\,(a + bv) = w_H\,(a, a + b)\,.$$

The definition of the weight immediately leads to a Gray map from R to F_q^2 which can be extended to $(F_q + vF_q)^n$:

$$\phi : R \to F_q^2,\ \phi\,(a + bv) = (a, a + b)\,.$$

This map is a distance preserving map from R to F_q^2. Let $x = \sum_{g\in G} \alpha_g g$, and $y = \sum_{g\in G} \beta_g g$ be two elements in the group ring RG. Then, the inner product of x and y is given by term-by-term multiplication of the coefficients of x and y, namely $\langle x, y\rangle = \sum_{g\in G} \alpha_g \beta_g$.

Let $a_1 + b_1 v$ and $a_2 + b_2 v$ be any two elements in R. Then, we have

$$\begin{aligned} & (a_1 + b_1 v)\,(a_2 + b_2 v) \\ = \ & a_1 a_2 + (a_1 b_2 + b_1 a_2 + b_1 b_2)\,v = 0 \\ \Leftrightarrow \ & a_1 a_2 = 0 \text{ and } a_1 b_2 + b_1 a_2 + b_1 b_2 = 0. \end{aligned}$$

Hence, the following relation is valid.

$$\begin{aligned} & (a_1, a_1{+}b_1)\,(a_2, a_2{+}b_2) \\ = \ & a_1 a_2 + a_1 a_2 + a_1 b_2 + b_1 a_2 + b_1 b_2 = 0. \end{aligned}$$

As a result of this fact, the map ϕ is a orthogonality preserving map. The map

$$\theta : RG \to R^n,\ \theta\left(\sum_{i=1}^{n} \alpha_i g_i\right) = (\alpha_1, \alpha_2, \dots, \alpha_n) \tag{1}$$

is an isomorphism from RG to R^n. Thus every element in RG can be considered as an n-tuple in R^n.

A linear code C of length n over R, is a submodule of R^n. A linear code of length n, dimension k, and minimum (Hamming) distance d over R is termed as an $[n,k,d]_q$ code [12]. The algebraic structure of constacyclic codes is described in detail in [3, 1]. Let n be a positive integer and α be a unit element of R. A linear code C of length n over R is said to be $\alpha-$constacyclic if for any codeword $(c_0, c_1, \ldots, c_{n-1}) \in C$ we have that $(\alpha c_{n-1}, c_0, c_1, \ldots, c_{n-2}) \in C$. If we take α as -1, then the code is called negacyclic.

The cyclic codes of length m are ideals in the quotient ring $R[x]/\langle x^m - 1\rangle$. Further, for a cyclic group C_m of order m we have $R[x]/\langle x^m - 1\rangle \cong RC_m$. Similar to the cyclic codes $e-$constacylic codes of length m can be viewed as ideals in the quotient ring $R[x]/\langle x^m - e\rangle$. Our aim is to construct a group G such that the isomorphism $R[x]/\langle x^m - e\rangle \cong RG$ exist, where $e \in F_q$ and $e \neq 0, 1$. For $n = 2p^k$, where p is an odd prime and k is an integer we show that the set of all doubled elements $G = 2\mathbb{Z}_n^*$ in $\mathbb{Z}_n^*$, is a multiplicative cyclic group of order $m = \varphi(n)$ (φ is the Euler totient function) with identity element $e = p^k + 1$, such that $e \neq 1$. Afterwards, we obtain self dual and self orthogonal $(p^k + 1)-$constacyclic codes of length $\varphi(n)$ over $(F_q + vF_q)$ by considering these codes as ideals in the group ring $(F_q + vF_q)G$.

Definition 1 *[7] Let u be a zero-divisor in RG, i.e. $uv = 0$ for some non-zero $v \in RG$. Let W be a submodule of RG with basis of group elements $S \subseteq G$. Then, a zero-divisor code is $C = \{ux | x \in W\} = uW$ or $C = \{xu | x \in W\} = Wu$.*

Definition 2 *[7] A zero-divisor u with rank$(U) = r$ is called a principal zero-divisor if and only if there exists a $v \in RG$ such that $uv = 0$ and rank$(V) = n - r$.*

Corollary 3 *[7] $C = \{xu | x \in W\}$ has a unique check element if and only if u is a principal zero divisor.*

The dual of a code with respect to the standard inner product forms a group ring encoding as well where the dual is defined by

$$C^\perp = \{y \in RG | \langle ux, y\rangle = 0, \forall x \in W\}.$$

Theorem 4 *[7] Let $u, v \in RG$ such that $uv = 0$. Let U and V be the RG matrices of u and v respectively, such that rank$(U) = r$ and rank$(V) = n-r$. Let W be a submodule over a basis $S \subset G$ of dimension r such that Su is linearly independent and $W^\perp$ denote the submodule over basis $G \backslash S$. Then, the dual code of $C = \{xu | x \in W\}$ is $C^\perp = \{xv^T | x \in W^\perp\} = \{y \in RG | yu^T = 0\}$.*

2. Constacyclic Codes over Group Ring $(F_q + vF_q)G$

In this section, we extend the notion of cyclic group ring codes to constacyclic group ring codes. Throughout this section, we assume p is an odd prime, $R = F_q + vF_q$ and $n = 2p^k$ under the restrictions $\gcd(q, \varphi(2p^k)) = 1$, and $p^k + 1 \neq 0, 1 \,(mod\ q)$.

Let $\mathbb{Z}_n$ be the set of integers modulo $n = 2p^k$. Let $G = 2\mathbb{Z}_n^* \subset \mathbb{Z}_n$ be the set of all doubled elements in $\mathbb{Z}_n^*$.

Theorem 5 *The set $G = 2\mathbb{Z}_n^*$, all doubled elements in $\mathbb{Z}_n^*$, is a cyclic multiplicative group with identity element $e = p^k + 1$.*

Corollary 6 *Let p be an odd prime and $n = 2p$. Then, $G = 2\mathbb{Z}_n^*$ the set of all doubled elements in $\mathbb{Z}_n^*$ is a cyclic multiplicative group with identity element $e \equiv p + 1$.*

Theorem 7 *Let G be the cyclic group given in Theorem 5 and $R = F_q + vF_q$ such that $\gcd(\varphi(p^k), q) = 1$. Also, let $u, v \in RG$ be principle zero divisors. Then, $(RG)u$ is an $e-$constacyclic code of length $\varphi(p^k)$ and dimension rank(u).*

Corollary 8 *The dual code of the code given in the Theorem 7 is a* $e^{-1}-$*constacyclic code of length* $\varphi\left(p^k\right)$ *and dimension* $rank\left(v\right)$.

3. Self Dual and Self Orthogonal Constacyclic Codes over $\left(F_q+vF_q\right)G$

This section is devoted to determining self dual and self orthogonal codes arising from constacyclic codes over group algebras.

Lemma 9 *Let* $C=\theta\left(\left(RG\right)u\right)$ *be an* $e-$*constacyclic code of length* $\varphi\left(p^k\right)$ *given in Theorem 7 with dual code* $C^{\perp}=\theta\left(\left(RG\right)v^T\right)$. *Then, the code* $C^{\perp}=\theta\left(\left(RG\right)v^T\right)$ *is also an* $e^{-1}-$*constacyclic code of length* $\varphi\left(p^k\right)$.

Theorem 10 *Let* $C=\theta\left(\left(RG\right)u\right)$ *be an* $e-$*constacyclic code of length* $\varphi\left(p^k\right)$ *given in Theorem 7 with dual code* $C^{\perp}=\theta\left(\left(RG\right)v^T\right)$. *Then,* C *is self dual if and only if* $e^2=1\left(mod\ q\right)$ *and* $u=v^T$.

Corollary 11 *Let* $C=\theta\left(\left(RG\right)u\right)$ *be an* $e-$*constacyclic code of length* $\varphi\left(p^k\right)$ *given in Theorem 7 with dual code* $C^{\perp}=\theta\left(\left(RG\right)v^T\right)$. *Then,* $p^k\equiv 2\left(mod\ q\right)$.

Theorem 12 *Let* $C=\theta\left(\left(RG\right)u\right)$ *be an* $e-$*consta cyclic code of length* $\varphi\left(p^k\right)$ *given in Theorem 7 with dual code* $C^{\perp}=\theta\left(\left(RG\right)v^T\right)$. *Then,* C *is self orthogonal if and only if* $e^2=1\left(mod\ q\right)$ *and for some* $w\in RG$ $u=wv^T$.

Corollary 13 *If* $C=\theta\left(\left(RG\right)u\right)$ *is an* $e-$*constacyclic code with parameters* $\left[\varphi\left(p^k\right),rank\left(u\right),d\right]_q$ *then,* $\phi\left(C\right)$ *is an* $e-$*constacyclic code with parameters* $\left[2\varphi\left(p^k\right),2rank\left(u\right),d\right]_q$.

4. Quantum Codes Obtained from Negacyclic Codes over $\left(F_q+vF_q\right)G$

The construction of quantum codes via classical codes over F_2 was first introduced by Calderbank and Shor [4] and Steane [13] in 1996. Later, construction quantum codes over different alphabets obtained from classical linear codes over F_q has been shown by Ketkar et al. in [10]. A quantum error correcting code Q is defined as follows:

Definition 14 *A* $q-$*ary quantum code* Q, *denoted by* $[[n,k,d]]_q$, *is a* q^k *dimensional subspace of the Hilbert space* $\mathbb{C}^{q^n}$ *and can correct all errors up to* $\lfloor\frac{d-1}{2}\rfloor$.

The following lemma is a method to get quantum error correcting codes via classical linear codes over finite fields.

Lemma 15 (CSS Code Construction) *[10] Let* C_1 *and* C_2 *denote two classical linear codes with parameters* $[n,k_1,d_1]_q$ *and* $[n,k_2,d_2]_q$ *such that* $C_2^{\perp}\leq C_1$. *Then there exists a* $[[n,k_1+k_2-n,d]]q$ *quantum code with minimum distance* $d=min\{wt(c)|c\in(C_1\backslash C_2^{\perp})\subset(C_2\backslash C_1^{\perp})\}$.

Corollary 16 *[10] If* C *is a classical linear* $[n,k,d]_q$ *code containing its dual,* $C^{\perp}\subset C$, *then there exists an* $[[n,2k-n,\geq d]]_q$ *quantum code.*

For further and detailed information readers can refer to the references [4, 5, 13]. We will use Corollary 16 to derive quantum error-correcting codes based on self dual and self orthogonal constacyclic codes over group algebras given in Theorem 10 and 12. All the computations are done using MAGMA [3].

Table 1. Some parameters of quantum codes obtained from self dual (self orthogonal) 2−constacyclic codes of length 40 over $\mathbb{F}_3$.

v^T	u	$\phi(C^\perp)$	$\phi(C)$	Q
$g^{10}+2g^9+g^8+g^7+g^6+g^4+2g^3+2g+2$	$2g^{10}+g^9+2g^8+2g^7+2g^6+2g^4+g^3+g+1$	$[40,20,6]_3$	$[40,20,6]_3$	$[[40,0,\geq 6]]_3$
$g^{12}+g^{10}+g^8+g^4+2g^2+1$	$g^8+g^6+2g^2+1$	$[40,16,6]_3$	$[40,24,4]_3$	$[[40,8,\geq 4]]_3$
$g^{14}+2g^{13}+g^{11}+2g^{10}+g^9+g^7+2g^5+g^4+g^3+2g^2+g+2$	$2g^6+2g^5+g^2+g+1$	$[40,12,9]_3$	$[40,28,4]_3$	$[[40,16,\geq 4]]_3$
$g^{16}+2g^{14}+g^{13}+g^{11}+2g^{10}+g^9+2g^8+g^5+g^4+2g^3+g+1$	$g^4+2g^3+g^2+1$	$[40,8,12]_3$	$[40,32,3]_3$	$[[40,24,\geq 3]]_3$
$g^{18}+2g^{17}+2g^{16}+2g^{14}+g^{13}+g^{12}+g^{10}+2g^9+2g^8+2g^6+g^5+g^4+g^2+2g+2$	$2g^2+g+1$	$[40,4,15]_3$	$[40,36,2]_3$	$[[40,32,\geq 2]]_3$

Table 2. Some parameters of quantum codes obtained from self dual (self orthogonal) 4−constacyclic codes of length 44 over $\mathbb{F}_5$.

u	v^T	$\phi(C^\perp)$	$\phi(C)$	Q
$g^{11}+3g^{10}+3g^9+4g^8+4g^7+2g^6+4g^5+2g^4+g^3+3g^2+g+3$	$g^{11}+3g^{10}+3g^9+4g^8+4g^7+2g^6+4g^5+2g^4+g^3+3g^2+g+3$	$[44,22,6]_5$	$[44,22,6]_5$	$[[44,0,6]]_5$
$g^{16}+g^{15}+2g^{14}+3g^{13}+3g^{12}+4g^{11}+2g^{10}+4g^9+3g^8+g^7+2g^6+g^5+3g^2+1$	$g^6+2g^4+4g^2+4g+1$	$[44,12,12]_5$	$[44,32,4]_5$	$[[44,20,\geq 4]]_5$
$g^{12}+2g^{10}+3g^6+2g^4+4g^2+1$	$g^{10}+g^8+4g^6+4g^4+3g^2+1$	$[44,20,6]_5$	$[44,24,5]_5$	$[[44,4,\geq 5]]_5$
$g^{17}+g^{16}+g^{14}+2g^{13}+2g^{12}+4g^{11}+3g^6+3g^5+3g^3+g^2+g+2$	$3g^5+g^4+3g^3+g^2+4g+1$	$[44,10,12]_5$	$[44,34,2]_5$	$[[44,24,\geq 2]]_5$

Example 17 *Let* $R = F_3 + vF_3$ *and*

$$G = \{2,4,6,8,12,14,16,18,22,24,26,28,32,34,36,38,42,44,46,48\} \subset \mathbb{Z}_{50},$$

be the multiplicative cyclic group mentioned in Corollary 6. Also, let $u = g^{18}+g^{17}+2g^{16}+2g^{14}+2g^{13}+g^{12}+g^{10}+g^9+2g^8+2g^6+2g^5+g^4+g^2+g+2$ *and* $v^T = 2g^2+2g+1$ *be two zero divisors in the group algebra* R *such that* $rank\,(u) = 2$ *and* $rank\,(v) = 18$*. Then, the two sided ideal* $(RG)\,u = \{xu|\, x \in RG\} \subset RG$ *is a* 2−*constacyclic code (negacyclic code) of parameters* $[20,2,15]_3$*. The generator matrix of this code can be computed as*

$$G = \begin{pmatrix} 1 & 0 & 1 & 2 & 2 & 0 & 2 & 1 & 1 & 0 & 1 & 2 & 2 & 0 & 2 & 1 & 1 & 0 & 1 & 2 \\ 0 & 1 & 2 & 2 & 0 & 2 & 1 & 1 & 0 & 1 & 2 & 2 & 0 & 2 & 1 & 1 & 0 & 1 & 2 & 2 \end{pmatrix}.$$

So, $\phi(\theta((RG)\,u))$ *is a* 2−*constacyclic code (negacyclic code) of parameters* $[40,4,15]_3$*. The dual code of this code is the two sided ideal* $(RG)\,v^T = \left\{xv^T \middle|\, x \in RG\right\} \subset RG$ *with parameters* $[20,18,2]_3$*. Further,* $\phi\left(\theta\left((RG)\,v^T\right)\right)$ *is a* 2−*constacyclic code (negacyclic code) of parameters* $[40,36,2]_3$*. By using Corollary 16 we have quantum code* Q *of parameters* $[[40,32,\geq 2]]_3$.

5. Conclusion

In this work, we determine self dual and self orthogonal codes arising from constacyclic codes of length $\varphi\left(p^k\right)$ over group ring $(F_q + vF_q)\,G$. Further, we obtained some parameters for quantum

Table 3. Some parameters of quantum codes obtained from self dual (self orthogonal) 6−constacyclic codes of length 36 over $\mathbb{F}_7$.

u	v^T	$\phi(C^\perp)$	$\phi(C)$	Q
$g^{10}+5g^8+4g^6+2g^4+3g^2+1$	$g^8+4g^6+2g^2+1$	$[36,16,3]_7$	$[36,20,3]_7$	$[[36,4,\geq 3]]_7$
$g^{16}+3g^{14}+2g^{12}+6g^{10}$ $+4g^8+5g^6+g^4+3g^2+2$	$4g^2+1$	$[36,4,9]_7$	$[36,32,2]_7$	$[[36,28,\geq 2]]_7$
$g^{12}+3g^6+2$	$4g^6+1$	$[36,12,3]_7$	$[36,24,2]_7$	$[[36,12,\geq 2]]_7$

codes derived from self dual and self orthogonal codes arising from these codes. This family of codes awaits further studies since most of obtained codes are near optimal codes.

References

[1] Aydin N Siap I and Ray-Chaudhuri D K 2001 *Design Code Cryptogr* **24** 313-326
[2] Berlekamp E R 2015 *World Scientific*
[3] Bosma W Cannon J and Playoust C 1997 *J. Symbolic Comput* **24** 235-265
[4] Calderbank A R and Shor P W 1996 *Phys. Rev. A* **54** 1098
[5] Calderbank A R Rains E M Shor P W and Sloane N J A 1998 *IEEE Trans. Inform. Theory* **44** 1369
[6] Hurley T 2006 *Int. J. Pure Appl. Math* **31** 319-335
[7] Hurley P and Hurley T 2009 *Int. J. of Inform. and Coding Theory* **1** 57-87
[8] Kai X and Zhu S 2013 *IEEE Trans. Inform. Theory* **59** 1193-1197
[9] Kai X Zhu S and Li P 2014 *IEEE Trans. Inform. Theory* **60** 2080-2086
[10] Ketkar A Klappenecker A Kumar S and Sarvepalli P K 2006 *IEEE Trans. Inform. Theory* **52** 4892-4914
[11] Milies C P and Sehgal S K 2002 *Springer*
[12] Ling S and Xing C 2004 *Cambridge University Press*
[13] Steane A M 1996 *Phys. Rev. A* **54** 4741
[14] Xiaoyan L 2004 *IEEE Trans. Inform. Theory* **50** 547-549

Journal of Physics: Conference Series **766** (2016) 012020 doi:10.1088/1742-6596/766/1/012020

Quantum codes from cyclic codes over $\mathbb{F}_3 + u\mathbb{F}_3 + v\mathbb{F}_3 + uv\mathbb{F}_3$

Mehmet Özen[1], **N. Tuğba Özzaim**[1] **and Halit İnce**[1]

[1] Department of Mathematics, Sakarya University, Sakarya, Turkey

E-mail: ozen@sakarya.edu.tr, tugbaozzaim@gmail.com, ince@sakarya.edu.tr

Abstract. In this paper, it will be argued that structure of cyclic codes over $\mathbb{F}_3 + u\mathbb{F}_3 + v\mathbb{F}_3 + uv\mathbb{F}_3$ where $u^2 = 1, v^2 = 1$ and $uv = vu$ for arbitrary length n. We define a new Gray map which is a distance preserving map. By using decomposition theory, we find generator polynomials of cyclic codes over $\mathbb{F}_3 + u\mathbb{F}_3 + v\mathbb{F}_3 + uv\mathbb{F}_3$ and obtain the result that cyclic codes over $\mathbb{F}_3 + u\mathbb{F}_3 + v\mathbb{F}_3 + uv\mathbb{F}_3$ are principally generated. Further, using these results we determine the parameters of quantum codes which constructed from cyclic codes over $\mathbb{F}_3+u\mathbb{F}_3+v\mathbb{F}_3+uv\mathbb{F}_3$. We present some of the results of computer search.

1. Introduction

Quantum error-correction plays an important role in quantum computing since it is used to correct any errors in quantum data due to decoherence and other quantum noise. Firstly, the existence of quantum error correcting code was shown by Shor [1] and independently by Steane [2]. In 1998, Calderbank et. al. published a paper in which they developed the theory to construct quantum codes by using classical error correcting codes. In recent years, a considerable literature has grown up around the quantum error correcting codes. Some authors constructed quantum codes by using the Gray image of cyclic codes over some finite rings. For example, in [7], a new method constructing quantum codes from cyclic codes over finite ring $\mathbb{F}_2 + v\mathbb{F}_2$ where $v^2 = v$ is given by Qian. In [4], Kai and Zhu construct quantum codes from Hermitian self orthogonal codes over $\mathbb{F}_4$ as Gray images of linear and cyclic codes over $\mathbb{F}_4 + u\mathbb{F}_4$ where $u^2 = 0$. In [5], Yin and Ma give an existence condition of quantum codes which are obtained from cyclic codes over $\mathbb{F}_2 + u\mathbb{F}_2 + u^2\mathbb{F}_2$ with Lee metric. In [6], some nonbinary quantum codes using classical codes over Gaussian integers are obtained by Ozen et. al. In [9], Guenda et. al. extend the CSS construction to finite commutative Frobenius rings. Dertli et. al. obtain quantum codes from cyclic codes over $\mathbb{F}_2 + u\mathbb{F}_2 + v\mathbb{F}_2 + uv\mathbb{F}_2$ in [10].

In [8], Ashraf and Mohammad give a construction for quantum codes from cyclic codes over $\mathbb{F}_3 + v\mathbb{F}_3$ where $v^2 = 1$. This ring is motivation to construct a new ring $\mathbb{F}_3 + u\mathbb{F}_3 + v\mathbb{F}_3 + uv\mathbb{F}_3$ where $u^2 = 1$ and $v^2 = 1$ because an advantage of the ring $\mathbb{F}_3 + u\mathbb{F}_3 + v\mathbb{F}_3 + uv\mathbb{F}_3$ is that it allows to find more parameters of quantum codes over $\mathbb{F}_3$.

This paper has been organised in the following way. The second section is concerned with the preliminaries used for this study. The third section presents a new Gray map from $\mathbb{F}_3+u\mathbb{F}_3+v\mathbb{F}_3+uv\mathbb{F}_3$ to $\mathbb{F}_3^4$ and linear codes over $\mathbb{F}_3+u\mathbb{F}_3+v\mathbb{F}_3+uv\mathbb{F}_3$ and their properties. In Section 4, we investigate the structure of cyclic codes over $\mathbb{F}_3+u\mathbb{F}_3+v\mathbb{F}_3+uv\mathbb{F}_3$. In Section 5, we give condition for cyclic codes over $\mathbb{F}_3+u\mathbb{F}_3+v\mathbb{F}_3+uv\mathbb{F}_3$ to contain their dual and determine the parameters of quantum codes which constructed from cyclic codes over $\mathbb{F}_3 + u\mathbb{F}_3 + v\mathbb{F}_3 + uv\mathbb{F}_3$. At the end of the paper, we give an example and a table to illustrate the results.

2. Preliminaries

Let $\mathbb{F}_3 = \{0, 1, 2\}$ be a finite field and R denotes the ring $\mathbb{F}_3+u\mathbb{F}_3+v\mathbb{F}_3+uv\mathbb{F}_3 = \{a+ub+vc+uvd : a, b, c, d \in \mathbb{F}_3\}$ where $u^2 = 1, v^2 = 1$ and $uv = vu$. This ring is a principal ideal ring but not a chain ring and the ideals $< 2 + u + 2v >, < 2 + v + uv >, < u + v + 2uv >, < 2u + 2v + 2uv >$ are maximal ideals of $\mathbb{F}_3 + u\mathbb{F}_3 + v\mathbb{F}_3 + uv\mathbb{F}_3$.

Let C be a nonempty subset of R^n. If C is a R-submodule of R^n, then C is called a linear code of length n over R. For any $a = (a_0, a_1, \ldots, a_{n-1})$ and $b = (b_0, b_1, \ldots, b_{n-1}) \in \mathbb{F}_3$ the inner product is given by $a.b = a_0b_0 + a_1b_1 + \cdots + a_{n-1}b_{n-1}$. If $a.b = 0$ then a and b are said to be orthogonal. Let C be a linear code of length n over R. Then the dual code of C is $C^{\perp} = \{a \in R | a.b = 0, \forall b \in C\}$. C is said to be self orthogonal if $C \subseteq C^{\perp}$ and self dual if $C = C^{\perp}$. A linear code C of length n is called cyclic if $\tau(c) = (c_{n-1}, c_0, \ldots, c_{n-2}) \in C$,where τ is cyclic shift operator, whenever $c = (c_0, c_1, ..., c_{n-1}) \in C$. Let R_n denote the quotient ring $R[x]/\langle x^n - 1\rangle$. Then we consider the following correspondence:

$$\begin{aligned} \phi : R^n &\longrightarrow R_n \\ c = (c_0, c_1, ..., c_{n-1}) &\longrightarrow c_0 + c_1x + ... + c_{n-1}x^{n-1} = c(x). \end{aligned}$$

It is well known that C is a cyclic code if and only if $\phi(C)$ is an ideal of R_n. Let $c = (c_0, c_1, ..., c_{n-1})$ be a codeword in C. The Hamming weight of $c = (c_0, c_1, ..., c_{n-1})$ is the number of nonzero coordinates in c and is denoted by $w_H(c)$. The Hamming distance between two elements $c_1, c_2 \in C$ is defined as $d_H(c_1, c_2) = \omega_H(c_1 - c_2)$.

Let $\alpha \in (\mathbb{F}_3^n)^{\ell}$ with $\alpha = (\alpha_1, \alpha_2, \ldots, \alpha_{n\ell}) = (\alpha^{(1)}|\alpha^{(2)}| \ldots |\alpha^{(\ell)})$. Let φ_ℓ be the quasi cyclic shift on $(\mathbb{F}_3^n)^{\ell}$ given by $\varphi_\ell(\alpha^{(1)}|\alpha^{(2)}| \ldots |\alpha^{(\ell)}) = (\tau(\alpha^{(1)})|\tau(\alpha^{(2)})| \ldots |\tau(\alpha^{(\ell)}))$. A code of length $n\ell$ over $\mathbb{F}_3$ is called as quasi cyclic code of index ℓ if $\varphi_\ell(C) = C$.

3. Gray map and linear codes over $\mathbb{F}_3 + u\mathbb{F}_3 + v\mathbb{F}_3 + uv\mathbb{F}_3$

3.1. *Gray map*

Any element of R can be written as $a + vb + uc + uvd \in R$, where $a, b, c, d \in \mathbb{F}_3$. The Gray map from R to $\mathbb{F}_3^4$ is defined as follows:

$$\begin{aligned} \psi : R &\longrightarrow \mathbb{F}_3^4 \\ a + vb + uc + uvd &\longrightarrow (a + b + c + d, a - b + c - d, a + b - c - d, a - b - c + d). \end{aligned}$$

This map can be naturally extended from R^n to $\mathbb{F}_3^{4n}$. We define the Lee weight of any $x \in R$ as $\omega_L(x) = \omega_H(\psi(x))$. For any elements $x, y \in R$, the Lee distance is given by $d_L(x, y) = \omega_L(x - y)$.

Theorem 3.1. *The Gray map is an isometry from (R^n, Lee distance) to ($\mathbb{F}_3^{4n}$, Hamming distance).*

Proof. For any $x_1, x_2 \in R$ and $\lambda \in \mathbb{F}_3$, $\psi(x_1 + x_2) = \psi(x_1) + \psi(x_2)$ and $\psi(\lambda x_1) = \lambda\psi(x_1)$. So ψ is linear. Now we show that ψ is distance preserving map. From the definition, we have $d_L(x_1, x_2) = \omega_L(x_1 - x_2) = \omega_H(\psi(x_1 - x_2)) = \omega_H(\psi(x_1) - \psi(x_2)) = d_H(\psi(x_1) - \psi(x_2))$. ☐

Theorem 3.2. *If C is a linear codes of length n over $\mathbb{F}_3 + u\mathbb{F}_3 + v\mathbb{F}_3 + uv\mathbb{F}_3$ with $|C| = 3^k$ and minimum Lee weight d_L then $\psi(C)$ is $[4n, k, d_H]$ ternary linear codes over $\mathbb{F}_3$.*

Theorem 3.3. *If C is self orthogonal then $\psi(C)$ is self orthogonal.*

Proof. Let C be self orthogonal and $\alpha = a + vb + uc + uvd$, $\beta = x + vy + uz + uvt \in C$ where $a, b, c, d, x, y, z, t \in \mathbb{F}_3$. Since C is self orthogonal, we have $\alpha.\beta = ax + by + cz + dt + v(ay + bx + ct + dz) + u(az + bt + cx + dy) + uv(at + bz + cy + dx) = 0$. So we have $ax + by + cz + dt = (ay + bx + ct + dz) = (az + bt + cx + dy) = (at + bz + cy + dx) = 0$.

Then the other side $\psi(\alpha).\psi(\beta) = (a+b+c+d, a-b+c-d, a+b-c-d, a-b-c+d)(x+y+z+t, x-y+z-t, x+y-z-t, x-y-z+t) = 0$. Hence $\psi(C)$ is self orthogonal. □

3.2. Linear codes over $\mathbb{F}_3 + u\mathbb{F}_3 + v\mathbb{F}_3 + uv\mathbb{F}_3$

Let A_1, A_2, A_3, A_4 are linear codes, then we denote that $A_1 \oplus A_2 \oplus A_3 \oplus A_4 = \{a_1+a_2+a_3+a_4 : a_i \in A_i; 1 \le i \le 4\}$ and $A_1 \otimes A_2 \otimes A_3 \otimes A_4 = \{(a_1, a_2, a_3, a_4) : a_i \in A_i; 1 \le i \le 4\}$.

Definition 3.4. *Let C be a linear code of length n over R. Define*

$$\begin{aligned} C_1 &= \{x+y+z+t \in \mathbb{F}_3 | x + vy + uz + uvt \in C\} \\ C_2 &= \{x-y+z-t \in \mathbb{F}_3 | x + vy + uz + uvt \in C\} \\ C_3 &= \{x+y-z-t \in \mathbb{F}_3 | x + vy + uz + uvt \in C\} \\ C_4 &= \{x-y-z+t \in \mathbb{F}_3 | x + vy + uz + uvt \in C\} \end{aligned}$$

Theorem 3.5. *Let C be a linear code of length n over R. Then $\psi(C) = C_1 \otimes C_2 \otimes C_3 \otimes C_4$ and $|C| = |C_1|.|C_2|.|C_3|.|C_4|$.*

Proof. For any $(x_1, \dots, x_n, y_1, \dots, y_n, z_1, \dots, z_n, t_1, \dots, t_n) \in \psi(C)$, let $c_i = x_i + y_i + z_i + t_i + v(x_i - y_i + z_i - t_i) + u(x_i + y_i - z_i - t_i) + uv(x_i - y_i - z_i + t_i)$; $1 \le i \le n$. Since ψ is bijection $c = (c_1, c_2, \dots, c_n) \in C$. By the definition of C_1, C_2, C_3, C_4 we have $(x_1, x_2 \dots, x_n) \in C_1, (y_1, y_2, \dots, y_n) \in C_2, (z_1, z_2, \dots, z_n) \in C_3, (t_1, t_2, \dots, t_n) \in C_4$. So, $(x_1, \dots, x_n, y_1, \dots, y_n, z_1, \dots, z_n, t_1, \dots, t_n) \in C_1 \otimes C_2 \otimes C_3 \otimes C_4$. It means that $\psi(C) \subseteq C_1 \otimes C_2 \otimes C_3 \otimes C_4$.

On the other hand, let $(x_1, \dots, x_n, y_1, \dots, y_n, z_1, \dots, z_n, t_1, \dots, t_n) \in C_1 \otimes C_2 \otimes C_3 \otimes C_4$ where $x = (x_1, x_2 \dots, x_n) \in C_1, y = (y_1, y_2, \dots, y_n) \in C_2, z = (z_1, z_2, \dots, z_n) \in C_3, t = (t_1, t_2, \dots, t_n) \in C_4$. There are $a = (a_1, a_2, \dots, a_n), b = (b_1, b_2, \dots, b_n), c = (c_1, c_2, \dots, c_n), d = (d_1, d_2, \dots, d_n) \in C$ such that $a_i = x_i + (1+2v+u+2uv)p_i$, $b_i = y_i + (1+v+u+uv)q_i$, $c_i = z_i + (1+2v+2u+uv)r_i$, $d_i = t_i + (1+v+2u+2uv)s_i$ where $p_i, q_i, r_i, s_i \in \mathbb{F}_3$. Since C is linear then $c = (1+v+u+uv)a + (1+2v+u+2uv)b + (1+v+2u+2uv)c + (1+2v+2u+uv)d = x + y + z + t + (x - y + z - t)v + (x + y - z - t)u + (x - y - z + t)uv$. It follows that $\psi(C) = (x, y, z, t) \in \psi(C)$. So $C_1 \otimes C_2 \otimes C_3 \otimes C_4 \subseteq \psi(C)$. Therefore $\psi(C) = C_1 \otimes C_2 \otimes C_3 \otimes C_4$. Moreover since $\psi(C)$ is bijection, $|C| = |\psi(C)| = |C_1 \otimes C_2 \otimes C_3 \otimes C_4| = |C_1|.|C_2|.|C_3|.|C_4|$. □

The rest of the paper, the terms λ_1, λ_2, λ_3, λ_4 will refer to $(1+v+u+uv)$, $(1+2v+u+2uv)$, $(1+v+2u+2uv)$, $(1+2v+2u+uv)$ respectively.

Lemma 3.6. *For $i = 1, 2, 3, 4$, if G_i is generator matrix of ternary linear codes C_i, then the generator matrix of C is* $\begin{pmatrix} \lambda_1 G_1 \\ \lambda_2 G_2 \\ \lambda_3 G_3 \\ \lambda_4 G_4 \end{pmatrix}$.

Proof. If G_i is generator matrix of ternary linear code C_i then the generator matrix of $\psi(C) = C_1 \otimes C_2 \otimes C_3 \otimes C_4$ is $\begin{pmatrix} G_1 & 0 & 0 & 0 \\ 0 & G_2 & 0 & 0 \\ 0 & 0 & G_3 & 0 \\ 0 & 0 & 0 & G_4 \end{pmatrix}$. By the Theorem 3.5, the generator matrix of C is $\begin{pmatrix} \lambda_1 G_1 \\ \lambda_2 G_2 \\ \lambda_3 G_3 \\ \lambda_4 G_4 \end{pmatrix}$.

□

Corollary 3.7. *If $\psi(C) = C_1 \otimes C_2 \otimes C_3 \otimes C_4$, then $C = \lambda_1 C_1 \oplus \lambda_2 C_2 \oplus \lambda_3 C_3 \oplus \lambda_4 C_4$.*

Lemma 3.8. *Let d_H and d_L denote the Hamming and Lee distance respectievly. Then $d_H = d_L = min\{d(C_1), d(C_2), d(C_3), d(C_4)\}$.*

Proof. Since ψ distance preserving map we obtain $d_L(C) = d_H(\psi(C)) = d_H(C_1 \otimes C_2 \otimes C_3 \otimes C_4) = min\{d(C_1), d(C_2), d(C_3), d(C_4)\}$. □

4. Cyclic codes over $\mathbb{F}_3 + u\mathbb{F}_3 + v\mathbb{F}_3 + uv\mathbb{F}_3$

Theorem 4.1. *Let $C = \lambda_1 C_1 \oplus \lambda_2 C_2 \oplus \lambda_3 C_3 \oplus \lambda_4 C_4$ be a linear code of length n over R. Then C is a cyclic code of length n over R if and only if for $i = 1, 2, 3, 4$, C_i is a cyclic code of length n over $\mathbb{F}_3$.*

Proof. Let $c = (c_1, c_2, \ldots, c_n) \in C$ where $c_j = \lambda_1 x_j + \lambda_2 y_j + \lambda_3 z_j + \lambda_4 t_i$ $j = 1, \ldots, n$ and $x = (x_1, x_2 \ldots, x_n) \in C_1, y = (y_1, y_2, \ldots, y_n) \in C_2, z = (z_1, z_2, \ldots, z_n) \in C_3, t = (t_1, t_2, \ldots, t_n) \in C_4$. Since for $i = 1, 2, 3, 4$, C_i is a cyclic code of length n over $\mathbb{F}_3$ we get $\tau(x) = (x_n, x_1, \ldots, x_{n-1}) \in C_1$, $\tau(y) = (y_n, y_1, \ldots, y_{n-1}) \in C_2$, $\tau(z) = (z_n, z_1, \ldots, z_{n-1}) \in C_3$, $\tau(t) = (t_n, t_1, \ldots, t_{n-1}) \in C_4$. Note that $\tau(c) = (c_n, c_1, \ldots, c_{n-1}) = \lambda_1 \tau(x) + \lambda_2 \tau(y) + \lambda_3 \tau(z) + \lambda_4 \tau(t) \in C$ which means that C is a cyclic code over R.

On the other side, let $x = (x_1, x_2 \ldots, x_n) \in C_1, y = (y_1, y_2, \ldots, y_n) \in C_2, z = (z_1, z_2, \ldots, z_n) \in C_3, t = (t_1, t_2, \ldots, t_n) \in C_4$. Assume that $c = (c_1, c_2, \ldots, c_n) \in C$ where $c_j = \lambda_1 x_j + \lambda_2 y_j + \lambda_3 z_j + \lambda_4 t_i$ for $j = 1, \ldots, n$. Since C is a cyclic code over R then we have $\tau(c) = (c_n, c_1, \ldots, c_{n-1}) \in C$. Note that again $(c_n, c_1, \ldots, c_{n-1}) = \lambda_1(x_n, \ldots, x_{n-1}) + \lambda_2(y_n, \ldots, y_{n-1}) + \lambda_3(z_n, \ldots, z_{n-1}) + \lambda_4(t_n, \ldots, t_{n-1}) \in C$. Hence $\tau(x) \in C_1$ $\tau(y) \in C_2$ $\tau(z) \in C_3$ $\tau(t) \in C_4$. That means for $i = 1, 2, 3, 4$, C_i is a cyclic code of length n over $\mathbb{F}_3$. □

Now we investigate generator polynomial of cyclic codes of length n over R. For $i = 1, 2, 3, 4$ since C_i is a cyclic code over $\mathbb{F}_3$, we make use of the generators of cyclic codes over $\mathbb{F}_3$.

Theorem 4.2. *Let $C = \lambda_1 C_1 \oplus \lambda_2 C_2 \oplus \lambda_3 C_3 \oplus \lambda_4 C_4$ be a cyclic code of length n over R. Then there exist the polynomials $g_i(x)'s$,$1 \leq i \leq 4$, over $\mathbb{F}_3[x]$ such that $C = < \lambda_1 g_1(x), \lambda_2 g_2(x), \lambda_3 g_3(x), \lambda_4 g_4(x) >$ where $g_i(x)'s$ are generator polynomials of $C_i's$ for $i = 1, 2, 3, 4$. Moreover $|C| = 3^{4n - \sum_{i=1}^{4} \deg(g_i(x))}$.*

Proof. Let $C_i = < g_i(x) >$ and $C = \lambda_1 C_1 \oplus \lambda_2 C_2 \oplus \lambda_3 C_3 \oplus \lambda_4 C_4$ then

$$C = \{c(x) = \lambda_1 g_1 s_1 + \lambda_2 g_2 s_2 + \lambda_3 g_3 s_3 + \lambda_4 g_4 s_4 | s_i \in \mathbb{F}_3[x]\}.$$

It is easily seen that $C \subseteq < \lambda_1 g_1(x), \lambda_2 g_2(x), \lambda_3 g_3(x), \lambda_4 g_4(x) >$.

On the other hand, let $d(x) \in < \lambda_1 g_1(x), \lambda_2 g_2(x), \lambda_3 g_3(x), \lambda_4 g_4(x) >$ then for $1 \leq i \leq 4$, there is $k_i(x) \in R[x]$ such that

$$d(x) = \lambda_1 g_1(x) k_1(x) + \lambda_2 g_2(x) k_2(x) + \lambda_3 g_3(x) k_3(x) + \lambda_4 g_4(x) k_4(x).$$

Then there exists polynomial $s_i(x) \in \mathbb{F}_3[x]$ such that $\lambda_1 s_1(x) = \lambda_1 k_1(x)$, $\lambda_2 s_2(x) = \lambda_2 k_2(x)$, $\lambda_3 s_3(x) = \lambda_3 k_3(x)$, $\lambda_4 s_4(x) = \lambda_4 k_4(x)$. Hence we have $< \lambda_1 g_1(x), \lambda_2 g_2(x), \lambda_3 g_3(x), \lambda_4 g_4(x) > \subseteq C$. This gives that $C = < \lambda_1 g_1(x), \lambda_2 g_2(x), \lambda_3 g_3(x), \lambda_4 g_4(x) >$. From Theorem 3.5 we know that $|C| = |C_1|.|C_2|.|C_3|.|C_4|$. Thus we get $|C| = 3^{4n - \sum_{i=1}^{4} \deg(g_i(x))}$. □

Journal of Physics: Conference Series **766** (2016) 012020 doi:10.1088/1742-6596/766/1/012020

Theorem 4.3. *Let C be a cyclic code of length n over R, then there is a polynomial $g(x)$ such that $C = < g(x) >$ where $g(x) = \lambda_1 g_1(x) + \lambda_2 g_2(x) + \lambda_3 g_3(x) + \lambda_4 g_4(x)$ and $g(x)|x^n - 1$.*

Proof. By Theorem 4.2, we know that $C = < \lambda_1 g_1(x), \lambda_2 g_2(x), \lambda_3 g_3(x), \lambda_4 g_4(x) >$ where $g_i(x)$ is generator polynomial of C_i for $i = 1, 2, 3, 4$. Let $g(x) = \lambda_1 g_1(x) + \lambda_2 g_2(x) + \lambda_3 g_3(x) + \lambda_4 g_4(x)$, obviously $< g(x) > \subseteq C$.

On the other side, note that $\lambda_1 g(x) = \lambda_1 g_1(x)$, $\lambda_2 g(x) = \lambda_2 g_2(x)$, $\lambda_3 g(x) = \lambda_3 g_3(x)$, $\lambda_1 g(x) = \lambda_1 g_1(x)$ then this gives $C \subseteq < g(x) >$. Hence $C = < g(x) >$. Since $g_i(x)|x^n - 1$ there exists $r_i(x) \in R[x]/x^n - 1$ such that $x^n - 1 = g_1(x)r_1(x) = g_2(x)r_2(x) = g_3(x)r_3(x) = g_4(x)r_4(x)$. It follows that $x^n - 1 = g(x)[\lambda_1 r_1(x) + \lambda_2 r_2(x)\lambda_3 g_3(x)r_3(x) + \lambda_4 g_4(x)r_4(x)]$, hence $g(x)|x^n - 1$. □

Corollary 4.4. *Every ideal of $R[x]/x^n - 1$ is principal.*

Theorem 4.5. *Let ψ be the gray map and φ_4 and τ be quasi and cyclic shift operator as in the preliminaries. Then $\psi\tau = \varphi_4\psi$.*

Proof. The proof is directly from the definition of quasi and cyclic codes. □

Theorem 4.6. *C is a cyclic code of length n over R if and only if $\psi(C)$ is a quasi cyclic code of index 4 over $\mathbb{F}_3$ with length $4n$.*

Proof. Let C be a cyclic code of length n over R i.e. $\tau(C) = C$. Using Theorem 4.5 we have $\varphi_4(\psi(C)) = \varphi(\tau(C)) = \psi(C)$. Hence $\psi(C)$ is a quasi cyclic code of index 4 over $\mathbb{F}_3$.

On the other hand, let $\psi(C)$ is a quasi cyclic codes of index 4 over $\mathbb{F}_3$ i.e. $\varphi_4(\psi(C)) = \psi(C)$ then using Theorem 4.5 again we get $\psi(\tau(C)) = \varphi_4(\psi(C)) = \psi(C)$. Note that ψ is injective so $\tau(C) = C$. □

5. Quantum codes from cyclic codes over $\mathbb{F}_3 + u\mathbb{F}_3 + v\mathbb{F}_3 + uv\mathbb{F}_3$

Theorem 5.1. *Let C and $\hat{C}$ be two linear codes with parameters $[n, k, d]$ and $[n, \hat{k}, \hat{d}]$ such that $C^\perp \subseteq \hat{C}$ then there exist a quantum code with parametre $[n, k + \hat{k}, min\{d, \hat{d}\}]$. Especially if $C^\perp \subseteq C$ then there exists an [[n,2k-n,d]] quantum code.*

Theorem 5.2. *Let $C = \lambda_1 C_1 \oplus \lambda_2 C_2 \oplus \lambda_3 C_3 \oplus \lambda_4 C_4$ be a cyclic code of length n over R. Then*

$$C^\perp =< \lambda_1 h_1^*(x) + \lambda_2 h_2^*(x) + \lambda_3 h_3^*(x) + \lambda_4 h_4^*(x) >$$

and $|C^\perp| = 3^{\deg g_1(x) + \deg g_2(x) + \deg g_3(x) + \deg g_4(x)}$ where for $i = 1, 2, 3, 4$, $h_i^*(x)$ are reciprocal polynomials of $h_i(x)$ i.e. $h_i(x) = x^n - 1/g_i(x)$, $h_i^*(x) = x^{\deg h_i(x)} h_i(x^{-1})$.

Lemma 5.3. *A cyclic code C with generator polynomial $g(x)$ contains its dual if and only if*

$$x^n - 1 \equiv 0 \ (mod\ g(x)g^*(x))$$

where $g^*(x)$ is reciprocal polynomial of $g(x)$.

Theorem 5.4. *$C = < \lambda_1 g_1(x), \lambda_2 g_2(x), \lambda_3 g_3(x), \lambda_4 g_4(x) >$ be a cyclic code of length n over R. Then $C^\perp \subseteq C$ if and only if $x^n - 1 \equiv 0 \ (mod\ g_i g_i^*)$ for $i = 1, 2, 3, 4$.*

Proof. For $i = 1, 2, 3, 4$, let $x^n - 1 \equiv 0 \pmod{g_i g_i^*}$ then from Lemma 5.3, we have $C_i^\perp \subseteq C_i$. This implies that $\lambda_i C_i^\perp \subseteq \lambda_i C_i$. Hence

$$< \lambda_1 h_1^*(x) + \lambda_2 h_2^*(x) + \lambda_3 h_3^*(x) + \lambda_4 h_4^*(x) > \subseteq < \lambda_1 g_1(x), \lambda_2 g_2(x), \lambda_3 g_3(x), \lambda_4 g_4(x) > .$$

Thus $C^\perp \subseteq C$. Conversely, if $C^\perp \subseteq C$ then

$$\lambda_1 C_1^\perp \oplus \lambda_2 C_2^\perp \oplus \lambda_3 C_3^\perp \oplus \lambda_4 C_4^\perp \subseteq \lambda_1 C_1 \oplus \lambda_2 C_2 \oplus \lambda_3 C_3 \oplus \lambda_4 C_4.$$

Since C_i is a ternary code such that $\lambda_i C_i$ is equal to $C \mod \lambda_i$. Thus $C_i^\perp \subseteq C_i$ and $x^n - 1 \equiv 0 \pmod{g_i g_i^*}$ for $i = 1, 2, 3, 4$. □

Corollary 5.5. *Let $C = \lambda_1 C_1 \oplus \lambda_2 C_2 \oplus (\lambda_3 C_3 \oplus \lambda_4 C_4$ be a cyclic code of length n over R. Then $C^{\perp} \subseteq C$ if and only if $C_i^{\perp} \subseteq C_i$ for $i = 1, 2, 3, 4$.*

By Theorem 5.1 and Corollary 5.5, the quantum codes can be constructed.

Theorem 5.6. *Let $C = \lambda_1 C_1 \oplus \lambda_2 C_2 \oplus \lambda_3 C_3 \oplus \lambda_4 C_4$ be a cyclic code of length n over R with size 3^k. If $C^{\perp} \subseteq C$ and there exist a quantum code with parameters $[[4n, 2k - 4n, d_L]]$.*

Example 5.7. *Let $n = 18$. Then $x^{18} - 1 = (x+1)^9(x+2)^9$ in $\mathbb{F}_3$. Let $g(x) = \lambda_1 g_1(x) + \lambda_2 g_2(x) + \lambda_3 g_3(x) + \lambda_4 g_4(x)$ where $g_1(x) = g_2(x) = g_3(x) = x^5 + 2x^3 + 2x^2 + 1$ and $g_4(x) = x^6 + x^5 + 2x^4 + x^3 + 2x^2 + x + 1$ so that $g_1^*(x) = g_2^*(x) = g_3^*(x) = x^5 + 2x^3 + 2x^2 + 1$ $g_4^*(x) = x^6 + x^5 + 2x^4 + x^3 + 2x^2 + x + 1$. Let $C = < g(x) >$ be a cyclic code over R. Clearly $x^{18} - 1$ is divisible by $g_i g_i^*$ for $i = 1, 2, 3, 4$. Hence by Theorem 5.5 we have $C^{\perp} \subseteq C$. Then a quantum code with parameters $[[72, 30, 3]]$ can be obtained.*

Table 1. The parameters of quantum code.

n	Generator polynomials	$[[N, K, D]]$
6	$g_1 = g_2 = x + 1$ and $g_3 = g_4 = x + 2$	$[[24, 16, 2]]$
9	$g_1 = g_2 = g_3 = g_4 = x + 2$	$[[36, 28, 2]]$
11	$g_1 = g_2 = x^5 + x^4 + 2x^3 + x^2 + 2$ and $g_3 = g_4 = x^5 + 2x^3 + x^2 + 2x + 2$	$[[44, 4, 5]]$
15	$g_1 = g_2 = g_3 = g_4 = x + 2$	$[[60, 52, 2]]$
18	$g_1 = g_2 = g_3 = g_4 = x + 2$	$[[72, 64, 2]]$
21	$g_1 = g_2 = g_3 = x + 2$ and $g_4 = x + 1$	$[[84, 76, 2]]$
22	$g_1 = g_2 = g_3 = g_4 = x^{10} + x^9 + x^8 + 2x^7 + 2x^6 + 2x^5 + x^4 + 2x^3 + 2x^2 + x + 2$	$[[88, 8, 7]]$
23	$g_1 = g_2 = g_3 = g_4 = x^{11} + x^{10} + x^9 + 2x^8 + 2x^7 + x^5 + x^3 + 2$	$[[92, 4, 8]]$
24	$g_1 = g_2 = g_3 = g_4 = x^5 + x^3 + 2x + 2$	$[[96, 56, 3]]$
30	$g_1 = g_2 = g_3 = x + 2$ and $g_4 = x^2 + 2$	$[[120, 110, 2]]$
35	$g_1 = g_2 = g_3 = g_4 = x^{12} + 2x^{11} + 2x^{10} + x^9 + 2x^8 + x^7 + x^5 + 2x^4 + x^2 + 1$	$[[140, 44, 5]]$

References

[1] P. W. Shor, Scheme for reducing decoherence in quantum computer memory, Phys. Rev. A, 52 (1995), 2493-2496.
[2] A. M. Steane, Simple quantum error correcting codes, Phys. Rev. Lett., 77 (1996), 793 - 797.
[3] A. R. Calderbank, E.M.Rains, P.M.Shor, N.J.A.Sloane, Quantum error correction via codes over $GF(4)$, IEEE Trans. Inf. Theory, 44 (1998) , 1369 - 1387.
[4] X.Kai,S.Zhu, Quaternary construction of quantum codes from cyclic codes over $\mathbb{F}_4 + v\mathbb{F}_4$, Int. J. Quantum Inform., 9 (2011) , 689 - 700.
[5] X. Yin, W. Ma, Gray Map And Quantum Codes Over The Ring $\mathbb{F}_2 + u\mathbb{F}_2 + u^2\mathbb{F}_2$; International Joint Conferences of IEEE TrustCom-11, (2011).
[6] M. Ozen, M. Guzeltepe, Quantum Codes From Codes over Gaussian Integers with Respect to The Mannheim Metric ,Quantum Information and Computation , 12 (2012) , 813 - 819.
[7] J.Qian, Quantum codes from cyclic codes over $\mathbb{F}_2 + v\mathbb{F}_2$, Journal of Inform. Computational Science 10 : 6 (2013) , 1715 - 1722.
[8] M. Ashraf, G. Mohammad, Quantum codes from cyclic codes over $\mathbb{F}_3 + v\mathbb{F}_3$, International Journal of Quantum Information, vol. 12, No. 6(2014) 1450042.
[9] K. Guenda , T. A. Gulliver , Quantum codes over rings, Int. J. Quantum Inform, 12, (2014), 1450020.
[10] A. Dertli, Y. Cengellenmis, S. Eren, On quantum codes obtained from cyclic codes over A_2, Int. J. Quantum Inform., vol. 13, 2(2015) 1550031.

International Conference on Quantum Science and Applications (ICQSA-2016) IOP Publishing
Journal of Physics: Conference Series **766** (2016) 012021 doi:10.1088/1742-6596/766/1/012021

Multi-Partied Quantum Digital Signature Scheme Without Assumptions On Quantum Channel Security

Mustafa ŞAHİN[1],**İhsan YILMAZ**[2]
[1,2] Çanakkale Onsekiz Mart University, Computer Eng. Dept., Çanakkale, Turkey
E-mail: msahin@comu.edu.tr,iyilmaz@comu.edu.tr

Abstract. In this study, we generalized quantum digital signature scheme for three-partied[5] to the multi-partied. Also we analyzed this scheme for repudiation, forgery and colluding adversaries.

1. Introduction

Digital signatures are like signatures on paper. The main purpose is to identify the signatory and to determine any falsification made on the document. Available methods of digital signature are based on difficult mathematical problems [8]. However, to solve this problem several quantum algorithms are proposed [1]. Gottesman and Chuang[7] has proposed concept of quantum digital signature. The method of Gottesman and Chuang[7] becomes unsuccessful as in the states of swap-test $|\psi\rangle_B|\phi\rangle_C + |\phi\rangle_B|\psi\rangle_C$. Complete-symmetry test in Buhrman et.al[2] is used to solve this problem. Andersson et al.[9] showed that coherent states and linear optical elements can be used instead of swap-test experimentally.Dunjko at al [10] showed that QDS can be generated without quantum memory. They used optic and photon detector. Walden et all .[11] showed that available usable/sell-able quantum key distribution devices can be used for the solution of the some problems. Juan Miguel Arrazola et all.[3] generalized the schema of Dunjko and Wallden [10] to the multiuser schema. Hua-Lei Yin et all.[4] show that it is not necessary authenticated quantum channel. Amiri et all .[5, 6] developed quantum signature without needing all security assumptions on quantum channel.

In this study, we generalized quantum digital signature scheme for three-partied[5] to the multi-partied. We don't have any assumptions about trust of quantum channels between participants. And all participants creates keys and send part of it to the message sender so that sender can not create any entangled keys and sends them to the participants to repudiate them. And, also we analyzed this scheme for repudiation, forgery and colluding adversaries.

2. Key Generation Stage

In this study, to create correlated but not the same keys between participants, we have used quantum key distribution(QKD)[13, 14] via phase randomized weak-coherent states and decoy states[15]. We don't do post-processing stage of QKD so that these keys have some little mismatches. Asymmetric coding of BB84[17] was used. $\{X, Z\}$ basis are chosen with different probabilities $\{p_X, p_Z\}$. $p_X \geq 1/2$, $p_Z = 1-p_X$ and also intensities$\{\mu_1, \mu_2, \mu_3\}$ of laser pulses can be chosen with different probabilities $\{p_{\mu_1}, p_{\mu_2}, p_{\mu_3}\}$.Where $\mu_1 > \mu_2 + \mu_3$ and $\mu_2, \mu_3 \geq 0$. The intensities $\{\mu_2, \mu_3\}$ are used for decoy states. Secret keys will be created from the measurement results which Alice and Bob choose X basis and same intensities.

Following the same procedure in[5] Bob is sender and Alice is receiver to prevent against Alice who can create entangled states. Also she can send the same key to the other participants. After measurement, Bob selects $L + n_{cor}$ bits from X_μ and randomly divide L bits into two part with length $L/2$, $(K^B_{keep}, K^B_{forward})$. $K^B_{forward}$ bits will be used to forward to the other participants. n_{cor} bits, Cor^B_X, from X basis measurements will be shared via classical channel to calculate correlation between Alice's and Bob's keys $\{\tilde{K}^A_m, K^B_m\}$. Cor^B_X wont be used into protocol. Z_μ will be used to estimate Eve's information about Bob's key. To achieve this we will use the smooth min-entropy. We would like to bound Eve's success rate about signature creation which includes less errors from certain number of errors. So we will calculate the smooth min entropy of Eve about Bob's key K^B_{keep}.

Bob has following key set $\{Cor^B_X, Z^B, K^B_{keep}, K^B_{forward}\}$. Lengths of the keys are $\{|Cor^B_X| = n_{cor}, |Z^B| = \eta^Z, |K^B_{keep}| = |K^B_{forward}| = L/2\}$. Alice also has these keys with the same manner. As mentioned, Cor^B_X will be used to calculate correlation between $\tilde{K}^A$ and K^B. Z^B can be used to gain information about eavesdropper existence. In the protocol $K^B_{forward}$ will be sent to other participants to protect himself against repudiation from Alice.

Since Eve can get information about the measurement of $\{Z^A, Z^B\}$ on quantum channel, and about Cor^B_X, Cor^A_X, and also about $basis^B, basis^A$. We can calculate information of Eve about K^B_{keep} by evaluating correlations of $\{Z^A, Z^B\}$. Also we assume that Eve is a participant in the protocol, so Eve has $(K^B_{forward}, Cor^B_X, Cor^A_X, basis^B, basis^A)$. Eve's all information system can be denoted as $\mathcal{H}_E$

The number of mismatches with receiver's key is important for quantum signature protocols. We will show how we can bound the forger's mismatch count for our protocol. Thus if we provide bigger mismatch count for forger's (message,signature) pair then any receiver wont accept the message with small probability. Our main goal is to achieve that $d(\tilde{K}^A_m, K^B_m) < d(\tilde{K}^{Eve}_m, K^B_m)$. $\tilde{K}^{Eve}_m$ is the Eve's guess about Bob's key by using all collected information about keys. Where $d(.,.)$ represents Hamming distance between the two states. Min and Smooth entropy [19] are used to calculate information of states.

Lim et al.[12] and Ma et al. [16] gives bounds for Eve's information about Bob's key in the context of QKD. We will use same ways to bound Eve's mismatch count about Bob's key. But Eve has extra information $K^B_{forward}$.

By extracting error correction and privacy amplification steps in [12], we can write following [5]:

$$H^{\epsilon}_{min}(X^B_{keep}|E) \gtrsim S_{X,0} + S_{X,1} - S_{X,1}h(\phi_{X,1}). \tag{1}$$

Where $S_{X,0}, S_{X,1}$ corresponds respectively number of vacuum events and number of single photon events in X basis measurements. We discard double photon events and never use in the protocol due to photon-splitting attacks. Where $\phi_{X,1} = c_{X,1}/S_{X,1}$ is the phase error rate in X measurements where $c_{X,1}$ is the number of phase errors. And $h(.)$ is binary entropy[16]. As shown and proofed in [5]:

$$< p_r > \leq \sum_{k=0}^{r} \binom{L/2}{k} 2^{-H^{\epsilon}_{min}(X^B_{keep}|E)_{\rho_{X_{keep}E}}} + \epsilon \tag{2}$$

This shows Eve's probability of making at most r mistakes by using best strategies to guess X_{keep} with all information he/she has. In this equation $\rho_{X_{keep}E}$ is $\epsilon - close$ to near optimal $\rho'_{X_{keep}E}$. As $L/2$ grows binom coefficient can be written as $\binom{L/2}{k} \approx 2^{(L/2)h(\frac{r}{L/2})}$. By using this approximation and Eq.(1), Eq.(2) can be written as:

$$< p_r > \leq \ 2^{-(S_{X,0}+S_{X,1}-S_{X,1}h(\phi_{X,1}))+(L/2)h(\frac{r}{L/2})} + \epsilon \tag{3}$$

We can get $L/2$ parenthesis as follows:

$$< p_r > \leq \ 2^{-(L/2)(\frac{S_{X,0}}{L/2}+\frac{S_{X,1}}{L/2}-\frac{S_{X,1}}{L/2}h(\phi_{X,1})-h(\frac{r}{L/2}))} + \epsilon \tag{4}$$

If we have neglected ϵ in Eq.(4), we get the following equation.

$$\frac{S_{X,0}}{L/2} + \frac{S_{X,1}}{L/2} - \frac{S_{X,1}}{L/2}h(\phi_{X,1}) - h(\frac{r}{L/2}) > 0 \tag{5}$$

Which determines Eve's probability of making fewer than r errors. To bound this probability we can define minimum error rate ER_{Eve} of Eve as follows:

$$h(ER_{Eve}) = (\frac{S_{X,0}}{L/2} + \frac{S_{X,1}}{L/2} - \frac{S_{X,1}}{L/2}h(\phi_{X,1})) \tag{6}$$

If we achieve error rate between Alice-Bob lesser than error rate of Eve about Bob's key X^B_{keep} then the protocol will be secure. The condition of $ER_{Eve} > ER_{AB}$ must be held. Alice and Bob can use k length of Cor^B_X and Cor^A_X keys to calculate the estimate of error rate about keys $\tilde{K}^A$ and K^B which are used in the protocol. By using Cor^B_X and Cor^A_X samples with Serfling[20] inequality we can bound $\tilde{K}^A, K^B$ error rate as follows:

$$ER_{AB} \leq \tilde{ER}_{AB} + \varepsilon_{ER_{AB}} \tag{7}$$

Where $\tilde{ER}_{AB}$ is observed from $\{Cor^B_X, Cor^A_X\}$ keys and $\varepsilon_{ER_{AB}}$ is calculated from Serfling inequality as follows [20]:

$$\epsilon_{ER_{AB}} \leq \sqrt{\frac{ln(P_k(\epsilon))}{2k}\left(1-\left(\frac{k-1}{L/2}\right)\right)} \tag{8}$$

If we select acceptance threshold value $t_{v_1} > \tilde{ER}_{AB}$ in the protocol then we can bound ER_{Eve} with small probability like in Eq.(8).

3. Multi-Partied Quantum Signature Protocol with Majority Vote Dispute Resolution

Arrazola et.al.[18] defined properties of the multi-party unconditionally secure signature(USS) protocols to achieve their security goals as Non-Forgery,Non-repudiation and Transferability.

Quantum Signature Protocol has been formed with the following sets and functions[18]:

(i) $P = \{P^0, P^1, P^2...P^{N-1}\}$ are N different participants. We assumed that P^0 is sender and $P^i, i = 1..N-1$ are the other possible receivers and dispute resolvers.

(ii) X is possible messages set and Σ is the possible signature set of the messages X.

(iii) $Sign(x)$ is a signature function that maps $X \to \Sigma$. That $Sign(x)$ function gets x value and returns $\sigma \in \Sigma$.

(iv) $l = \{-1, 0, 1, 2, 3, ..l_{max}\}$ is the set of verification levels that l verification level means that how many times a signed message can be transferred to the other participants. $l_{max} \leq N-1$. $l = 0$ guarantees that this message comes from signer but does not guarantees it can be forwarded other participants. $l = -1$ level will be used in Majority-Vote dispute resolution so $l = -1$ level is used to prevent repudiation from sender.

(v) $Ver(X, \Sigma, P, l) \rightarrow \{True, False\}$ gets message, signature, participant and verification level and returns $\{True, False\}$ which means this message and signature will be accepted by P_i participant at level l.

All participants have imperfect quantum channels and authenticated classical channels between them to make post-processing. We can outline the protocol for single bit value $m = \{0, 1\}$. To generalize to the multi-bit message, the protocol can be repeated. The protocol can be outlined as follows:

(i) **Key Generation And Distribution Stage:** $P^i, i = 1..N-1$ participants uses individually Key Generation Stage to create and share L length key pairs $K_m(P^1), \tilde{K}_m(P^1 \rightarrow P^0), K_m(P^2), \tilde{K}_m(P^2 \rightarrow P^0), .., K_m(P^{N-1}), \tilde{K}_m(P^{N-1} \rightarrow P^0)$ with P_0. $K_m(P^i), i = 1..N-1$ are receiver's keys and $\tilde{K}_m(P^i \rightarrow P^0), i = 1..N-1$ are sender's keys which are received from quantum channel with noise and has little changes. All participants send their key pair's part to P_0. P_0 collects all keys $\tilde{K}_m(P^i \rightarrow P^0), i = 1..N-1$ which are sent to her from other participants to transfer message to them later. P_0 use signature $Sig_m = (\tilde{K}_m(P^i \rightarrow P^0), i = 1..N-1)$ to transfer message m to the other participants. Note that only P_0 knows all keys which were sent by other participants that will protect the protocol against forgery.

(ii) **Key Sharing Stage:** $P^i, i = 1..N-1$ participants randomly choose $\lfloor |L|/2 \rfloor$ of bits from their keys $K_m(P^i), i = 1..N-1$ and then send them with the positions to the other participants except P^0 via authenticated classical channel. $P^i, i = 1..N-1$ save the positions and values of the forwarded bits. $P^i, i = 1..N-1$ will firstly and importantly use the unsent bits and bits which come from other parties to check the validity of the signature that will be sent from P^0. All participants except P^0 shares half of the her/his keys to protect him/her selfs against repudiation from P^0. Since P^0 does not know which positions of the keys are sent to the other participants, P^0 can not deceive $P^i, i = 1..N-1$ about signature. Secondly, they use forwarded part of the keys to protect him/her selfs against collective adversaries. We can show their new keys as $S_m^{P^1}, S_m^{P^2}, .., S_m^{P^{N-1}}$ that formed from unsent bits and other forwarded keys from other participants. The length of the new keys is $(N-1)L/2$. $P^i, i = 1..N-1$ knows all bits of $K_m^{P^i}$ and half of the $\{K_m^{P^j}, j \neq i, j = 1..N-1\}$. If say P^1 is dishonest, he knows only half bits of the $\{K_m^{P^j}, j = 2..N-1\}$, so P^1 cannot forge the other participants.

(iii) **Messaging Stage:** In this stage, P^0 wants to send one bit message $m = \{0, 1\}$ to P^1 or any other participant. Then receiver checks the validity of the signature and then forwards message and signature to the other participants.

(a) P^0 sends (m, Sig_m) one bit message to P^1 (or any P^i).

(b) P^1 checks the Sig_m with his unsent and forwarded parts of $K_m^{P^1}$, and keys which are sent by other participants. Then P^1 saves separately unmatched bit counts. P^1 accepts the message if the following mismatch count rules are hold.

1. The mismatch count of unsent part of the key $K_m^{P^1}$, and the parts of the $S_m^{P^1}$ which are received from other participants is less than $t_{v_1} L/2$. Where $t_{v_1} < 1/2$ is a acceptance threshold that is determined by the security level and settings of the protocol. Our protocols verification function can be defined as follows:

$$Ver(m, Sig_m, P^i, l = -1) = \begin{cases} \text{True} & |Ver(m, Sig_m, P^j, l = 0) = True| > \frac{N-1}{2} \\ \text{False} & otherwise \end{cases} \tag{9}$$

$$Ver(m, Sig_m, P^i, l = 0) = \begin{cases} \text{True} & MC(Sig_m^{Pj}) < t_{v_i} \frac{L}{2} \\ \text{False} & MC(Sig_m^{Pj}) >= t_{v_i} \frac{L}{2} \end{cases} \tag{10}$$

Where $MC(.)$ is mismatch count in the parts of the signature. If we select t_{v_j} as in Eq.(12) then $Ver(m, Sig_m, P^i, l \geq 1)$ will be hold.

2. The mismatch count of sent part of the key K_m^{P1} is less than $t_{v_1}L/2$.

(c) P^1 forwards the data (m, Sig_m) to the one of the other participant $P^j, j = 2..N-1$, say P^2.

(d) P^2 checks the unmatched count in the same manner with P^1 but he/she uses different threshold value t_{v_2}. Where $0 < t_{v_1} < t_{v_2} < 1/2$. Every participant must use different acceptance threshold for the (m, Sig_m) which is sent from P^0 or forwarded by other participant. The number of forward determines the threshold, so every forward process needs bigger threshold. This is necessary of all unconditionally secure signature protocols [18]. P^2 then forwards (m, Sig_m) to the other participant.

(e) All participants can use the same manner except with bigger threshold values.

Majority-Vote Dispute Resolution: This method is invoked whenever there is a disagreement on a message's signature is valid for all users. When this method returns $False$ all participants accept that signature is invalid otherwise valid. If more than half of the users get $Ver(x, \sigma, i, -1) = True$ than signature is valid for all users otherwise invalid.

4. Security Analysis

We can define this protocol's security analysis as follows:

(i) **Non-forgery:** Since all participants has half keys of the other participants, they can try to forge any other participant. We assume P_i wants to forge P_j, $j \neq i$, $i, j >= 1..N-1$. To achieve forging P_i must send (m, Sig_m) to P_j with fewer than $t_{v_j}L/2$ mismatch count. Due to P_i knows only half key of the P_j, K^{P_j}, then P_i cannot achieve $t_{v_j}L/2$ threshold condition for the other part of P_j, K^{P_j}.
Let $ER^{MAX} = max\ ER_{P_i P_j}, i, j = 0..N-1$ is the maximum of the error rates observed between $\{Cor_X^{P_i}$ and $Cor_X^{P_j}\}$ with k size. By calculating Eq.(6) for all key generation stages between P^0 and $P^j, j = 1..N-1$ then we can get $ER_{Eve}^{MIN} = min\ ER_{Eve(P^0,P^j),j=1..N-1}$. If we select t_{v_j} between $0 \leq ER^{MAX} < t_{v_j} < ER_{Eve}^{MIN} \leq 1/2$ then P_j participant will accept the legitimate (m, Sig_m) from P_0 and also will reject forged (m, Sig_m) from the participant P_i, $i \neq j, i, j = 1..N-1$. Because all verification functions must be different we can choose acceptance threshold values as follows:

$$h = \frac{|ER_{Eve}^{MIN} - ER^{MAX}|}{N-1} \tag{11}$$

$$v_j = ER^{MAX} + hj \tag{12}$$

Every threshold value is different from the other at least h. Then threshold values holds $ER^{MAX} < t_{v_1} < t_{v_2} < ... < t_{v_{N-1}} < ER_{Eve}^{MIN}$. By increasing L this protocol will be much secure. If at least $(N-1)/2$ participants are dishonest in the protocol, non-forgery condition cannot be achieved.

(ii) **Non-repudiation:** P_0 wants that $P_i, i = 1..N-1$ accepts validity of (m, Sig_m). When P_i forwards (m, Sig_m) to the other participants P_j, $j \neq i, j = 1..N-1$, they rejects the same (m, Sig_m). In order that P_j participant reject (m, Sig_m), he/she has bigger mismatch count than $t_{v_j}L/2$ mismatch count some part of Sig_m. If P_i accepts the message which is sent by P_0, then P_i has lesser mismatch count than $t_{v_j}L/2$ mismatch count for all parts of Sig_m. Note that P_i has half parts of other P_j participants' exact keys K^{P_j} due to protocol. To repudiate P_0 must inject more errors P_j's key part but it is impossible due to key sharing part of protocol.

Journal of Physics: Conference Series **766** (2016) 012021 doi:10.1088/1742-6596/766/1/012021

5. Conclusion

In this study we have try to develop multi-participant quantum digital signature protocol by extending Amiri[5]. We generalized this protocol to the multi-participant case. We used majority vote dispute resolution and bounded-ordered acceptance thresholds for this protocol to make secure against repudiation and forgery. We also used decoy states against to the photon number splitting attack. Also majority vote dispute resolution method has advances according to using arbiter. Since the acceptance mechanism of the majority vote method is distributed to the all participants, this method is much more secure than arbiter.

References

[1] Shor. P. W., Polynomial-Time Algorithms for Prime Factorization and Discrete Logarithms on a Quantum Computer, SIAM Journal on Computing, 26(5),pp. 1484-1509, (1997).
[2] Buhrman H., Cleve R., Watrous J., and Wolf R. Quantum Fingerprinting. Physical Review Letters, vol:87, num:16, (2001).
[3] Arrazola J. M, Walden P., and Andersson E. Multiparty Quantum Signature Schemes, arXiv:quanth-ph/1505.07509v1 (2015).
[4] Yin H. Y., Yao F., Chen Z. B. Practical Quantum Digital Signature, arXiv:quanth-ph/1507.03333v3 (2016).
[5] Amiri R., Wallden P., Kent A., and Andersson E. Secure Quantum Signatures Using Insecure Quantum Channels. arXiv:quanth-ph/1507.02975v1 (2015).
[6] Amiri R.,and Andersson E. Unconditionally Secure Quantum Signatures. Entropy, 17, 5635-5659, (2015).
[7] Gottesman, D. and Chuang, I. Quantum Digital Signatures:arXiv:quanth-ph/0105032 (2001).
[8] Lamport L. Constructing digital signatures from a one-way function. Technical Report CSL-98-, SRI International (October 1979).
[9] Andersson, E., Curty, M., and Jex, I. (2006). Experimentally realizable quantum comparison of coherent states and its applications. Physical Review A, 74(2), 022304-1. 10.1103/PhysRevA.74.022304
[10] Dunjko, V., Wallden, P. and Andersson, E. Quantum digital signatures without Quantum Memory. Phys. Rev. Lett. 112, 040502 (2014).
[11] Wallden, P., Dunjko, V., Kent. A., Andersson, E. Quantum digital signatures with quantum-key-distribution components. Phys. Rev. A 91, 042304 (2015).
[12] C. C. W. Lim, M. Curty, N. Walenta, F. Xu, and H. Zbinden, Concise security bounds for practical decoy-state quantum key distribution, Phys. Rev. A 89, 022307 (2014).
[13] C. H. Bennett and G. Brassard, in Proceedings of IEEE International Conference on Computers, Systems, and Signal Processing, Bangalore, India (IEEE, New York, 1984), pp. 175–179.
[14] V. Scarani, H. Bechmann-Pasquinucci, N. J. Cerf, M. Dusek, N. Lütkenhaus, and M. Peev, ibid. 81, 1301 (2009).
[15] W. Y. Hwang, Quantum Key Distribution with High Loss: Toward Global Secure Communication, Phys. Rev. Lett. 91, 057901 (2003).
[16] Ma X., Qi B., Zhao Y., and Lo H-K., Practical decoy state for quantum key distribution, Physical Review A, 72, 012326 (2005).
[17] H.-K. Lo, H. F. Chau, and M. Ardehali,Decoy-State Quantum Key Distribution, J. Cryptol. 18, 133 (2005).
[18] J. M. Arrazola, P. Wallden, and E. Andersson, Multiparty quantum signature schemes, Quantum Inf. Comput. 6, 0435 (2016).
[19] Renner R. Security of quantum key distribution PhD Thesis Swiss Federal Institute of Technology (2005), arXiv:quant-ph/0512258v2.
[20] R. J. Serfling, Probability inequalities for the sum in sampling without replacement, Ann. Stat. 2, 39 (1974).

International Conference on Quantum Science and Applications (ICQSA-2016) IOP Publishing
Journal of Physics: Conference Series **766** (2016) 012022 doi:10.1088/1742-6596/766/1/012022

Quantum Codes From Cyclic Codes Over The Ring R_2

Alev Altınel, Murat Güzeltepe

Department of Mathematics, Sakarya University, TR54187 Sakarya, Turkey
Department of Mathematics, Sakarya University, TR54187 Sakarya, Turkey

E-mail: alevaltinell@gmail.com

E-mail: mguzeltepe@sakarya.edu.tr

Abstract. Let R_2 denotes the ring $F_2 + uF_2 + v_2 + uvF_2 + wF_2 + uwF_2 + vwF_2 + uvwF_2$. In this study, we construct quantum codes from cyclic codes over the ring R_2, for arbitrary length n, with the restrictions $u^2 = 0$, $v^2 = 0$, $w^2 = 0$, $uv = vu$, $uw = wu$, $vw = wv$ and $u(vw) = (uv)w$. Also, we give a necessary and sufficient condition for cyclic codes over R_2 that contains its dual. As a final point, we obtain the parameters of quantum error-correcting codes from cyclic codes over R_2 and we give an example of quantum error-correcting codes form cyclic codes over R_2.

1. Introduction

Quantum error-correcting codes play a prominent role both quantum communication and quantum computation. The first error-correcting code was discovered by Shor in [1] and independently by Steane in [2]. Calderbank *et al.* gave a systematic way to construct quantum error-correcting code using classical error-correcting code, which was the problem of finding quantum error-correcting codes was transformed into the problem of finding additive self-orthogonal codes over the field F_4 in [3]. Also many quantum error-correcting codes have been constructed by using classical error-correcting codes over finite field F_q in [2, 4]. In addition, quantum codes were constructed from cyclic codes of odd length over the finite ring $F_2 + uF_2$ in [5]. In the light of the paper, quaternary construction of quantum codes from cyclic codes over $F_4 + uF_4$ was studied in [6]. Besides, quantum codes from cyclic codes of odd length n over finite commutative ring $F_2 + vF_2$ was constructed in [7]. We inspired the paper in [7] and construct quantum codes from cyclic codes over the ring R_2.

This paper is organized as follows. In Section 2, we give some basic definitions and the structure of the ring R_2. Also, we define a Gray map and construct linear and cyclic codes over the ring R_2. In Section 3, we give fundamental theorem for quantum error-correcting codes and we obtain the parameters of quantum error-correcting codes from cyclic codes over R_2 and give some examples.

2. Preliminaries

The ring R_2 is defined as a characteristic 2 ring, which depends on the restrictions $u^2 = v^2 = w^2 = 0$, $uv = vu$, $uw = wu$, $vw = wv$ and $u(vw) = (uv)w$. In addition, it is clear that R_2 is

Table 1. Ideals and The number of the elements of the ideals

The ideals	The number of the elements	The ideals	The number of the elements
$I_0 = \{0\}$	$\lvert I_0\rvert = 1,$		
I_{uvw}	$\lvert I_{uvw}\rvert = 2,$	$I_{u,vw}$	$\lvert I_{u,vw}\rvert = 32,$
I_{uv}	$\lvert I_{uv}\rvert = 4,$	$I_{v,uw}$	$\lvert I_{v,uw}\rvert = 32,$
I_{uw}	$\lvert I_{uw}\rvert = 4,$	$I_{w,uv}$	$\lvert I_{w,uv}\rvert = 32,$
I_{vw}	$\lvert I_{vw}\rvert = 4,$	$I_{uv,u+w}$	$\lvert I_{uv,u+w}\rvert = 32,$
I_{uv+uw}	$\lvert I_{uv+uw}\rvert = 4,$	$I_{uv,v+w}$	$\lvert I_{uv,v+w}\rvert = 32,$
I_{uv+vw}	$\lvert I_{uv+vw}\rvert = 4,$	$I_{uv,u+v+w}$	$\lvert I_{uv,u+v+w}\rvert = 32,$
I_{uw+vw}	$\lvert I_{uw+vw}\rvert = 4,$	$I_{uw,u+v}$	$\lvert I_{uw,u+v}\rvert = 32,$
$I_{uv+uw+vw}$	$\lvert I_{uv+uw+vw}\rvert = 4,$	$I_{uv,uw,v+w}$	$\lvert I_{uv,uw,v+w}\rvert = 32,$
$I_{uv,uw} = I_{uv} \oplus I_{uw}$	$\lvert I_{uv,uw}\rvert = 8,$	$I_{uv,vw,u+w}$	$\lvert I_{uv,vw,u+w}\rvert = 32,$
$I_{uv,vw}$	$\lvert I_{uv,vw}\rvert = 8,$	$I_{uv,vw,u+v+w}$	$\lvert I_{uv,vw,u+v+w}\rvert = 32,$
$I_{uv,uw+vw}$	$\lvert I_{uv,uw+vw}\rvert = 8,$	$I_{u+v,uv+uw}$	$\lvert I_{u+v,uv+uw}\rvert = 32,$
$I_{uw,vw}$	$\lvert I_{uw,vw}\rvert = 8,$	$I_{u,v}$	$\lvert I_{u,v}\rvert = 64,$
$I_{uw,uv+vw}$	$\lvert I_{uw,uv+vw}\rvert = 8,$	$I_{u,w}$	$\lvert I_{u,w}\rvert = 64,$
$I_{vw,uv+uw}$	$\lvert I_{vw,uv+uw}\rvert = 8,$	$I_{v,w}$	$\lvert I_{v,w}\rvert = 64,$
$I_{uv+uw,uv+vw}$	$\lvert I_{uv+uw,uv+vw}\rvert = 8,$	$I_{u,v+w}$	$\lvert I_{u,v+w}\rvert = 64,$
I_u	$\lvert I_u\rvert = 16,$	$I_{v,u+w}$	$\lvert I_{v,u+w}\rvert = 64,$
I_v	$\lvert I_v\rvert = 16,$	$I_{w,u+v}$	$\lvert I_{w,u+v}\rvert = 64,$
I_w	$\lvert I_w\rvert = 16,$	$I_{u+v,u+w}$	$\lvert I_{u+v,u+w}\rvert = 64,$
I_{u+v}	$\lvert I_{u+v}\rvert = 16,$	$I_{u+v,u+vw}$	$\lvert I_{u+v,u+vw}\rvert = 64,$
I_{u+w}	$\lvert I_{u+w}\rvert = 16,$	$I_{u+v,w+uv}$	$\lvert I_{u+v,w+uv}\rvert = 64,$
I_{u+vw}	$\lvert I_{u+vw}\rvert = 16,$	$I_{u+w,v+w}$	$\lvert I_{u+v,v+w}\rvert = 64,$
I_{v+w}	$\lvert I_{v+w}\rvert = 16,$	$I_{u+w,u+vw}$	$\lvert I_{u+w,u+vw}\rvert = 64,$
I_{v+uw}	$\lvert I_{v+uw}\rvert = 16,$	$I_{v+w,v+uw}$	$\lvert I_{v+w,v+uw}\rvert = 64,$
I_{w+uv}	$\lvert I_{w+uv}\rvert = 16,$	$I_{u+vw,v+uw}$	$\lvert I_{u+vw,v+uw}\rvert = 64,$
I_{u+v+w}	$\lvert I_{u+v+w}\rvert = 16,$	$I_{u+vw,w+uv}$	$\lvert I_{u+vw,w+uv}\rvert = 64,$
I_{u+v+uw}	$\lvert I_{u+v+uw}\rvert = 16,$	$I_{v+uw,w+uv}$	$\lvert I_{v+uw,w+uv}\rvert = 64,$
I_{u+v+uv}	$\lvert I_{u+v+uv}\rvert = 16,$	$I_{uv,u+w,v+w}$	$\lvert I_{uv,u+w,v+w}\rvert = 64,$
I_{v+w+uv}	$\lvert I_{v+w+uv}\rvert = 16,$	$I_{vw,u+v,u+w}$	$\lvert I_{vw,u+v,u+w}\rvert = 64,$
$I_{u+v+w+uv}$	$\lvert I_{u+v+w+uv}\rvert = 16,$	$I_{u,v,w}$	$\lvert I_{u,v,w}\rvert = 128,$
$I_{uv,vw,uw}$	$\lvert I_{uv,vw,uw}\rvert = 16,$	$I_1 = R_2$	$\lvert I_1\rvert = 256.$

isomorphic to

$$F_2\left[X,Y,Z\right]/\left\langle X^2,Y^2,Z^2,XY-YX,XZ-ZX,YZ-ZY\right\rangle.$$

We list all the ideals of R_2 as can be seen at Table 1. Due to the structure of the ideals, we can see the ideals are not all principal just as $I_{u,v,w}$ is not a principal ideal. The ideal $I_{u,v,w} = M$ is a maximal ideal of R_2 and R_2 is a local ring since it has a only one maximal ideal. And it is known that if a ring R is a local ring with maximal ideal M , then an element in $R - M$ is a unit in R. So, we can determine the units of R_2 as $R_2^* = R_2 - I_{u,v,w}$.

To define the dual code of a code in R_2^n, we give an inner product which is, in fact, the usual inner product as:

$$\langle (x_1, x_2, \ldots, x_n), (y_1, y_2, \ldots, y_n) \rangle = x_1y_1 + x_2y_2 + \ldots + x_ny_n, \tag{1}$$

where the operations belong to the ring R_2 . Now we define the inner product for the ring R_2 at the same way concerning the dot product in F_2 as follows :

$$\begin{aligned}&\langle \bar{a}_0 + u\bar{a}_1 + v\bar{a}_2 + uv\bar{a}_3 + w\bar{a}_4 + uw\bar{a}_5 + vw\bar{a}_6 + uvw\bar{a}_7, \bar{b}_0 + u\bar{b}_1 + v\bar{b}_2 + uv\bar{b}_3 + \\ &w\bar{b}_4 + uw\bar{b}_5 + vw\bar{b}_6 + uvw\bar{b}_7\rangle = \bar{a}_0 \cdot \bar{b}_0 + u(\bar{a}_0 \cdot \bar{b}_1 + \bar{a}_1 \cdot \bar{b}_0) + v(\bar{a}_0 \cdot \bar{b}_2 + \bar{a}_2 \cdot \bar{b}_0) + \\ &uv(\bar{a}_0 \cdot \bar{b}_3 + \bar{a}_1 \cdot \bar{b}_2 + \bar{a}_2 \cdot \bar{b}_1 + \bar{a}_3 \cdot \bar{b}_0) + w(\bar{a}_0 \cdot \bar{b}_4 + \bar{a}_4 \cdot \bar{b}_0) + uw(\bar{a}_0 \cdot \bar{b}_5 + \bar{a}_1 \cdot \bar{b}_4 + \bar{a}_4 \cdot \bar{b}_1 + \\ &\bar{a}_5 \cdot \bar{b}_0) + vw(\bar{a}_0 \cdot \bar{b}_6 + \bar{a}_2 \cdot \bar{b}_4 + \bar{a}_4 \cdot \bar{b}_2 + \bar{a}_6 \cdot \bar{b}_0) + uvw(\bar{a}_0 \cdot \bar{b}_7 + \bar{a}_1 \cdot \bar{b}_6 + \bar{a}_3 \cdot \bar{b}_4 + \bar{a}_4 \cdot \bar{b}_3 + \\ &\bar{a}_5 \cdot \bar{b}_2 + \bar{a}_6 \cdot \bar{b}_1 + \bar{a}_0 \cdot \bar{b}_7 + \bar{a}_7 \cdot \bar{b}_0),\end{aligned}$$

where the symbol $[\cdot]$ denotes the dot product in binary vectors.

Definition 1 *A code C of length n over the ring R_2 is a subset of R_2^n and C is linear if it is a submodule.*

Definition 2 *Let C be a linear code over R_2 of length n, then the dual of C is defined as*

$$C^{\perp} = \{\bar{y} \in R_2^n : \langle \bar{y}, \bar{x} \rangle = 0, \forall \bar{x} \in C\}.$$

$C^{\perp}$ is also a linear code over R_2 of length n resulting from the definition of the inner product in Eq. 1. A code C is self orthogonal if $C \subseteq C^{\perp}$ and self dual if $C = C^{\perp}$.

We denote that $A \oplus B = \{a + b : a \in A, b \in B\}$ and $A \otimes B = \{(a, b) : a \in A, b \in B\}$.

Definition 3 *For any ring R, a cyclic shift on R^n is a permutation σ such that $\sigma(q_0, q_1, \ldots, q_{n-1}) = (q_{n-1}, q_0, \ldots, q_{n-2})$. And a linear code C over a ring R that is invariant under the cyclic shift is called a cyclic code.*

Definition 4 *A Gray map ψ from R_2^n to F_2^{8n} can be defined as*

$$\psi(\bar{a}_0 + u\bar{a}_1 + v\bar{a}_2 + uv\bar{a}_3 + w\bar{a}_4 + uw\bar{a}_5 + vw\bar{a}_6 + uvw\bar{a}_7) =$$

$$\left(\bar{a}_7, \sum_{i=6,7} \bar{a}_i, \sum_{i=5,7} \bar{a}_i, \sum_{i=4,5,6,7} \bar{a}_i, \sum_{i=3,7} \bar{a}_i, \sum_{i=2,3,6,7} \bar{a}_i, \sum_{i=1,3,5,7} \bar{a}_i, \sum_{i=0}^{7} \bar{a}_i\right).$$

The Gray map ψ is a weight preserving map from R_2^n (Lee weight) to F_2^{8n} (Hamming weight). And if C is self orthogonal, so is $\psi(C)$.

Definition 5 *For any $q \in R_2$, we define the Lee weight as*

$$w_L(q) = w_H(\psi(q)),$$

where w_H denotes well-known Hamming weight for binary codes.

Let C be a linear code of length n over R_2. Define

$$\begin{aligned}C_1 &= \{a_7 \in F_2^n : \exists a_i \in F_2^n, 0 \le i \le 6, q \in C\} \\ C_2 &= \{a_6 + a_7 \in F_2^n : \exists a_i \in F_2^n, 0 \le i \le 5, q \in C\} \\ C_3 &= \{a_5 + a_7 \in F_2^n : \exists a_0, a_1, a_2, a_3, a_4, a_6 \in F_2^n, q \in C\} \\ C_4 &= \{a_4 + a_5 + a_6 + a_7 \in F_2^n : \exists a_i \in F_2^n, 0 \le i \le 3, q \in C\} \\ C_5 &= \{a_3 + a_7 \in F_2^n : \exists a_0, a_1, a_2, a_4, a_5, a_6 \in F_2^n, q \in C\} \\ C_6 &= \{a_2 + a_3 + a_6 + a_7 \in F_2^n : \exists a_0, a_1, a_4, a_5 \in F_2^n, q \in C\} \\ C_7 &= \{a_1 + a_3 + a_5 + a_7 \in F_2^n : \exists a_0, a_2, a_4, a_6 \in F_2^n, q \in C\} \\ C_8 &= \left\{\sum_{i=0}^{7} a_i \in F_2^n : q \in C\right\},\end{aligned}$$

where $q = a_0 + ua_1 + va_2 + uva_3 + wa_4 + uwa_5 + vwa_6 + uvwa_7$. Then $C_i's$ are binary codes at length n for every $i = 1, 2, \ldots, 8$. Furthermore, the linear code C is uniquely defined as

$$C = (1 + u + v + uv + w + uw + vw + uvw)\, C_1 \oplus (u + uv + uw + uvw)\, C_2 \oplus$$

$$(v + uv + vw + uvw)\, C_3 \oplus (w + uw + vw + uvw)\, C_4 \oplus (uv + uvw)\, C_5 \oplus$$

$$(uw + uvw)\, C_6 \oplus (vw + uvw)\, C_7 \oplus (uvw)\, C_8.$$

Theorem 1 *Let C be a linear code of length n over R_2, then $\psi(C) = \bigotimes_{i=1}^{8} C_i$ and $|C| = \prod_{i=1}^{8} |C_i|$.*

Here,

$$|C| = \prod_{i=1}^{8} |C_i| = 2^{n-\deg f_1} \cdot 2^{n-\deg f_2} \cdot \ldots \cdot 2^{n-\deg f_8} = 2^{8n-\sum_{i=1}^{8} \deg f_i},$$

where $f_i's$ are generator polynomials of $C's$ respectively for $i = 1, 2, \ldots, 8$.

$d_L = \min\{d_H(C_i) : i = 1, 2, \ldots, 8\}$ where d_L denotes the minimum Lee weight of linear code C over R_2 and $d_H(C_i)$ denotes the minimum Hamming weights of binary codes $C_i's$ respectively.

Proposition 1 *Let*

$$C = (1 + u + v + uv + w + uw + vw + uvw)\, C_1 \oplus (u + uv + uw + uvw)\, C_2 \oplus$$

$$(v + uv + vw + uvw)\, C_3 \oplus (w + uw + vw + uvw)\, C_4 \oplus (uv + uvw)\, C_5 \oplus$$

$$(uw + uvw)\, C_6 \oplus (vw + uvw)\, C_7 \oplus (uvw)\, C_8$$

be a linear code over R_2. Then C is a cyclic code over R_2 if and only if $C_i{}'s$ are binary cyclic codes.

3. Quantum Codes From Cyclic Codes Over R_2

We start to this section by giving well-known theorem on quantum codes:

Theorem 2 *(CSS Construction) Let C and C' be two binary codes with parameters $[n, k_1, d_1]$ and $[n, k_2, d_2]$, respectively. If $C^\perp \subseteq C'$, then an $[[n, k_1 + k_2 - n, \min\{d_1, d_2\}]]$ quantum code can be constructed. Especially, $C^\perp \subseteq C$, then there exists an $[[n, 2k_1 - n, d_1]]$ quantum code.*

Proposition 2 *Suppose*

$$C = (1 + u + v + uv + w + uw + vw + uvw)\, C_1 \oplus (u + uv + uw + uvw)\, C_2 \oplus$$

$$(v + uv + vw + uvw)\, C_3 \oplus (w + uw + vw + uvw)\, C_4 \oplus (uv + uvw)\, C_5 \oplus$$

$$(uw + uvw)\, C_6 \oplus (vw + uvw)\, C_7 \oplus (uvw)\, C_8$$

is cyclic code of length n over R_2. Then

$$C = \langle (1 + u + v + uv + w + uw + vw + uvw)\, f_1, (u + uv + uw + uvw)\, f_2, (v + uv + vw + uvw)\, f_3, (w + uw + vw + uvw)\, f_4, (uv + uvw)\, f_5, (uw + uvw)\, f_6, (vw + uvw)\, f_7, (uvw)\, f_8 \rangle$$

and

$$|C| = 2^{8n-\sum_{i=1}^{8} \deg f_i},$$

where f_i's are generators polynomials of C_i's respectively. Also,

$$\begin{aligned}C^{\perp} = \langle & (1+u+v+uv+w+uw+vw+uvw)\, h_1^* + (u+uv+uw+uvw)\, h_2^* + \\ & (v+uv+vw+uvw)\, h_3^* + (w+uw+vw+uvw)\, h_4^* + (uv+uvw)\, h_5^* + \\ & (uw+uvw)\, h_6^* + (vw+uvw)\, h_7^* + (uvw)\, h_8^* \rangle\end{aligned}$$

and

$$\left|C^{\perp}\right| = 2^{\sum_{i=1}^{8} \deg f_i},$$

where for $i = 1, 2, \ldots, 8$, h_i^* are the reciprocal polynomials of h_i i.e., $h_i(x) = (x^n - 1)/f_i(x)$, $h_i^*(x) = x^{\deg h_i} h_i\left(\frac{1}{x}\right)$ for $i = 1, 2, \ldots, 8$.

Proposition 3 *Suppose C is a cyclic code of length n over R_2, then there is a unique polynomial $f(x)$ such that $C = \langle f(x)\rangle$ and $f(x)/(x^n - 1)$, where*

$$\begin{aligned}f(x) = & (1+u+v+uv+w+uw+vw+uvw)\, f_1(x) + (u+uv+uw+uvw)\, f_2(x) + \\ & (v+uv+vw+uvw)\, f_3(x) + (w+uw+vw+uvw)\, f_4(x) + (uv+uvw)\, f_5(x) + \\ & (uw+uvw)\, f_6(x) + (vw+uvw)\, f_7(x) + (uvw)\, f_8(x).\end{aligned}$$

Lemma 1 *[3] A binary cyclic code C with generator polynomial $f(x)$ contains its dual code if and only if $x^n - 1 \equiv 0 \,(\mathrm{mod}\, ff^*)$ where f^* is the reciprocal polynomial of f.*

Theorem 3 *Let $C = \langle f(x)\rangle$ be a cyclic code of length n over R_2 and $f(x)/(x^n - 1)$, where*

$$\begin{aligned}f(x) = & (1+u+v+uv+w+uw+vw+uvw)\, f_1(x) + (u+uv+uw+uvw)\, f_2(x) + \\ & (v+uv+vw+uvw)\, f_3(x) + (w+uw+vw+uvw)\, f_4(x) + (uv+uvw)\, f_5(x) + \\ & (uw+uvw)\, f_6(x) + (vw+uvw)\, f_7(x) + (uvw)\, f_8(x).\end{aligned}$$

Then $C^{\perp} \subseteq C$ if and only if $x^n - 1 \equiv 0 \,(\mathrm{mod}\, f_i f_i^*)$ for $i = 1, 2, \ldots, 8$.

Corollary 1 *Let*

$$\begin{aligned}C = & (1+u+v+uv+w+uw+vw+uvw)\, C_1 \oplus (u+uv+uw+uvw)\, C_2 \oplus \\ & (v+uv+vw+uvw)\, C_3 \oplus (w+uw+vw+uvw)\, C_4 \oplus (uv+uvw)\, C_5 \oplus \\ & (uw+uvw)\, C_6 \oplus (vw+uvw)\, C_7 \oplus (uvw)\, C_8\end{aligned}$$

is a cyclic code of length n over R_2. Then $C^{\perp} \subseteq C$ if and only if $C_i^{\perp} \subseteq C_i$ for $i = 1, 2, \ldots, 8$.

Example 1 *For $n = 63$,*

$$\begin{aligned}x^{63} - 1 = & (x+1)\left(x^2+x+1\right)\left(x^3+x+1\right)\left(x^3+x^2+1\right)\left(x^6+x+1\right) \\ & \left(x^6+x^3+1\right)\left(x^6+x^4+x^2+x+1\right)\left(x^6+x^4+x^3+x+1\right) \\ & \left(x^6+x^5+1\right)\left(x^6+x^5+x^2+x+1\right)\left(x^6+x^5+x^3+x^2+1\right) \\ & \left(x^6+x^5+x^4+x+1\right)\left(x^6+x^5+x^4+x^2+1\right) \\ = & f_1 f_2 f_3 f_4 f_5 f_6 f_7 f_8 f_9 f_{10} f_{11} f_{12} f_{13}\end{aligned}$$

in $F_2[x]$. Hence, $f_1^* = f_1$, $f_2^* = f_3$, $f_3^* = f_2$. So, $x^{63} - 1$ is divisibly by $f_i f_i^*$, $i = 3, 4, 5, 7, 8, 9, 10, 11, 12, 13$. Thus we have a binary cyclic code over R_2

$$\begin{aligned}C = \langle & (1+u+v+uv+w+uw+vw+uvw)\, f_3, (u+uv+uw+uvw)\, f_4, \\ & (v+uv+vw+uvw)\, f_5, (w+uw+vw+uvw)\, f_7, (uv+uvw)\, f_8, \\ & (uw+uvw)\, f_9, (vw+uvw)\, f_{10}, (uvw)\, f_{11}\rangle\end{aligned}$$

and its dual code is generated as following over R_2 by Proposition 2:

$$C^{\perp} = \{(1+u+v+uv+w+uw+vw+uvw)\, f_3^* + (u+uv+uw+uvw)\, f_4^* + (v+uv+vw+uvw)\, f_5^* + (w+uw+vw+uvw)\, f_7^* + (uv+uvw)\, f_8^* + (uw+uvw)\, f_9^* + (vw+uvw)\, f_{10}^* + (uvw)\, f_{11}^*\}$$

Thus, the code C contains its dual code $C^{\perp}$.

By using Theorem 2 and Theorem 3 we can construct the parameters of quantum codes.

Theorem 4 *Let*

$$C = (1+u+v+uv+w+uw+vw+uvw)\, C_1 \oplus (u+uv+uw+uvw)\, C_2 \oplus (v+uv+vw+uvw)\, C_3 \oplus (w+uw+vw+uvw)\, C_4 \oplus (uv+uvw)\, C_5 \oplus (uw+uvw)\, C_6 \oplus (vw+uvw)\, C_7 \oplus (uvw)\, C_8$$

be a cyclic code of arbitrary length n over R_2 with size

$$256^{k_1} 128^{k_2} 64^{k_3} 32^{k_4} 16^{k_5} 8^{k_6} 4^{k_7} 2^{k_8}.$$

If $C_i^{\perp} \subseteq C_i$ where $i = 1, 2, \ldots, 8$ then $C^{\perp} \subseteq C$ and there exists a quantum error-correcting code with parameters

$$[[8n, 8k_1 + 7k_2 + 6k_3 + 5k_4 + 4k_5 + 3k_6 + 2k_7 + k_8 - 8n, d_H]].$$

Example 2 *For $n = 4$, in $F_2[x]$*

$$\left(x^4 - 1\right) = (x+1)^2 \left(x^2 + 1\right).$$

Let $C = \langle (vw+uvw)\, f_1(x), (uvw)\, f_2(x) \rangle$, where $f_1(x) = (x+1) = f_1^*(x)$ and $f_2(x) = (x^2+1) = f_2^*(x)$. In the light of Theorem 1, we know that C is linear cyclic codes with the parameters $[32, 3, 16]$. Also, with the help of Proposition 2, $C^{\perp} = \langle uvw\,(x^3) + vw\,(x^2) + uvw\,(x) + vw \rangle$ is a linear codes with the parameters $[32, 29, 2]$. We can clearly see $C^{\perp} \subseteq C$. Thus, we obtain a quantum code with parameters $[[16, 13, 2]]$.

References

[1] Shor P. W. 1995 *Phys. Rev.* A Scheme for reducing decoherence in quantum memory **52** 2493-2896
[2] Steane A. M. 1996 *Phys. Rev.* A Simple quantum error-correcting codes **54** 4741-51
[3] Calderbank A. R., Rains, E. M., Shor, P. W., Sloane, N. J. A. 1998 *IEEE Trans. Inform. Theory* Quantum error correction via codes over $GF(4)$ **44** 1369-87
[4] A Ketkar *et al.* 2006 *IEEE Trans. Inform. Theory* Nonbinary quantum stabiliter codes over finite fields **52** 4892-4914
[5] Qian J., Ma W. and Gua W. 2009 *Int. J. Quant. Inform.* Quantum codes from cyclic codes over finite ring **7** 1277
[6] Kai X., Zhu S. 2011 *Int. J. Quant. Inform.* Quaternary construction of quantum codes from cyclic codes over $F_4 + uF_4$ **9:2** 689-700
[7] Qian J. 2013 *Journal of Information and Computational Science* Quantum Codes from Cyclic Codes over $F_2 + vF_2$ **10:6** 1715-22
[8] Guzeltepe M., Altinel A., On Some Codes Over The Ring R_2, Under review.

Journal of Physics: Conference Series **766** (2016) 012023 doi:10.1088/1742-6596/766/1/012023

On the solvability of p states quantum Rabi Model with Z_p-graded parity

Won Sang Chung[1]

Department of Physics and Research Institute of Natural Science, College of Natural Science, Gyeongsang National University, Jinju 660-701, Korea

E-mail: mimip4444@hanmail.net

Jae Yoon Kim

Department of Physics and Research Institute of Natural Science, College of Natural Science, Gyeongsang National University, Jinju 660-701, Korea

Abstract.
In this paper the p-level Rabi model with Z_p-graded symmetry is discussed. The p-level Rabi Hamiltonian is constructed by introducing the generalized Pauli matrices. The energy and wave function for the p-level Rabi equation are obtained by using the standard perturbation method.

1. Introduction

The quantum Rabi model(QRM) [1] is very useful in studying the interaction between atom and light. This model arises in various physical models such as quantum computation, condensed matter, atomic and molecular physics [2], nuclear and particle physics [3,4]. This model is described by the following Schrödinger equation:

$$H_R\psi = E\psi \tag{1}$$

where

$$H_R = w^a + \Delta\sigma_z + g(a^\dagger + a)\sigma_x, \tag{2}$$

and g is coupling strength and $a^\dagger, a$ are the creation and annihilation operator for the quantized field. The atomic ground state is $(0\ 1)^T$ and excited state is $(1\ 0)^T$. The energy scale is defined by setting $w = 1$.

The quantum Rabi model can be extended to Jaynes-Cummings model [6] by taking the interaction picture and using the rotating wave approximation which can be applied for the weak coupling case. For dimensionless coupling strength $\kappa = g/w \le 10^{-2}$, The quantum Rabi model is well captured by JCM which can be analytically solvable. The dynamics of the JCM have been studied in Refs. [7, 8].

Recently, solid-state semiconductor [7] and superconductor systems [8-10] have allowed the advent of the ultrastrong coupling regime, where the dimensionless coupling strength

[1] Talk given at the international conference on quantum science and applications, Eskisehir, Turkey (2016)

$\kappa = g/w \geq 0.1$ [11]. In this regime, rotating wave approximation cannot be adopted, so the related physics should be explained by QRM.

There has been many trials to obtain the exact solution of the Rabi Hamiltonian until Braak [12] recently succeeded in doing it. He solved the quantum Rabi equation explicitly without the famous rotating wave approximation. The solution for $\Delta = 0$ is given by Charlier polynomial, but for $\Delta \neq 0$ it is not shown to be related to any kind of orthogonal function. So for this case Braak used the numerical method to obtain the energy levels for this model.

In this paper we discuss the p-level Rabi model with Z_p-graded symmetry, which is a generalization of standard two-level Rabi model. This paper is organized as follows: In section II we obtain the p-level Rabi Hamiltonian by introducing the generalized Pauli matrices. We also diagonalize the p-level Rabi Hamiltonian with a help of two kinds of unitary transformations. In section III we obtain the energy and wave function for the p-level Rabi equation by using the standard perturbation method.

2. p-level Rabi model with Z_p-graded symmetry

Now let us start with the p-level Rabi model with Z_p-graded symmetry, whose Hamiltonian is given by

$$H = \sum_{k=0}^{p-1} \Delta_k \sigma_{p-1}^k + wa^\dagger a + g(a\sigma_0 + a^\dagger \sigma_0^{p-1}) \tag{3}$$

where

$$\sum_{k=0}^{N-1} \Delta_k q^{ki} = E_i, \quad or \quad \Delta_i = \frac{1}{p}\sum_{k=0}^{p-1} q^{-ki} E_k \tag{4}$$

and the generalized Pauli matrices are

$$\sigma_i = \sum_{k=0}^{p-1} q^{ik} e_{k,k+p-1}, \quad (i = 0, 1, \cdots, p-2) \tag{5}$$

$$\sigma_{p-1} = \sum_{k=0}^{p-1} q^k e_{k,k} \tag{6}$$

and $e_{ij} = |i\rangle\langle j|$, $q = e^{2\pi i/p}$. These matrices are idempotent because they obey

$$\sigma_i^p = I, \quad (i = 0, 1, \cdots, p-1) \tag{7}$$

and

$$\sigma_i^\dagger = \sigma_i^{p-1} \tag{8}$$

These matrices are unitary matrices:

$$\sigma_i \sigma_i^\dagger = \sigma_i^\dagger \sigma_i = I, \quad (i = 0, 1, \cdots, p-1) \tag{9}$$

These Z_p-graded matrices satisfy

$$\{\sigma_{i_1}, \sigma_{i_2}, \cdots, \sigma_{i_p}\} = p!\delta_{i_1 i_2 \cdots i_p} I, \tag{10}$$

where

$$\{\sigma_{i_1}, \sigma_{i_2}, \cdots, \sigma_{i_p}\} = \sum_{\{j_1,\cdots,j_p\}\in \mathrm{perm}\{i_1,\cdots,i_p\}} \sigma_{j_1}\sigma_{j_2}, \cdots, \sigma_{j_p} \tag{11}$$

$$\delta_{i_1 i_2 \cdots i_p} = \begin{cases} 1 & (\text{if } i_1 = i_2 = \cdots = i_p) \\ 0 & (\text{otherwise}) \end{cases} \tag{12}$$

and perm$\{i_1, \cdots, i_p\}$ implies set of all permutations of $\{i_1, \cdots, i_p\}$. The proof of the eq.(11) is given in the Appendix.

Now let us introduce the Z_p-graded Walsh-Hadamard matrix as follows:

$$W = \frac{1}{\sqrt{p}} \begin{pmatrix} 1 & 1 & 1 & \cdots & 1 \\ 1 & q & q^2 & \cdots & q^{p-1} \\ 1 & q^2 & (q^2)^2 & \cdots & (q^2)^{p-1} \\ \vdots & & & & \\ 1 & q^{p-1} & (q^{p-1})^2 & \cdots & (q^{p-1})^{p-1} \end{pmatrix} \tag{13}$$

Indeed this matrix is unitary:

$$WW^\dagger = W^\dagger W = I \tag{14}$$

The Z_p-graded Walsh-Hadamard matrix transforms one of Z_p-graded Pauli matrix into another one:

$$W\sigma_{p-1}W^\dagger = \sigma_0^{p-1} \tag{15}$$

$$W\sigma_0 W^\dagger = \sigma_{p-1} \tag{16}$$

Thus, under this unitary transform, the p-level Rabi Hamiltonian is changed into

$$H' = WHW^\dagger = \Delta_0 + \sum_{k=1}^{p-1} \Delta_{p-k}\sigma_0^k + wa^\dagger a + g(a\sigma_2 + a^\dagger \sigma_2^2) \tag{17}$$

Now let us consider the case that $\Delta_0 = 0$ and $\Delta_1 = \Delta_2 = \cdots = \Delta_{p-1} = \Delta$ is real. The Hamiltonian can be written in a matrix form, acting in the space spanned by $|0\rangle, |1\rangle, \cdots, |p-1\rangle$,

$$H = \begin{pmatrix} a^\dagger a + g(a + a^\dagger) & \Delta & \Delta & \cdots & \Delta \\ \Delta & a^\dagger a + g(qa + q^{-1}a^\dagger) & \Delta & \cdots & \Delta \\ \Delta & \Delta & a^\dagger a + g(q^2 a + q^{-2}a^\dagger) & \cdots & \Delta \\ \vdots & & & & \\ \Delta & \Delta & \Delta & \cdots & a^\dagger a + g(q^{p-1}a + q^{-(p-1)}a^\dagger) \end{pmatrix} \tag{18}$$

Now let us diagonalize the Hamiltonian H' by means of the unitary matrix

$$U = \frac{1}{\sqrt{p}} \begin{pmatrix} 1 & R & R^2 & \cdots & R^{p-1} \\ 1 & qR & q^2R^2 & \cdots & q^{p-1}R^{p-1} \\ 1 & q^2R & (q^2)^2R^2 & \cdots & (q^2)^{p-1}R^{p-1} \\ \vdots & & & & \\ 1 & q^{p-1}R & (q^{p-1})^2R^2 & \cdots & (q^{p-1})^{p-1}R^{p-1} \end{pmatrix} \tag{19}$$

where $R = q^{a^\dagger a}$. Using the eq.(19) we have

$$H'' = UH'U^\dagger = wa^\dagger a + g(a + a^\dagger) + \Delta \sum_{k=1}^{p-1} R^k \sigma_{p-1}^k \tag{20}$$

The Z_p-symmetry of the chiral three states Rabi model can be used to eliminate the discrete degree of freedom from the problem. Each Z_p-parity-chain $\mathcal{H}_r, r = 0, 1, 2, \cdots, p-1$ is isomorphic to the Hilbert space of photon number states $\mathcal{H}$. In each $\mathcal{H}_r$, the Hamiltonian (20) reads

$$H_r = wa^\dagger a + g(a + a^\dagger) + \Delta \sum_{k=1}^{p-1} q^{kr} R^k \tag{21}$$

The complication of this reduced Hamiltonian comes from the last two terms. On the other hand, these terms is instrumental for the analytical solution of the model. To elucidate its meaning, it is convenient to represent the eq. (21) in Bargmanns space of analytical functions which is isometrically isomorphic to the Bargmann space for photon number states $\mathcal{B}$, where $a, a^\dagger$ are realized by ∂_z, z, respectively. The space $\mathcal{B}$ is spanned by functions $f(z)$ of a complex variable z. Using $Rf(z) = f(qz)$, in each Z_p-parity-chain $\mathcal{H}_r$, we can write the Schrödinger equation as

$$z\frac{d}{dz}\psi_r(z) + g\left(\frac{d}{dz} + z\right)\psi_r(z) = \epsilon_r\psi_r(z) - \Delta \sum_{k=1}^{p-1} q^{kr}\psi_r(q^k z) \tag{22}$$

where $r = 0, 1, 2, \cdots, p-1$.

Now let us consider the case $r = 0$. In this case the Hamiltonian H_0 reads in $\mathcal{B}$,

$$H_0 = z\frac{d}{dz} + g\left(\frac{d}{dz} + z\right) + \Delta \sum_{k=1}^{p-1} R^k \sigma_{p-1}^k \tag{23}$$

The Schroödinger equation $(H_0 - \epsilon_0)\psi(z) = 0$ corresponds to a linear but non-local differential equation in the complex domain,

$$z\frac{d}{dz}\psi(z) + g\left(\frac{d}{dz} + z\right)\psi(z) = \epsilon_0\psi(z) - \Delta \sum_{k=1}^{p-1} \psi(q^k z) \tag{24}$$

The theory of these equations initiated by Riemann and Fuchs [13] can now be applied to the above equation. First, if we set

$$\psi(z) = \phi_0(z), \psi(q^{p-1}z) = \phi_1(z), \psi(q^{p-2}z) = \phi_2(z), \cdots, \psi(qz) = \phi_{p_1}(z), \tag{25}$$

we obtain the following equations:

$$(z + q^r g)\frac{d}{dz}\phi_r + (gq^{-r}z - \epsilon_0)\phi_r = -\Delta \sum_{s\neq r} \phi_s \tag{26}$$

This system has p regular singular points at $z = -gq^k, k = 0, 1, \cdots, p-1$ and an irregular singular point at $z = \infty$. From the eq.(25), we can easily find that

$$\phi_0(z) = \phi_1(qz) = \phi_2(q^2 z) = \cdots \phi_{p-1}(q^{p-1}z) \tag{27}$$

If ϵ_0 belongs to the spectrum of H_0, all ϕ_r's must be analytic at points $z = -q^r g, r = 0, 1, \cdots, p-1$.

The eq.(26) is a system of p differential equations of Fuchsian type. In general it has p independent solutions. The required solution must be analytic in the whole complex plane, i.e. is entire, in order for E_r to belong to the spectrum of the system. The singular points of the system are at $z = -gq^k$. When $\Delta = 0$, each differential equation behaves like $(z + gq^k)^\rho$ in the vicinity of each $z = -gq^k$, so we are lead to the following indicial equation:

$$\rho - E_r - g^2 = 0$$

So for all p independent solutions $\phi_{r,k}(z)$ to be analytic at the singular points, we need

$$E_r = N - g^2, \quad (N = 0, 1, 2, \cdots)$$

This gives the exact isolated energies of the p states Rabi model. These energies have the Rabi-like form but are p-fold degenerate.

3. Pertubative solution of p states quantum Rabi equation

In this section we solve the p states quantum Rabi Model by standard pertubation method. If we set $\phi_{r,k}(z) = \psi_r(q^k z)$ in the eq.(26), we obtain the following equations:

$$(z + gq^{-r})\frac{d}{dz}\phi_{r,s} + (gq^r z - E_r)\phi_{r,s} = -\Delta \sum_{k=1}^{p-1} q^{kr}\phi_{r,k+s} \tag{28}$$

Now let us apply the perturbation method for small value of Δ by putting

$$\phi_{r,k} = \sum_{i=0}^{\infty} \phi_{r,k}^{(i)}\Delta^i, \quad E_r = \sum_{i=0}^{\infty} E_r^{(i)}\Delta^i \tag{29}$$

Inserting eq.(29) into the eq.(28), we have

$$(z + q^{-s}g)\frac{d}{dz}\phi_{r,s}^{(i)} + (gq^s z - E_r^{(0)})\phi_{r,s}^{(i)} = \sum_{m=0}^{i-1} E_r^{(i-m)}\phi_{r,s}^{(m)} - \sum_{k=1}^{p-1} q^{kr}\phi_{r,s+k}^{(i-1)}$$

where we set $\phi_{r,k+pN}^{(i)} \equiv \phi_{r,k}^{(i)}, k = 0, 1, 2, \cdots, p-1$ for any integer N. If we replace $\phi_{r,s}^{(i)} = e^{-q^s g y_s}\bar{\phi}_{r,s}^{(i)}$ with $y_s = z + q^{-s}g$, we have

$$y_s \partial_{y_s} \bar{\phi}_{r,s}^{(i)} = \sum_{m=0}^{i-1} E_r^{(i-m)}\bar{\phi}_{r,s}^{(m)} - \sum_{k=1}^{p-1} q^{kr} e^{g^2(q^k-1)} e^{g(q^s - q^{k+s})y_s} \bar{\phi}_{r,k+s}^{(i-1)} \tag{30}$$

Solving the eq.(30) for $i = 0$, we have

$$\bar{\phi}_{r,s}^{(0)} = 1 \tag{31}$$

where we set $E_r^{(0)} = -g^2$. Solving the eq.(30) for $i = 1$, we have

$$\bar{\phi}_{r,s}^{(1)} = -\sum_{k=1}^{p-1} q^{kr} e^{g^2(q^k-1)}(q^s - q^{k+s})gy_s\ {}_2F_2(1, 1; 2, 2; (q^s - q^{k+s})gy_s)$$

and

$$E_r^{(1)} = \sum_{k=1}^{p-1} q^{kr} e^{g^2(q^k-1)}$$

For $p = 2M$, we get

$$E_r^{(1)} = (-1)^r e^{-2g^2} + 2\sum_{k=1}^{M-1} e^{g^2(\cos\frac{k}{M}\pi - 1)} \cos\left(\frac{kr}{M}\pi - g^2 \sin\frac{k}{M}\pi\right)$$

For $p = 2M + 1$, we get

$$E_r^{(1)} = 2\sum_{k=1}^{M} e^{g^2\left(\cos\frac{2k}{2M+1}\pi - 1\right)} \cos\left(\frac{2kr}{2M+1}\pi - g^2 \sin\frac{2k}{2M+1}\pi\right)$$

For $i = 2$, we have

$$E_r^{(2)} = -g^2 \sum_{k=1}^{p-1}\sum_{m=1}^{p-1} q^{(k+m)r} e^{g^2(q^k+q^m-2)}(1-q^k)(1-q^m)$$
$$\times {}_2F_2(1,1;2,2;g^2(1-q^k)(1-q^m)) \tag{32}$$

and

$$\bar{\phi}_{r,s}^{(2)} = -E_r^{(1)} \sum_{k=1}^{p-1} q^{kr} e^{g^2(q^k-1)} q^s(1-q^k) g y_s {}_3F_3(1,1,1;2,2,2;-q^s(1-q^k) g y_s)$$
$$+\sum_{k=1}^{p-1}\sum_{m=1}^{p-1} q^{(k+m)r} e^{g^2(q^k+q^m-2)} {}_2F_2(1,1;2,2;g^2(1-q^k)(1-q^m))$$
$$\times q^s(1-q^k) g y_s {}_2F_2(1,1;2,2;q^s(1-q^k) g y_s)$$
$$+\sum_{n=1}^{\infty}\sum_{k=1}^{p-1}\sum_{m=1}^{p-1}\sum_{l=1}^{n} q^{(k+m)r} e^{g^2(q^k+q^m-2)} \binom{n}{l} \frac{1}{lnn!}(q^s(1-q^k)g)^{n-l}$$
$$\times {}_1F_1(l;l+1;g^2(1-q^k)(1-q^m)) y_s^n \tag{33}$$

4. Conclusion

In this paper we discuss the the p-level Rabi model with Z_p-graded symmetry, which is a generalization of standard two-level Rabi model. We obtained the p-level Rabi Hamiltonian by introducing the generalized Pauli matrices. We investigated some properties of the generalized Pauli matrices. We diagonalized the p-level Rabi Hamiltonian with a help of two kinds of unitary transformations. For this Hamiltonian, we eliminated the discrete degree of freedom by introducing Z_p-parity-chain $\mathcal{H}_r, r = 0, 1, 2, \cdots, p-1$ which is isomorphic to the Hilbert space of photon number states $\mathcal{H}$. In each $\mathcal{H}_r$, we rewrote the Hamiltonian in terms of the Bargmann varibles. For this Hamiltonian, we obtained the energy and wave function by using the standard perturbation method. In this process, we obtained the explicit form of the energy and wave function up to a second order in Δ.

Acknowledgement

One of author (W.S Chung) gives many thanks to Professor Algin and his students for kind hospitality during stay in Eskisehir. This work was supported by the National Research Foundation of Korea Grant funded by the Korean Government (NRF-2015R1D1A1A01057792) and by the Gyeongsang National University Fund for Professors on Sabbatical Leave, 2016.

Appendix

When $i_1 = i_2 = \cdots = i_p$ dose not hold, we have

$$\{\sigma_{i_1}, \sigma_{i_2}, \cdots, \sigma_{i_p}\}$$

$$= \sum_{m=1}^{p-1} \sum_{\mathrm{perm}\{i_1,\cdots,i_{p-1}\}} \sigma_{i_1}\sigma_{i_2}\cdots\sigma_{i_{m-1}}\sigma_{p-1}\sigma_{i_m}\cdots\sigma_{i_{p-1}}$$

$$= \sum_{m=1}^{p-1} \sum_{\mathrm{perm}\{i_1,\cdots,i_{p-1}\}} \sum_{l_1=0}^{p-1} q^{i_1 l_1} e_{l_1,l_1+p-1} \cdots \sum_{l_{m-1}=0}^{p-1} q^{i_{m-1}l_{m-1}} e_{l_{m-1},l_{m-1}+p-1} \sum_{k=0}^{p-1} q^k e_{k,k}$$

$$d \times \sum_{l_{m+1}=0}^{p-1} q^{i_{m+1}l_{m+1}} e_{l_{m+1},l_{m+1}+p-1} \cdots \sum_{l_{p-1}=0}^{p-1} q^{i_{p-1}l_{p-1}} e_{l_{p-1},l_{p-1}+p-1}$$

$$= \sum_{m=1}^{p-1} \sum_{\mathrm{perm}\{i_1,\cdots,i_{p-1}\}} \sum_{l_1=0}^{p-1} q^{\sum_{n=0}^{p-1}(i_n(l_1+(p-1)(n-1))+l_1+(m-1)(p-1)} e_{l_1+(p-1)(m-1),l_1+1}$$

$$= 0$$

because

$$\sum_{m=1}^{p-1} q^{(m-1)(p-1)} = 0$$

References

[1] I. Rabi, Phys. Rev. 49, 324 (1936); Phys. Rev. 51, 652 (1937).
[2] C. Cohen-Tannoudji, B. Diu, F. Laloe, Quantum Mechanics, (Wiley-Interscience 2006).
[3] E. Sassaroli, Am. J. Phys 67, 869 (1999).
[4] W. Haxton and B. Holstein, Am. J. Phys 68, 15 (2000).
[5] E. Jaynes and F. Cummings, Proc. IEEE 51, 89 (1963).
[6] B. Shore and P. Knight J. Mod. Opt. 40, 1195 (1993).
[7] G. Khitrova, H. M. Gibbs, M. Kira, S. W. Koch, and A. Scherer, Nat. Phys. 2, 81 (2006).
[8] J. Bourassa, J. M. Gambetta, A. A. Abdumalikov, Jr., O. Astafiev, Y. Nakamura, and A. Blais, Phys. Rev. A 80, 032109 (2009).
[9] P. Forn-Daz, J. Lisenfeld, D. Marcos, J. J. Garca-Ripoll, E. Solano, C. J. P. M. Harmans, and J. E. Mooij, Phys. Rev. Lett. 105, 237001 (2010).
[10] T. Niemczyk et al, Nature Phys. 6, 772 (2010).
[11] C. Ciuti and I. Carusotto, Phys. Rev. A 74, 033811 (2006).
[12] D. Braak, Phys. Rev. Lett 107, 100401 (2011).
[13] E. Ince, Ordinary Differential Equations, Dover, New York (1956)

Active Free Surface Density Maps

S Çelen

Ege University, Faculty of Engineering, Mechanical Engineering Department, Bornova, Izmir, 35100-TURKEY

E-mail: seracelen@hotmail.com, serap.celen@ege.edu.tr

Abstract. Percolation problems were occupied to many physical problems after their establishment in 1957 by Broadbent and Hammersley. They can be used to solve complex systems such as bone remodeling. Volume fraction method was adopted to set some algorithms in the literature. However, different rate of osteoporosis could be observed for different micro-structures which have the same mass density, mechanical stimuli, hormonal stimuli and nutrition. Thus it was emphasized that the bone might have identical porosity with different specific surfaces. Active free surface density of bone refers the used total area for its effective free surface. The purpose of this manuscript is to consolidate a mathematical approach which can be called as "active free surface density maps" for different surface patterns and derive their formulations. Active free surface density ratios were calculated for different Archimedean lattice models according to Helmholtz free energy and they were compared with their site and bond percolation thresholds from the background studies to derive their potential probability for bone remodeling.

1. From Historical Bone Investigations to Novel Formulation

In 1957, Broadbent and Hammersley established the percolation theory to explain the random properties of the flow through a medium. After then the theory was used to solve many physical problems [1]. It may be a beneficial tool to investigate complex systems such as bone remodeling. In the literature, some equations were proposed which were based on volume fraction method [2, 3]. However, different osteoporosis rates were reported for different micro-structures which have the same environmental conditions. Thus it was emphasized that the bone had identical porosity although it had different specific surfaces [4-7]. "Mass and the mechanostat" concept was suggested to describe osteoblast and osteoclast tasks for the bone by Frost [8]. In fact, Wolff's law was caused an awareness for the changeability of bone architecture according to the applied external stress in 1892 [9]. However, Frost, Cowin, Huiskes et al. reported that the strain value was important [8,10,11]. Huiskes et al. investigated the strain energy density (SED) to conduct mechanical signals for remodeling of the bone structure [11]. Skerry et al. proposed that the osteocyte should be a mechanosensor of bone [12]. It was reported that it delivered mechanical signals to activate osteoblasts and osteoclasts [12, 13]. Thus small-strain approach was developed [2-14]. The free surface density of bone was explained as the total surface area which was in contact with soft tissue (Figure 1) [4-7]. Martin reported the importance of specific surfaces of bone on osteoblast and osteoclast activity. His reports were culminated in a conclusion that the higher specific surface area, the greater bone remodeling potential. Thus the potential of bone remodeling should be related with morphology and dimensions of the specific surface [4-7].

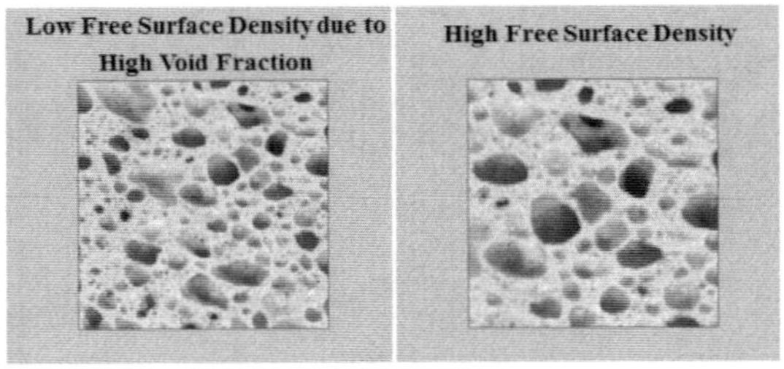

Figure 1. Representation of two bone samples which have different free surface density.

Martin proposed an empirical formula for effective free surface density [4-6]. After then Rouhi et al. redeveloped it for elliptical and different hypothetical micro-structural geometries [7]. Their arguments need to be updated in several respects. There is a research gap for a novel theory which includes random networks between Archimedean lattice models for bone remodeling. The most important development should be "the description of a dispersion error factor". In the present paper, I report my theoretical study as "Active Free Surface Density (AFSD) Maps" to derive a novel equation for bone remodeling. AFSD ratios were calculated for different Archimedean lattice models according to Helmholtz free energy and they were compared with their site and bond percolation thresholds from the background studies to derive their potential probability for bone remodeling.

2. Material and Method

2.1. Field Equations of Background Studies

Helmholtz free energy was given according to the free surface density of bone (Equation 1-3) [4-7,10, 20, 21, 26]. Related nomenclature was given in Table 1.

$$-\gamma v \dot{\psi} - \gamma v \eta \dot{\theta} + T_{ij} L_{ij} - \frac{1}{\theta} q_i \theta_i + h \geq 0 \tag{1}$$

$$\psi = \psi(\theta, \theta_i, S_\lambda, E_{ij}) = 0 \tag{2}$$

$$\gamma v \psi - \gamma v \frac{\partial \psi}{\partial S_\lambda} \dot{S}_\lambda - \frac{1}{\theta} q_i \theta_i + h \geq 0 \tag{3}$$

Rouhi et.al and others assumed that zero strain reference state at constant temperature situation [4-7,10, 20, 21, 26].

$$T_{ij}(\theta_0, 0, S_\lambda, \delta_{ij}) = 0 \tag{4}$$

In the reports it was assumed that strains were small and Equation 5-10 were reported [3-7,10, 20, 21, 26, 27].

$$e = \xi - \xi_0 \tag{5}$$

$$E_{ij} = \frac{1}{2}(U_{i,j} + U_{j,i}) \tag{6}$$

$$C = C(S_\lambda, E) \tag{7}$$

$$\psi = \psi(S_\lambda, E) \tag{8}$$

$$\frac{\mathrm{de}}{\mathrm{dt}} = \phi(S_\lambda, E) \tag{9}$$

$$\phi(S_\lambda, E) = \frac{C(S_\lambda, E)\,(\det(1+2E))^{1/2}}{\gamma} \tag{10}$$

In the investigations $\phi(S_\lambda, E)$ and $\psi(S_\lambda, E)$ were approximated by neglecting higher order terms in Taylor Series expansion and they were reported following equations (Equation 11-14) [7].

$$\phi(S_\lambda, E) \cong a(S_\lambda) + A^*_{ij}(S_\lambda) E_{ij} + \frac{1}{2} B_{ijkm}(S_\lambda) E_{ij} E_{km} \tag{11}$$

$$a(S_\lambda) = \phi(S_\lambda, 0) \tag{12}$$

$$A^*_{ij}(S_\lambda) = \left[\frac{\partial \phi(S_\lambda, E_{ij})}{\partial E_{ij}}\right]_{Eij=0} \tag{13}$$

$$B_{ijkm}(S_\lambda) = \left[\frac{\partial^2 \phi(S_\lambda, E_{ij})}{\partial E_{ij} \partial E_{km}}\right]_{Eij=0} \tag{14}$$

They developed the derivative form of the difference of unstrained volume fraction before remodelling (Equation 15-19) [7].

$$\frac{de}{dt} = \{a(S_\lambda) + A_{ij}^*(S_\lambda)E_{ij}\} \tag{15}$$

$$\frac{de}{dt} = \{a_0 + a_1 S_\lambda + a_2 {S_\lambda}^2 + (A_{ij}^0 + A_{ij}S_\lambda)E_{ij}\} \tag{16}$$

$$A_{ij}^0 = A_{ij}^*(0, E_{ij}) \tag{17}$$

$$A_{ij} = \left[\frac{\partial A_{ij}^*(S_\lambda, E_{ij})}{\partial S_\lambda}\right]_{S_\lambda=0} \tag{18}$$

$$A_{ij} = \left[\frac{\partial^2 \phi(S_\lambda, E_{ij})}{\partial S_\lambda \partial E_{ij}}\right]_{E_{ij}=0}^{S_\lambda=0} \tag{19}$$

Equation 20 referred to the bone remodeling process [4-7,10, 20, 21, 26, 27].

$$\frac{de}{dt} = -\frac{dP_v}{dt} = A_{ij}S_\lambda(E_{ij} - {E_{ij}}^*) \tag{20}$$

Active free surface density (S_λ) of bone was reported in Equation 21 [7]. If the whole surface is effective for remodeling, thus λ equals to 1, and S_λ equals to S_B/V_t or S_v [4-7].

$$S_\lambda = \lambda\frac{S_B}{V_t} \tag{21}$$

Martin et al. reported that the specific surface increased with the rate of remodeling of bone. They gave Equation 22 and 23 to estimate the free surface density [4-7].

$$S_\lambda = 28.8P_v^5 - 101P_v^4 + 134P_v^3 - 93.9P_v^2 + 32.2P_v \tag{22}$$

$$\frac{de}{dt} = -\frac{dP_v}{dt} = A_{ij}(28.8P_v^5 - 101P_v^4 + 134P_v^3 - 93.9P_v^2 + 32.2P_v)(E_{ij} - {E_{ij}}^*) \tag{23}$$

Table 1. Nomenclature.

Physical Properties	Symbol
Helmholtz free energy	ψ
Bulk density	γ
Volume fraction	ν
Specific entropy	η
Absolute temperature, temperature	θ, θ_i
Stress tensor	T_{ij}
Kronoecker Delta	δ_{ij}
Velocity tensor (Gradient)	L_{ij}
Strain Tensor, Strain (Equilibrium)	E_{ij}, E_{ij}^*
Displacement vector	U
Displacement vector (Gradient)	U_{ij}
Surface Modeling Coefficient	C
Function of active free surface density variation with strain	ϕ
Heat flux vector	q_i
Entropy	h
Active free surface density	S_λ
Free surface density	S_B
Difference between the initial and strained volume fractions	e
Initial volume fraction (Before the strain)	ξ
Material constants of bone	$a, a_0, a_1, a_2, A_{ij}^* A_{ij}^0$ B_{ijkm}
Void Fraction (Porosity)	P_v
Scattering Factor	S_F

2.2. Current Model and Results

In the literature, the importance of micro-structural properties of bone was emphasized [2-32]. Some assumptions were made before the theoretical investigation. Wolff's law was valid thus the bone matrix's condition was depend on strain tensor history [9]. There was not any bone disease so blood plasma supplied the materials for the synthesis of bone matrix. Bone matrix established a suitable environment for the required mass transfer to the osteoblasts and osteoclasts. Proposal of Beaupre et al. was valid in the equilibrium condition, if $E_{ij}=E_{ij}*$ bone remodeling was interrupted [21]. Also we assumed that bone remodeling could only occur in the presence of an active surface area [4-7]. Morphological and functional characteristics of the bone surface allow us to evaluate its free energy. Five different Archimedean lattice models **(3^6, 4^4, 6^3, $4{,}8^2$, $3{,}12^2$)** which bond and site percolation thresholds were known from the literature were used [33, 38]. Because the bond percolation threshold allowed us to determine the potential lattice which might be promising for bone remodeling. Circumferential lengths were adjusted to the same size, thus all polygons had 404 μm equal perimeter. (Figure 2) Distances of the channels were fixed to 10 μm for whole polygons. To derive our novel equation for different surface patterns (on 1x1 mm^2 total planar area), it was assumed that polygons had been filled with bone, blood plasma was supposed to flow through micro-channels and they formed a porosity.

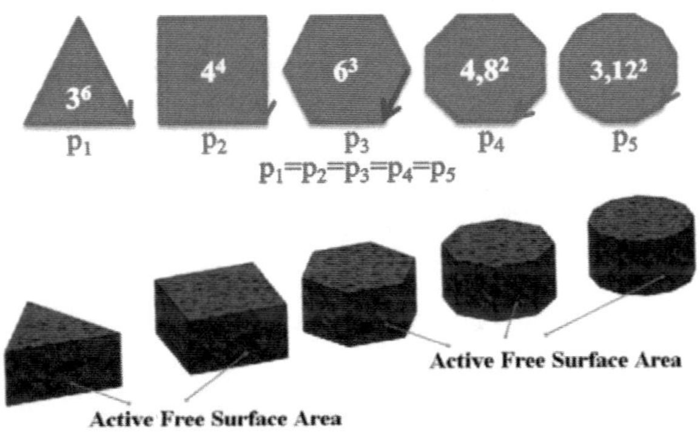

Figure 2. Schematical representation of perimeters for Archimedean lattice models.

AFSD ratios were estimated in the range of 0.71- 0.8498 for different Archimedean lattice models (Figure 3). The minimum void fraction was calculated about 0.18. Our AFSD results were found consistent with Lerebours et al.'s investigation [39], also our theory explained their scattered results. 3^6 lattice has a greater potential because of its high specific surface and minimum bond percolation threshold. Different AFSD ratios can be obtained with lattices which have the same void fraction such as 3^6 and 4^4 (Figure 3 right). AFDS ratio of 4^4 was calculated lower than 3^6. This situation reduces its potential for bone remodeling. This result can be attributed to the reduced micro-flow or signal transmission due to vertices of the porosity. A scattering (dispersion error) factor (S_F) should be added to calculate more accurate values according to the coordination number of the lattices (Figure 3 left). If coordination number equals to 6, Equation 24 should be used. If it is below from 6, Equation 25 should be used.

$$\frac{de}{dt} = -\frac{dP_v}{dt} = A_{ij}(18.355P_v^5 - 46.79P_v^4 + 33.85P_v^3 - 8.816P_v^2 - 1.5P_v)(\mathbf{1/S_F})(E_{ij} - E_{ij}^*) \quad (24)$$

$$\frac{de}{dt} = -\frac{dP_v}{dt} = A_{ij}\{(18.355P_v^5 - 46.79P_v^4 + 33.85P_v^3 - 8.816P_v^2 - 1.5P_v)(\mathbf{S_F})\}(E_{ij} - E_{ij}^*) \quad (25)$$

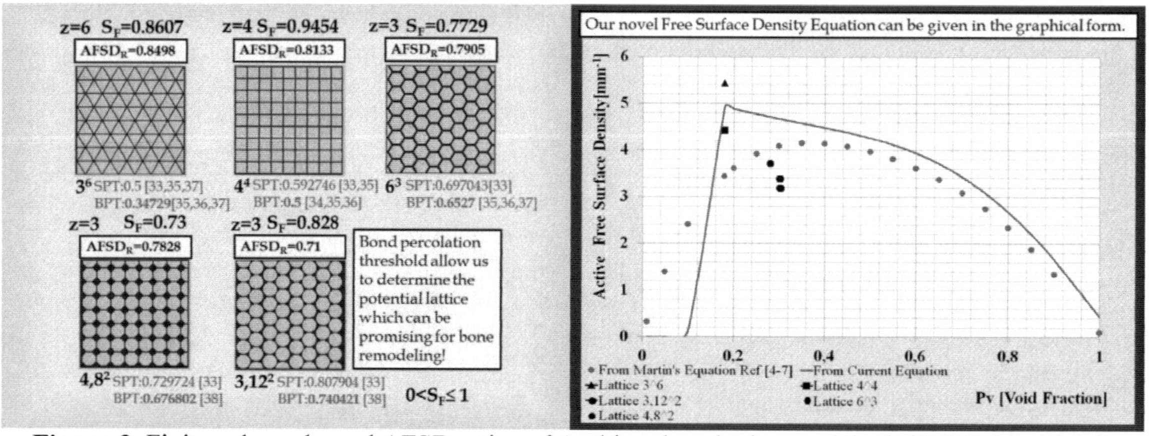

Figure 3. Finite subgraphs and AFSD ratios of Archimedean lattice models (left), Models (right).

3. Conclusions

In the scope of this paper, AFSD Maps were presented and a novel equation was proposed. $\mathbf{3^6}$ and $\mathbf{4^4}$ lattices were suitable for bone remodeling. Scattering factor will be beneficial for precise computation of AFSD values which have irregular dispersion. This study is a preliminary investigation and further studies are necessary to evaluate the maximum AFSD ratio of other three dimensional lattices.

Acknowledgements
The author expresses her sincere thanks to ICON Machinery.

References

[1] Broadbent S R and Hammersley J M 1957 Percolation processes *Math. Proc. Cambridge* **53** 03 629-41
[2] Cowin S C and Hegedus D H 1976 Bone remodeling I: theory of adaptive elasticity *J. Elasticity* **6** 3 313–26
[3] Hegedus D M and Cowin SC 1976 Bone Remodeling II: small strain adaptive elasticity *J. Elasticity* **6** 4 337-52
[4] Martin R B 1972 The effects of geometric feedback in the development of osteoporosis *J. Biomech.* **5** 5 447-55
[5] Martin R B 1974 The specific surface-porosity relationship in bone:refinement of the concept of geometric feedback Orthopaedic Research Society Meeting (Dallas)
[6] Martin R B 1984 Porosity and specific surface of bone *Crit. Rev. Biomed. Eng.* **10** 3 179-222
[7] Rouhi G, Herzog W, Sudak L, Firoozbakhsh K and Epstein M 2004 Free surface density instead of volume fraction in the bone remodeling equation: theoretical considerations *Forma* **19** 165-82
[8] Frost H M 1987 Bone "mass" and the "mechanostat":a proposal *The Anat. Rec.* **219** 1-9
[9] Wolff J 1892 *Das Gesetz der Transformation der Knochen* (Berlin: Verlag von August Hirschwald)
[10] Cowin S C 1987 Bone remodeling of diaphyseal surfaces by torsional loads:theoretical predictions *J. Biomech.* **20** 11-12 1111-20
[11] Huiskes R, Ruimerman R, van Lenthe G H and Janssen J D 2000 Effects of mechanical forces on maintenance and adaptation of form in trabecular bone *Nature* **405** 6787 704-6
[12] Skerry T M, Bitensky L, Chayen J and Lanyon L E 1989 Early strain-related changes in enzyme activity in osteocytes following bone loading in vivo *J. Bone Miner. Res.* **4** 5 783-8
[13] Burger E H, Klein-Nulend J and Smit T H 2003 Strain-derived canalicular fluid flow regulates osteoclast activity in a remodelling osteon-a proposal *J. Biomech.* **36** 1453-9
[14] Cowin S C 1993 Bone stress adaptation models *J. Biomech. Eng.* **115** 4B 528-33

[15] Bobyn J D, Pilliar R M, Cameron H U and Weatherly G C 1980 The optimum pore size for the fixation of porous-surfaced metal implants by the ingrowth of bone *Clin. Orthop. Relat. Res.* **150** 263-70

[16] Pilliar R M 1987 Porous-surfaced metallic implants for orthopedic applications *J. Biomed. Mater. Res.* **21** A1 1-33

[17] Cameron H U, Pilliar R M and Macnab I 1976 The rate of bone ingrowth into porous metal *J. Biomed. Mater. Res.* **10** 2 295-302

[18] Hulbert S F, Young F A, Mathews R S, Klawitter J J, Talbert C D and Stelling F H 1970 Potential of ceramic materials as permanently implantable skeletal prostheses *J. Biomed. Mater. Res.* **4** 433-56

[19] Itälä A I, Ylanen H O, Ekholm C, Karlsson K H and Aro H T 2001 Pore diameter of more than 100 microm is not requisite for bone ingrowth in rabbits *J. Biomed. Mater. Res.* **58** 6 679-83

[20] Bronner F and Worrell R V 1999 *Orthopaedics, principles of basic and clinical science* (Boca Raton, Florida: CRC Press)

[21] Beaupré G S, Orr T E and Carter D R 1990 An approach for time-dependent bone modeling and remodeling-application:a preliminary remodeling simulation *J. Orthop. Res.* **8** 5 662-70

[22] Carter D R, Wong M and Orr T E 1991 Musculoskelethal ontogeny, phylogeny and functional adaptation *J. Biomech.* **24** 1 3-16

[23] Mullender M G and Huiskes R 1995 Proposal for the regulatory mechanisms of Wolff's law *J. Orthop. Res.* **13** 4 503-12

[24] Prendergast P J and Huiskes R 1995 Mathematical modelling of microdamage in bone remodeling and adaptation In: *Bone structure and remodelling* ed A Odgaard and H Weinans (Singapore: World Scientific Publishers) pp 213-23

[25] McNamara L M and Prendergast P J 2007 Bone remodelling algorithms incorporating both strain and microdamage stimuli *J. Biomech.* **40** 1381-91

[26] Recker R R 1983 *Bone Histomorphometry: Techniques and Interpretation* (Boca Raton, Florida: CRC Press)

[27] Fritton S P, McLeod K J and Rubin C T 2000 Quantifying the strain history of bone: spatial uniformity and self similarity of low magnitude strains *J. Biomech.* **33** 317-25

[28] Çelen S, Efeoğlu C and Özden H 2011 *Phys. Procedia* **12** B 245–51

[29] Çelen S and Özden H 2012 *Appl. Surf. Sci.* **263** 579-85

[30] Çelen S 2015 *Opt. Laser Eng.* **70** 45-50

[31] Çelen S 2016 Mathematical principle of cellular accommodation:a schmitt trigger approach *Mat.-wiss. u. Werkstofftech.* **47** 4 301-6

[32] Weinans H, Huiskes R and Grootenboer H J 1992 The behavior of adaptive bone-remodeling simulation models *J. Biomech.* **25** 12 1425-41

[33] Suding P N and Ziff R M 1999 Site percolation thresholds for Archimedean lattices *Phys. Rev.* E **60** 1 275-83

[34] Jacobsen J L 2014 High-precision percolation thresholds and potts-model critical manifolds from graph polynomials *J. Phys.* A*: Math. Theor.* **47** 135001 1-78

[35] Christensen K 2002 *Percolation Theory* (London:Blackett Laboratory, Imperial College London) pp 1-39

[36] Sykes M F and Essam J W 1964 Exact critical percolation probabilities for site and bond problems in two dimensions *J. Math. Phys.* **5** 8 1117–27

[37] Wierman J C 2003 Pairs of graphs with site and bond percolation critical probabilities in opposite orders *Discrete Appl. Math.* **129** 545 – 48

[38] Parvianinen R 2007 Estimation of bond percolation thresholds on the Archimedean lattices *J.Phys.*A:*Math. Theor.* **40** 9253-58

[39] Lerebours C, Thomas C D L, Clement J G, Buenzli P R and Pivonka P 2015 The relationship between the porosity and specific surface in human cortical bone is subject specific *Bone* **72** 109-17

Differences observed in the surface morphology and microstructure of Ni-Fe-Cu ternary thin films electrochemically deposited at low and high applied current densities

U Sarac[1*], M Kaya[2] and M C Baykul[2]

[1]Department of Science Education, Bartın University, Bartın 74100, Turkey
[2]Department of Physics, Eskisehir Osmangazi University, Eskisehir 26480, Turkey

*Corresponding Author: usarac428@hotmail.com

Abstract. In this research, nanocrystalline Ni-Fe-Cu ternary thin films using electrochemical deposition technique were produced at low and high applied current densities onto Indium Tin Oxide (ITO) coated conducting glass substrates. Change of surface morphology and microstructural properties of the films were investigated. Energy dispersive X-ray spectroscopy (EDX) measurements showed that the Ni-Fe-Cu ternary thin films exhibit anomalous codeposition behaviour during the electrochemical deposition process. From the X-ray diffraction (XRD) analyses, it was revealed that there are two segregated phases such as Cu-rich and Ni-rich within the films. The crystallographic structure of the films was face-centered cubic (FCC). It was also observed that the film has lower lattice micro-strain and higher texture degree at high applied current density. Scanning electron microscopy (SEM) studies revealed that the films have rounded shape particles on the base part and cauliflower-like structures on the upper part. The film electrodeposited at high current density had considerably smaller rounded shape particles and cauliflower-like structures. From the atomic force microscopy (AFM) analyses, it was shown that the film deposited at high current density has smaller particle size and surface roughness than the film grown at low current density.

1. Introduction

Ni-Fe-Cu system has attractive properties such as low coercivity and high conductivity, therefore much attention has been paid both technologically and scientifically for the investigation of their properties [1–3]. It has been reported that the optimal compositions of the Ni-Fe-Cu thin films used for production of magnetic sensors include Ni between 45% - 80%, Fe between 5% - 20%, and Cu between 10% - 50% [4, 5]. There are different production methods of preparing such magnetic thin films. Among them, electrodeposition method has attracted significant attention for a long time in fabricating magnetic thin films owing to their unique advantages. However, it is well known that many electroplating conditions strongly affect the surface morphology and structural properties as well as magnetic properties of the electrochemically deposited magnetic thin films.

In this study, the differences observed in the microstructural and morphological properties of the electrochemically deposited Ni-Fe-Cu ternary thin films with respect to applied current density were presented and discussed. From the analyses, especially, it was revealed that a change in the current density applied during the electroplating process significantly affects the surface morphology and the texture degree of Ni-Fe-Cu ternary thin films grown onto ITO coated glass substrates by electrochemical deposition technique.

2. Experimental

Ni-Fe-Cu ternary thin films were electrochemically deposited from an electrolyte solution containing 0.07 M Ni sulfate, 0.004 M Cu sulfate, 0.05 M boric acid, and 0.008 M Fe sulfate at different applied current densities. Deionized water was used to prepare all of the electrolyte solutions. The electrolyte solutions used for the electrochemically deposition of Ni-Fe-Cu magnetic ternary thin films were freshly prepared before each deposition process. It was determined that the pH value and the temperature of freshly prepared electrolyte solutions were 4.2 ± 0.1 and 25 ± 1 °C, respectively. All electrochemical deposition experiments were carried out in a conventional three-electrode cell consisting of counter, working, and reference electrodes using a potentiostat/galvanostat (VersaSTAT 3). A platinum wire, ITO coated conducting glass substrates with a working area of about 1 cm^2 and a saturated calomel electrode (SCE) were served as the counter electrode, working electrode, and reference electrode, respectively. The films were prepared at different applied current densities such as low (-12 mA/cm^2) and high (-20 mA/cm^2) at ambient temperature and pressure without stirring. Before electrochemically deposition process, the surface of the substrates was cleaned in the medium for 5 min in acetone followed by 5 min in ethanol and finally rinsed ultrasonically in deionized water for 10 min. The film thickness was controlled by charge and it was kept constant at 500 nm by adjusting the electroplating time based on the Faraday's law for all applied current densities.

In order to obtain the crystal structure of the films, XRD measurements were performed with a Panalytical Empyrean XRD diffractometer (Cu Kα radiation, λ=0,154059 nm) by scanning in the 2θ=42°-54° range with 0.02 steps. The compositional analysis of the films was done by EDX consisted in the SEM. In order to investigate the effect of applied current density on the surface morphological structure of the films a SEM (Zeiss Supra 50Vp) and an AFM (Veeco, Multimode 5) were used.

3. Results and Discussion

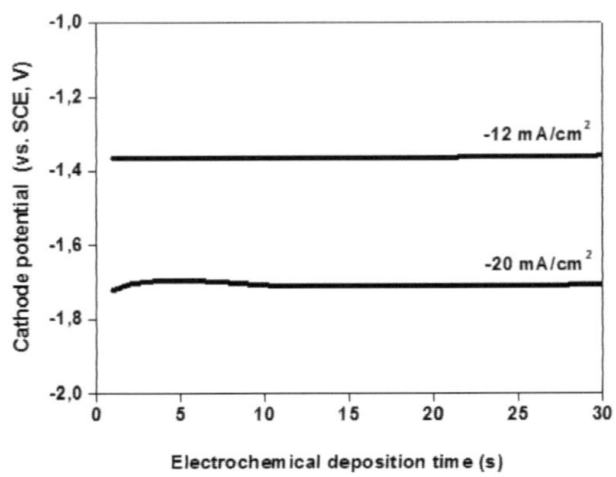

Figure 1. Galvanostatic potential–time transient curves of electrochemically deposited Ni-Fe-Cu ternary thin films with respect to applied current density.

In the present study, the morphological and microstructural properties of Ni-Fe-Cu ternary thin films electrochemically deposited at different applied current densities have been investigated. As a first step of this investigation, the galvanostatic potential–time transient curves are recorded with respect to electrochemical deposition time during the experiments. The galvanostatic potential-time transient curves recorded for the first 30 s of electrochemical deposition process as a function of applied current density are shown in Figure 1. It is observed that the cathode potential remains almost stable at each applied current density, reflecting that the Ni-Fe-Cu ternary thin films investigated in the present study can be produced properly onto ITO coated conducting glass substrates during the electrochemical deposition process. Furthermore, the cathode potential increases when the applied current density increases as seen from Figure 1.

From the compositional analyses carried out for the Ni-Fe-Cu ternary thin films grown at low (-12 mA/cm^2) and high (-20 mA/cm^2) current densities, the elements other than Ni, Fe, and Cu are also detected from the energy dispersion X-ray spectrometer (EDS) spectra shown in Figure 2. It is concluded that these detected elements (O, Si, In, Ca, and Sn) arise from the well-known ITO coated conducting glass substrates. However, we have only obtained the atomic percentages of Ni, Fe, and Cu within the films. In the analyses of the EDS spectra, same method was also carried out in our previous research [6] and the other earlier studies [7, 8]. The results obtained from the EDX analyses are showed that the film has 77 at.% Ni, 12.2 at.% Fe, and 10.8 at.% Cu at low applied current density, whereas the chemical composition of the film electrochemically deposited at high applied current density is 80 at.% Ni, 9.6 at.% Fe, and 10.4 at.% Cu, indicating that the preferential deposition of Fe is more pronounced at low current density. Therefore, by changing of applied current density, it is possible to prepare the Ni-Fe-Cu ternary thin films with different chemical compositions.

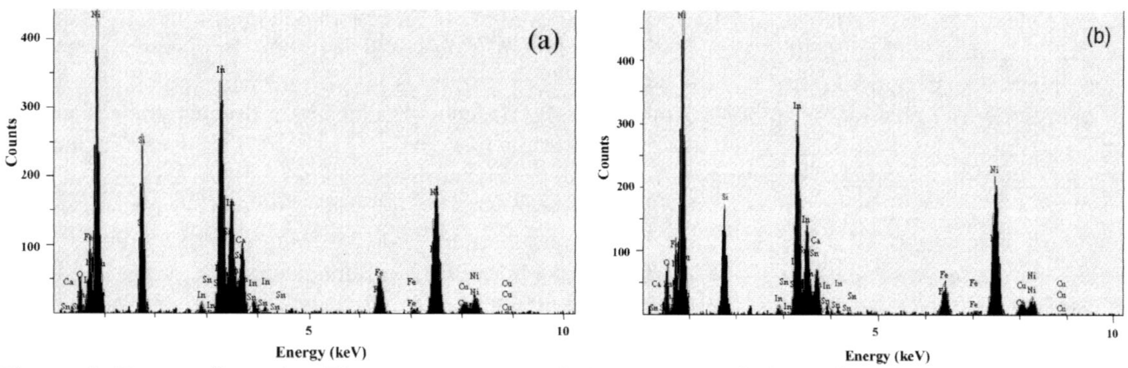

Figure 2. Energy dispersion X–ray spectrometer (EDS) spectra of electrodeposited Ni-Fe-Cu ternary thin films prepared at low and high current densities a) -12 mA/cm^2 and b) -20 mA/cm^2, respectively.

From the results of EDX measurements, there is clear evidence for the existence of anomalous codeposition behavior in Ni-Fe-Cu ternary thin films electrochemically deposited at low and high current densities, since the relative Fe concentration in the films is higher than the relative ion concentration of Fe^{2+} in the electrolyte solution. It was shown that the electrochemically deposited NiFe binary magnetic films prepared onto ITO coated glass substrates exhibit the anomalous codeposition behavior of two iron group elements of Fe and Ni [9]. Furthermore, the anomalous codeposition behavior was also observed in Ni-Cu-Fe ternary thin films electrochemically deposited onto ITO coated glass substrates in our previous study [10].

The XRD patterns revealed the crystallographic structure of the electrochemically deposited Ni-Fe-Cu ternary thin films grown at low and high applied current densities investigated in the present work and ITO coated glass substrate are shown in Figure 3. From the XRD patterns, the (111) and (200) diffraction peaks appeared at the angular positions of around 44° and 51°, respectively, which correspond to FCC structure, are clearly seen. However, the films exhibit two segregated peaks at the angular positions of around 44° and also 51°. According to the results obtained from the XRD analysis, the films are not single-phase. In agreement with previous reports [6, 10-16], we have been also find a phase separation consisting of Cu-rich and Ni-rich phases in electrodeposited Ni-Fe-Cu ternary thin films investigated in the present study. Besides, from the XRD patterns, a shift in the Ni-rich (111) and (200) peak positions towards higher 2θ angles is observed with increasing applied current density, implying that the lattice parameters of the Ni-rich (111) and (200) phases decrease. The lattice parameters of the Ni-rich (111) phase are found to be 0.3532 nm and 0.3526 nm for low and high applied current densities, respectively. These results are also in good agreement with those reported for the Ni-Fe films [17, 18] and Ni-Cu-Fe films [10] which have FCC crystallographic structure.

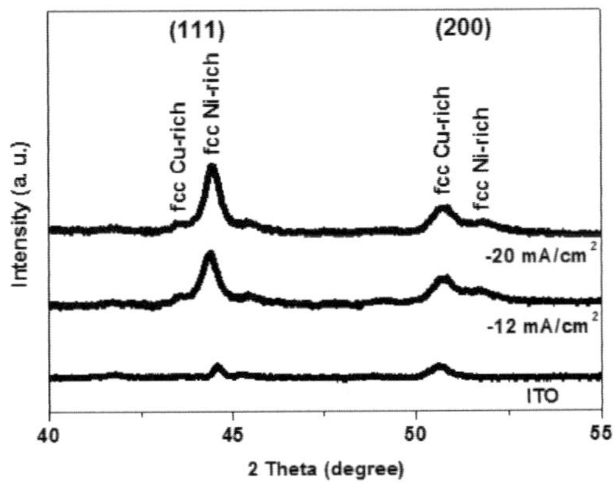

Figure 3. XRD patterns of electrochemically deposited Ni-Fe-Cu ternary thin films prepared at low and high current densities and ITO coated glass substrate.

The average crystallite sizes of the films are determined using a well-known Scherrer formula [19]. All of the films are found to have nanocrystalline structure and the average crystallite size is determined as 14.6 nm at low current density (-12 mA/cm^2) and it is 16.4 nm at high current density (-20 mA/cm^2), indicating that the crystallite size of the films increases with increasing applied current density. As can be seen from Figure 1, since the cathode potential increases as the applied current density increases, it is expected that the average crystallite sizes of the films decrease by increasing the applied current density. However, the variation of the average crystallite size with the applied current density shows an opposite behavior. It has been reported that the average crystallite size increases with increasing applied current density, which has been attributed to the co–deposition of hydrogen at the cathode electrolyte interface [20]. In the existence of hydrogen, the changes in the surface energy and growth mechanisms lead to an increase in the average crystallite size of films by increasing current density [20]. It was also reported in previous studies that an enhancement in the applied current density leads to an increment in the crystallite size, since it reduces the charge transfer resistance and the adsorption intensity of electrochemical active species [21-23].

The lattice micro-strain percentages of the films, ε, were also estimated from the XRD patterns using the following equation [12, 24].

$$\varepsilon = \left(\frac{\beta}{4 \times \mathrm{tg}\theta}\right) \times 100$$

where β represents the full width at half maximum (FWHM) of fcc Ni-rich (111) peak of the films and θ is the Bragg's angle. The lattice micro-strain percentages are found to be 0.62 % and 0.55 % for the films grown at low and high applied current densities, respectively. Thus, it is concluded that an increase in the applied current density induces a lower lattice micro-strain and larger crystallite size.

In order to obtain additional information on the microstructure, the evolution of integral intensities of the Ni-rich (111) and (200) peaks of the films as a function of applied current density are also analyzed by the peak profile fitting of XRD patterns. The integral intensity of the Ni-rich (111) peak increases and the integral intensity of the Ni-rich (200) peak decreases with increasing applied current density, indicating that the relative peak intensity of the Ni-rich (111) phase increases. Thus, we conclude that an increase in the applied current density gives rise to an increment in the texture degree of the Ni-rich (111) phase within the Ni-Fe-Cu ternary thin films.

The morphological structure of the films was investigated using a SEM. Figure 4(a) and (b) show the SEM images of the films electrodeposited at low and high applied current densities, respectively. SEM images revealed that the morphological structure is significantly influenced by the applied current density. As the film is prepared at low applied current density of -12 mA/cm^2, the film surface exhibits a morphology consisting of smaller rounded particles on the base part and larger cauliflower–like structures formed owing to the agglomeration of the smaller rounded particles on the upper part (Figure 4a). The same kind of surface morphology is observed as the film is electrodeposited at high applied current density of -20 mA/cm^2 (Figure 4b). However, the sizes of smaller rounded particles and larger cauliflower–like structures significantly decrease, while the number of them increases with increasing applied current density.

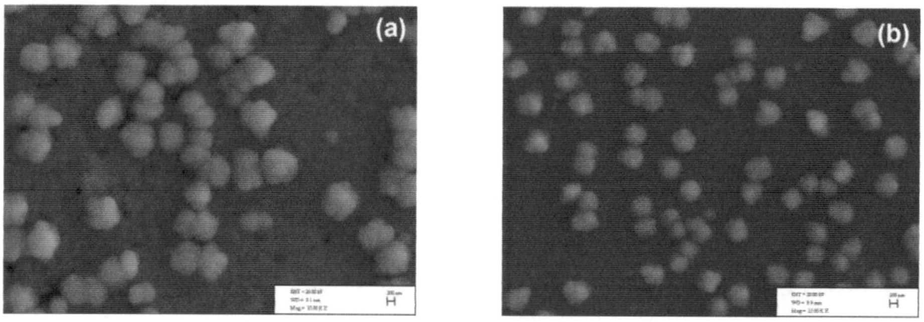

Figure 4. SEM images of the electrochemically deposited Ni-Fe-Cu ternary films grown at low and high current densities a) -12 mA/cm^2 and b) -20 mA/cm^2, respectively.

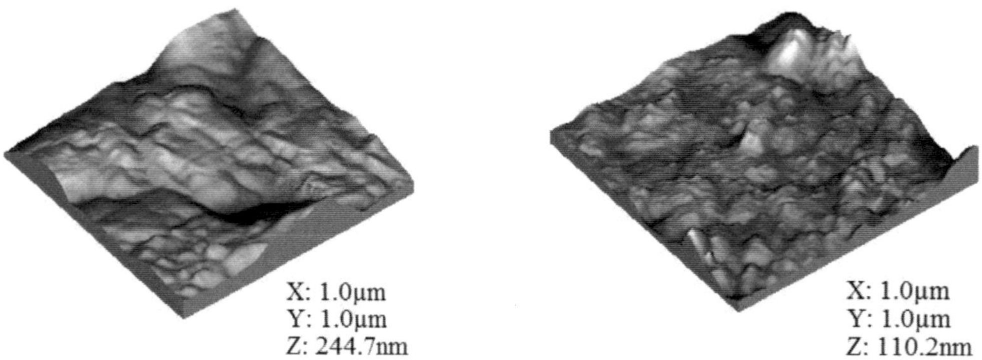

Figure 5. Three dimensional AFM images of the electrochemically deposited Ni-Fe-Cu ternary films produced at low and high current densities a) -12 mA/cm^2 and b) -20 mA/cm^2, respectively.

AFM analyses are also used to investigate of the surface morphology of Ni-Fe-Cu ternary thin films. From the three-dimensional AFM images shown in Figure 5, it is concluded that the surface of the films exhibits granular growth morphology irrespective of applied current density. A decrement in the surface roughness of the films is observed with increasing applied current density. The root mean square (RMS) surface roughness values are found to be 28.2 nm and 12.3 nm for the films grown at low and high applied current densities, respectively. An increase in the nucleation density leads to the formation of smaller particles on the ITO coated glass substrates at high applied current density during the electrochemical deposition process of the films, resulting more compact and smooth film. The differences observed in the particle size and/or surface morphology with the applied current density may be ascribed to the change of chemical composition of the films. It was also shown that the applied current density affects the particle size and/or surface morphology of the electrodeposited metallic thin films in previous studies [25-27].

4. Conclusions

We have been investigated the effect of applied current density on the crystallographic structure, crystallite size, texture degree, lattice strain, surface roughness, and grain size of electrochemically deposited Ni-Fe-Cu ternary thin films prepared onto ITO coated conducting glass substrates. Compositional analyses indicated that the Ni-Fe-Cu ternary thin films have the composition of 77-80 at.% Ni, 9.6-12.2 at.% Fe, and 10.4-10.8 at.% Cu. Electrochemically deposited Ni-Fe-Cu ternary thin films exhibit anomalous codeposition behavior irrespective of applied current density. The XRD analyses show that the Ni-Fe-Cu ternary thin films have two segregated Ni-rich and Cu-rich phases. Ni-Fe-Cu ternary thin films investigated in the present study are found to have FCC crystal structure. The film has larger crystallite size and lower lattice micro-strain at high applied current density. An increment in the applied current density induces an enhancement in the texture degree of Ni-rich (111) phase within the Ni-Fe-Cu ternary thin films. The films exhibit a morphology having of smaller rounded particles on the base part and larger cauliflower–like structures on the upper part regardless of applied current density. The film electrochemically deposited at high current density has smaller particles compared to those grown at low current density. The RMS surface roughness and particle size decrease when the film is electrochemically deposited at high current density. The differences observed in the properties of the films can be attributed to the change of the film composition affected by the applied current density.

Acknowledgements
The authors are thankful to K. O. Ay for EDX measurements and V. Simsek for providing XRD patterns. The authors are also grateful to Department of Materials Science and Engineering, Anadolu University, Turkey and M. Arikel for providing SEM and AFM images.

References
[1] Baykul M C, Sarac U and Alper M 2012 *J. Supercond. Nov. Magn.* **25** 2585
[2] Xu M, Luo G, Chai C, Mai Z, Lai W, Wu Z and Wang D 2000 *J. Cryst. Growth* **212** 291
[3] Hecker M, Tietjen D, Schneider C M, Cramer N, Malkinski L, Camley R E and Celinski Z 2002 *J. Appl. Phys.* **91** 7203
[4] Prioteasa I, Ciupină V, Oancea-Stănescu I M, Prodan G and Ştefanov C 2012 *Optoelectron. Adv. Mat.* **6** 597
[5] Zhang S and Levy P 1993 *J. Appl. Phys.* **73** 5315
[6] Sarac U and Baykul M C 2013 *J. Mater. Sci.: Mater. Electron.* **24** 2777
[7] Wang C Z, Meng G W, Fang Q Q , Peng X S, Wang Y W, Fang Q and Zhang L D 2002 *J. Phys. D Appl. Phys.* **35** 738
[8] Jia B and Wang L 2010 *BioResources* **5** 2248
[9] Su X and Qiang C 2012 *Bull. Mater. Sci.* **35** 183
[10] Sarac U and Baykul M C 2014 *J. Mater. Sci.: Mater. Electron.* **25** 2554
[11] Chang J K, Hsu S H, Sun I W and Tsai W T 2008 *J. Phys. Chem. C* **112** 1371
[12] Sarac U, Öksüzoğlu R M and Baykul M C 2012 *J. Mater. Sci.: Mater. Electron.* **23** 2110
[13] Sarac U and Baykul M C 2013 *J. Alloy. Compd.* **552** 195
[14] Sun L, Chien C L and Searson P C 2004 *Chem. Mater.* **16** 3125
[15] Kong D S, Wang J M, Shao H B, Zhang J Q and Cao C N 2011 *J. Alloy. Compd.* **509** 5611
[16] Fesharaki M J, Peter L, Schucknecht T, Rafaja D, Degi J, Pogany L, Neurohr K, Szeles E, Nabiyouni G and Bakonyi I 2012 *J. Electrochem. Soc.* **159** D162
[17] Li H and Ebrahimi F 2003 *Mat. Sci. Eng. A* **347** 93
[18] Leith S D, Ramli S and Schwartz D T 1999 *J. Electrochem. Soc.* **146** 1431
[19] Wilson A J C 1962 *Proc. Phys. Soc. Lond.* **80** 286
[20] Ebrahimi F and Ahmed Z 2003 *J. Appl. Electrochem.* **33** 733
[21] Hassani S H, Raeissi K and Golozar MA 2008 *J. Appl. Electrochem* **38** 689
[22] Farzaneh M A, Zamanzad-Ghavidel M R, Raeissi K, Golozar M A, Saatchi A and Kabi S 2011 *Appl. Surf. Sci.* **257** 5919
[23] Sarac U, Baykul M C and Uguz Y 2015 *J. Supercond. Nov. Magn.* **28** 1041
[24] Boubatra M, Azizi A, Schmerber G and Dinia A 2011 *J. Mater. Sci. Mater. Electron* **22** 1804
[25] Eom H, Jeon B, Kim D and Yoo B 2010 *Mater. Trans.* **51** 1842
[26] Sarac U and Baykul M C 2012 *J. Mater. Sci. Technol.* **28** 1004
[27] Baskaran I, Sankara Narayanan T S N and Stephen A 2006 *Mater. Lett.* **60** 1990

Journal of Physics: Conference Series **766** (2016) 012026 doi:10.1088/1742-6596/766/1/012026

Effect of Heat Treatment on the Structural Properties of TiO_2 Films Produced by Sol-Gel Spin Coating Technique

M. Nebi[1] and D. Peker[1]

[1] Physics Department, Eskisehir Osmangazi University, Eskisehir, TURKEY

E-mail: mnebi@ogu.edu.tr

Abstract. Due to have superior properties as fotocatalyst and have wide band gap, TiO_2 thin films often investigated by researchers and used by technological applications widely. In this study TiO_2 films were deposited on glass substrate by Sol-Gel Spin Coating Technic. TiO_2 films were deposited at different number of layer and then annealed at 400° C, 500° C, and 600° C in air. Effect of anneal temperature to structural properties were investigated by XRD analysis. It was observed by the light of XRD results that the structural properties of films had changed by anneal temperature.

1. Introduction

Titanium dioxide (TiO_2) is a large band gap (i.e., 3.20 eV for the anatase crystalline phase) semiconductor extensively studied due to its unique combination of chemical and physical properties used to develop various environment and energy related applications [1-3].

TiO_2 materials have thermal and chemical stability, good mechanical hardness and high UV photoactivity. TiO_2 is transparent to visible light and has high refractive index making it suitable for the development of various optical thin film applications [1-4].

It has been shown as one of the most important materials for applications in semiconductor electrochemistry, as gas sensor, as antireflective coatings, in electrochromic devices and solar energy conversion [5, 6].

When exposed to UV radiation with photon energy greater than the band gap energy of the material, the TiO_2 surface becomes super-hydrophilic, self-cleaning, self-sterilizing and antifogging. The application range of TiO_2 can be extended by improving the properties of these materials through doping with transition metal ions (Fe, Mo, Ag, Co, W, etc.) [7-17].

In this study TiO_2 films were deposited on glass substrate by Sol-Gel Spin Coating Technic. TiO_2 films were deposited at different number of layer and then annealed at 400° C, 500° C, and 600° C in air. Effect of anneal temperature to structural properties were investigated by XRD analysis.

2. Experimental

It is important that producing thin films by practical and economic way. There are various methods that using to produce semiconductive thin films. One of them which has wide use is Sol-Gel Spin Coating Technique.

This chemical method bases on preparing a solution and drop solution on very fast (ex; 1500-4000 rpm) spinning surface. This method uses to deposite film on hard and rough surfaces.

Advantages of Sol-Gel Spin Coating Technique:

- 1. No need to high temperature and vacuum
- 2. Easy to learn and use.
- 4. Deposited films are homogeneous at macro dimension.
- 5. Has easy control process
- 6. Can apply to surfaces which has different geometric shapes.
- 7. Energy conservation
- 8. Low cost and fast.

Spin Coating Method contains successive 4 steps;

- Dropping
- Spinning
- Drying
- Repeating same steps for multiple layers

Steps of Spin Coating Method are shown in Figure 2.1.

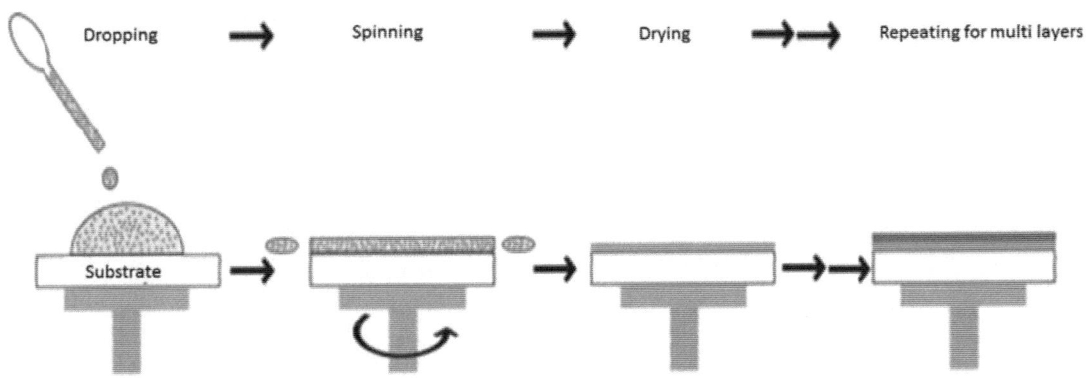

Figure 2.1. Schematic illustration of sol-gel spin coating method.

Prepared solution is drop on surface and surplus solution throw away from the sides of substrate by effect of spinning. At the same time, the film starts to deposit on substrate homogeneously. After substrates dried we can repeat this steps for tuning the film thickness of material.

2.1. Preparing Glass Substrates

In this study we used microscope glass for substrate. Microscope glasses cut as 1 cm square and then we applied following step for cleaning the glass substrates.

- Boil in a distilled water with little mass of detergent for 5 min.
- Wash with distilled water for 5 min.
- Wash with isopropyl alcohol water for 5 min.
- And then again wash with distilled water for 10 min.
- Dry at 70° C in a drying oven for an hour.

2.2. *Preparing TiO_2 Solution*

When 180 ml ethanol (C_2H_5OH) stay mixing in a beher cup we added respectively 4 ml acetylaceton ($C_5H_8O_2$), 2 ml distilled water and 14 ml titanium butoksit (Titanium (IV) n-butoxide, ($C_{16}H_{36}O_4Ti$) After chemicals added solution awaited for 24 hour under atmospheric conditions.

TiO_2 films were deposited at different number of layer 1-3-5 and then annealed at 400° C, 500° C, and 600° C in air. Then effect of anneal temperature to structural properties were investigated by XRD analysis.

3. Results

3.1. *XRD (X- Ray Diffractometer) Analysis*

The crystal structure and the preferred crystal orientation of TiO_2 films were investigated by X-ray diffraction (XRD) measurements. XRD measurements were performed by Panalytical Emprayan X-ray diffractometer using CuK_α (λ=1.5405 Å) radiation in the 2θ range 30° - 80° with a scanning speed of 2°/min. X-ray diffraction measurements were performed and the obtained results are shown in Figure 3.1 to 3.9.

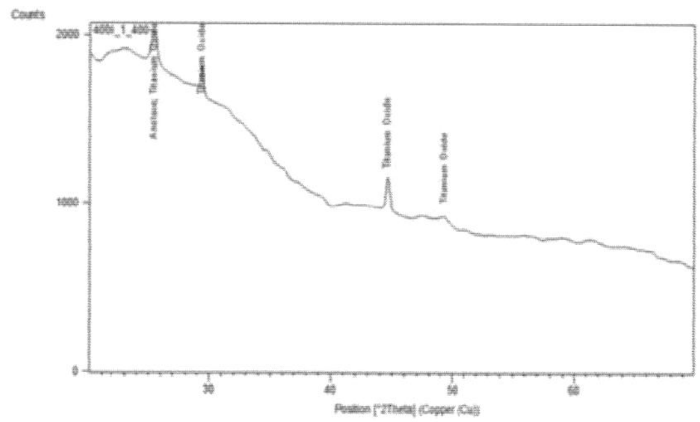

Figure 3.1. 4000 rpm, 1 layer, annealed at 400° C.

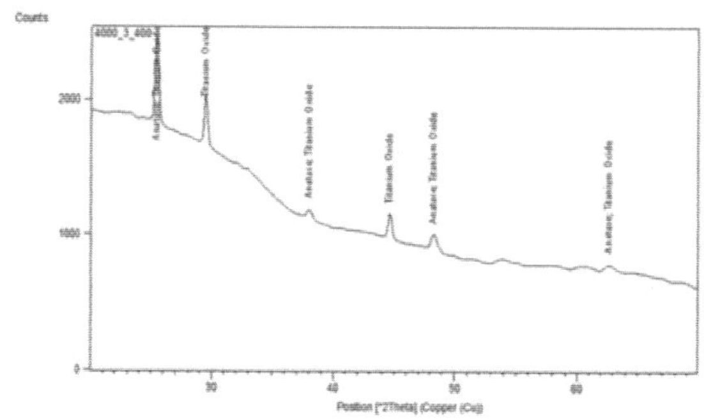

Figure 3.2. 4000 rpm, 3 layers, annealed at 400° C.

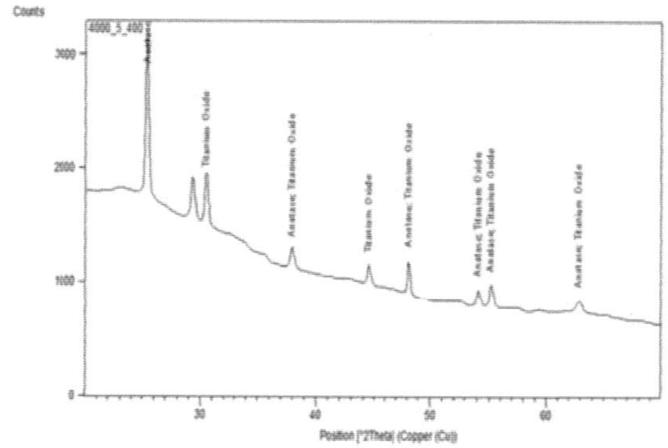

Figure 3.3. 4000 rpm, 5 layers, annealed at 400° C.

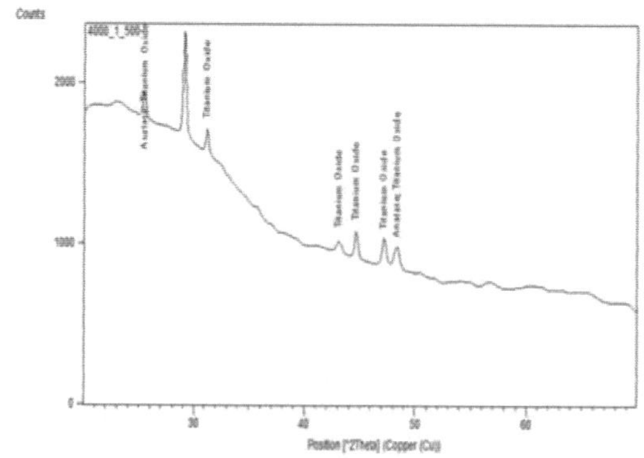

Figure 3.4. 4000 rpm, 1 layer, annealed at 500° C.

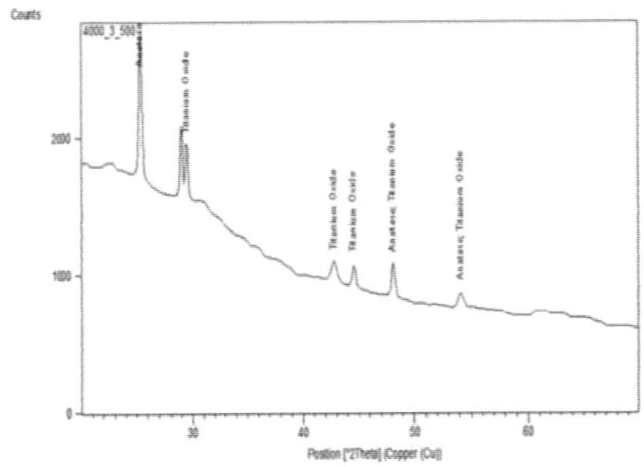

Figure 3.5. 4000 rpm, 3 layers, annealed at 500° C.

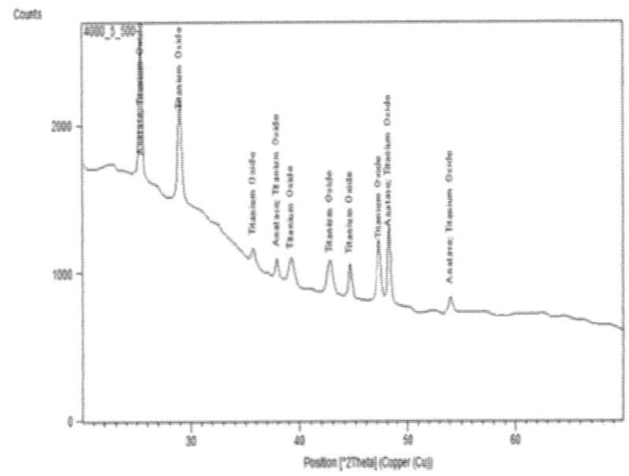

Figure 3.6. 4000 rpm, 5 layers, annealed at 500° C.

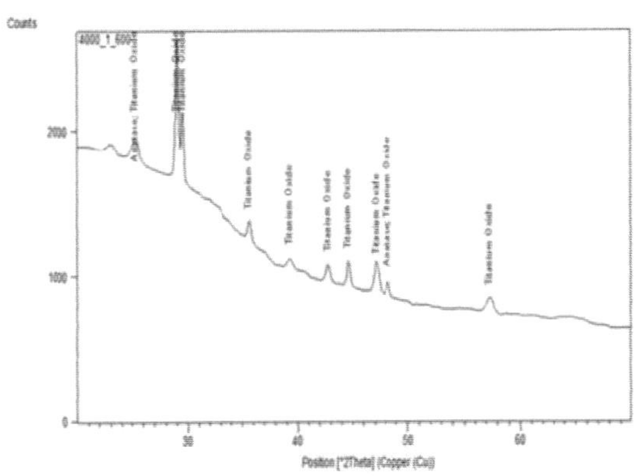

Figure 3.7. 4000 rpm, 1 layer, annealed at 600° C.

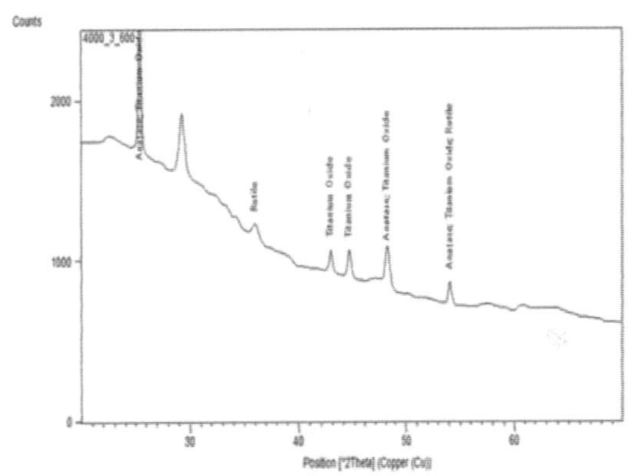

Figure 3.8. 4000 rpm, 3 layers, annealed at 600° C.

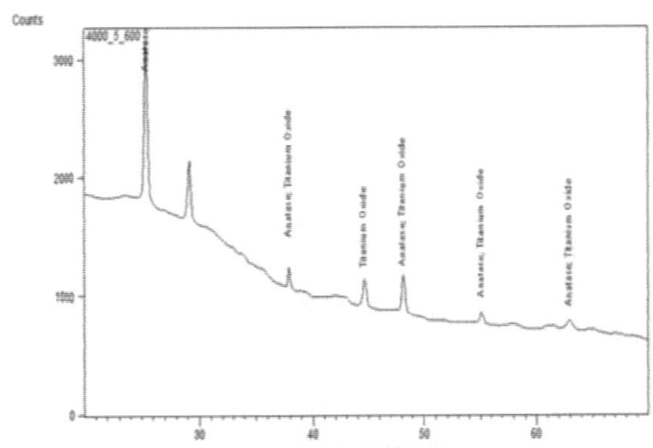

Figure 3.9. 4000 rpm, 5 layers, annealed at 600° C.

4. Conclusions

Effect of heat treatment on the structural properties of TiO_2 films produced by Sol-Gel Spin Coating Technique has been examined. It was observed by the light of XRD results that the structural properties of films, had changed by anneal temperature and number of layers. X-ray diffraction studies revealed that the crystalline structure of films which were annealed at 600° C in air are relatively better than the others.

References

[1] Fujishima A *et al* 2008 Surface Science Reports **63** 515
[2] Fujishima A *et al* 2000 Journal of Photochemistry and Photobiology C: Photochemistry Reviews **1** 1
[3] Manole A V *et al* 2013 Ceramics International **39** 4771
[4] Karunagaran B *et al* 2005 Physica B **369** 129
[5] Gratzel M and Kalyanasundaram K 1998 Coordination Chemistry Reviews **77** 347
[6] Subramanian M *et al* 2008 Thin Solid Films **516** 3776
[7] Thompson T L and Yates J T 2006 Surface Chemical Reviews **106** 4428
[8] Kubacka A *et al* 2009 Catalysis Today **143** 286
[9] Subasri R *et al* 2010 Materials Chemistry and Physics **124** 63
[10] Castro A L *et al* 2009 Journal of Solid State Chemistry **182** 1838
[11] Furubayashi Y *et al* 2006 Thin Solid Films **496** 157
[12] Hitosugi T *et al* 2008 Thin Solid Films **516** 5750
[13] Maghanga C M *et al* 2009 Thin Solid Films **518** 1254
[14] Sato Y *et al* 2008 Thin Solid Films **516** 5758
[15] Hung K H *et al* 2011 Journal of Alloys and Compounds **509** 10190
[16] Zhao L *et al* 2010 Journal of Sol–Gel Science and Technology **53** 475
[17] Liu J X *et al* 2011 Applied Surface Science **257** 10156

Journal of Physics: Conference Series **766** (2016) 012027 doi:10.1088/1742-6596/766/1/012027

Nuclear level densities and gamma-ray strength functions of ^{145,149,151}Nd isotopes

K O Ay[1], M Ozgur[1], E Algin[1], M Guttormsen[2], F L Bello Garrote[2], L Crespo Campo[2], A Görgen[2], T W Hagen[2], V W Ingeberg[2], B V Kheswa[2], M Klintefjord[2], A C Larsen[2], J E Midtbo[2], V Modamio[2], T Renstrom[2], S J Rose[2], E Sahin[2], S Siem[2], G M Tveten[2] and F Zeiser[2]

[1] *Department of Physics, Eskisehir Osmangazi University, 26480 Eskisehir, Turkey*
[2] *Department of Physics, University of Oslo, N-0316 Oslo, Norway*

E-mail: kursadosmanay@gmail.com

Abstract. The nuclear level densities and gamma-ray strength functions are the key elements for Hauser-Feshbach statistical model calculations to predict reaction cross sections which have many applications including astrophysics. The nuclear level densities and γ-ray strength functions have been determined for ^{145,149,151}Nd isotopes below the neutron separation energies using the Oslo method with the ^{144,148,150}Nd(*d,p*) reactions. The results from the first measurements as well as planned experiments at OCL will be presented.

1.Introduction

The study of nuclear level densities has been gaining a growing interest in the nuclear science community in recent years. Microscopic description of heavy nuclei up to high excitation energy has become possible with the introduction of new theoretical approaches and fast computers [1].

Currently, experimental level densities and γ-ray strength functions play an important role for testing of nuclear structure models and they are also important tools for various nuclear applications such as for calculation of reaction cross sections applied to astrophysical nucleosynthesis. They are important ingredients in designing nuclear power plants and also used for calculations in transmutation of nuclear waste.

The nuclear physics group at the University of Oslo has developed a method [2, 3] to determine the level density and the γ-ray strength function (γSF) simultaneously by using particle-γ coincidences. In this work, the Oslo method is applied to extract the level densities and γ-ray strength functions of the ^{145,149,151}Nd isotopes.

In the second section, the experimental procedure and the data analysis are described and in the third section the level densities and γ-ray strength functions extracted are presented.

2.Experimental Procedure and Data Analysis

Three experiments were performed at the MC-35 Scanditronix cyclotron of the Oslo Cyclotron Laboratory (OCL). The self-supporting ^{144}Nd (97% enrichment and 2 mg/cm^2 thickness), ^{148}Nd (95% enrichment and 1.9 mg/cm^2 thickness), and ^{150}Nd (97% enrichment and 1.9 mg/cm^2 thickness) targets were bombarded with a 13.5-MeV deuteron beam, and ^{144}Nd(*d,pγ*)^{145}Nd, ^{148}Nd(*d,pγ*)^{149}Nd and

^{150}Nd($d,p\gamma$)^{151}Nd reactions were studied. Particle-γ coincidences were measured with the SiRi particle telescope and the CACTUS γ-detector system [4, 5] which are shown in figures 1 and 2.

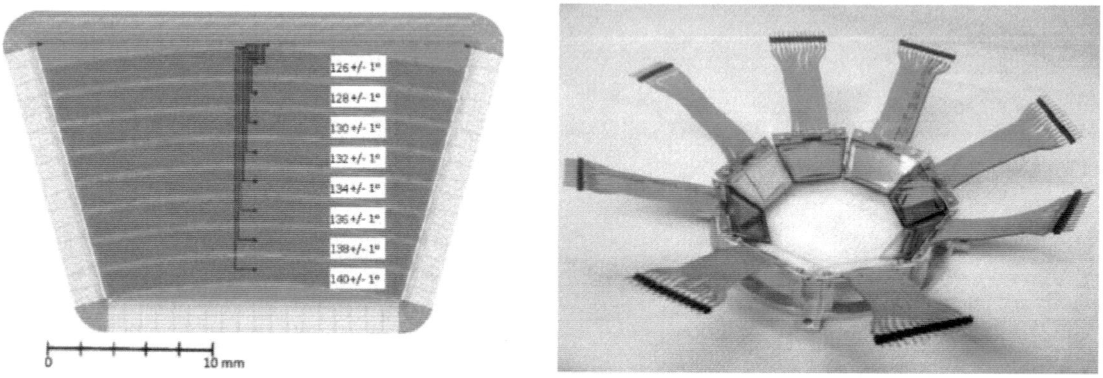

(a) (b)

Figure 1: SIRI particle-detector system. (a) The layout of one silicon chip with its eight ΔE detectors used for different reaction angles. (b) The whole detector system with the cables to read out the signal.

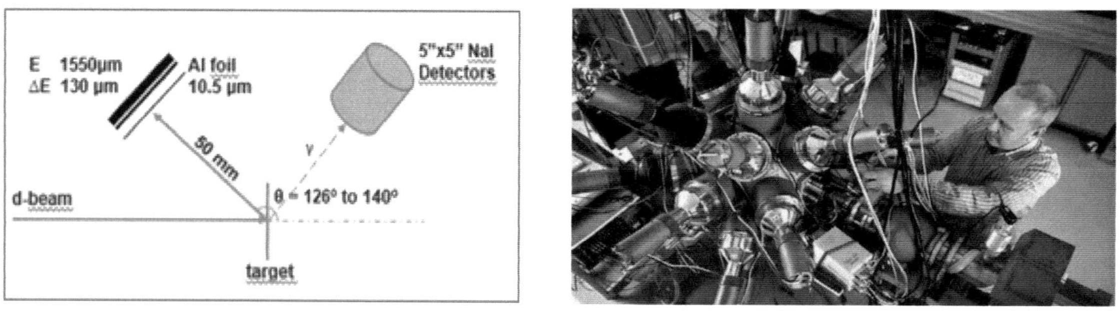

(a) (b)

Figure 2: Detector setup. (a) Positioning of the SiRi with respect to the target and beam (b) Complete setup of the CACTUS array.

In order to reduce the intense elastically scattered deuterons and to obtain a broad and higher spin distribution, the 64 SiRi telescopes are placed in backward direction and they cover eight angles relative to the beam axis. The front and back detectors have thicknesses of 130 μm and 1550 μm, respectively. The CACTUS array consists of 26 collimated 5" x 5" NaI(Tl) detectors with a total efficiency of 14.1% at E_γ = 1.33 MeV.

By using the reaction kinematics, the excitation energy of the compound nucleus is calculated from the measured ejectile energy. The particles and γ-rays are measured in coincidence, thus each γ-ray can be assigned to a certain initial excitation energy of the residual nucleus, by which the particle-γ ray coincidence matrix is obtained (Figure 3).

First main steps of the Oslo method are shown in Figure 4. After obtaining the raw matrix by sorting the data into a matrix of initial excitation energies E versus the NaI energy signal (a), this raw matrix is unfolded [6] by using the NaI response function for each excitation bin (b); and finally, by applying an iterative subtraction technique [7] to separate the distribution of the first generation γs from the total γ cascade, the first generation (primary) γ-ray matrix $P(E, E_\gamma)$ is obtained (c).

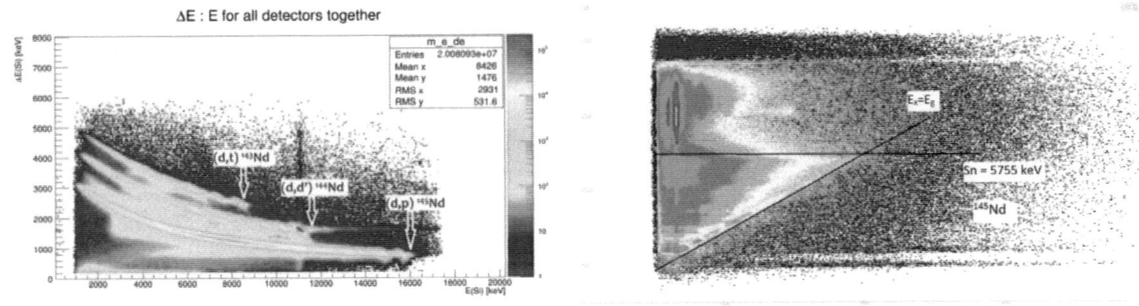

(a) (b)

Figure 3: (a) ΔE –E bananas for ^{144}Nd and (b) particle – γ ray coincidence matrix for ^{145}Nd.

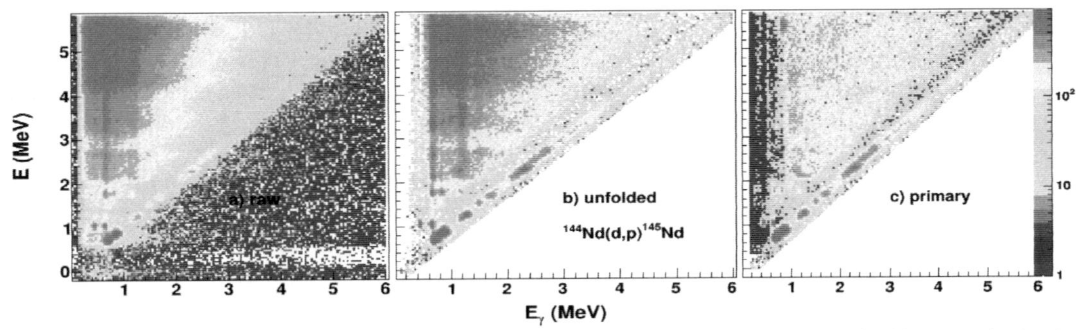

Figure 4: Initial excitation energy E versus γ-ray energy E_γ from particle-γ coincidences obtained via the ^{144}Nd($d,p\gamma$)^{145}Nd reaction. Main steps are: (a) obtaining a raw γ-ray spectra, (b) unfolding the spectra by using the NaI response function, and finally (c) extracting the primary (or first generation) γ-ray spectra $P(E, E_\gamma)$ as a function of excitation energy.

According to the Brink hypothesis [8] the γ-ray transmission coefficient $\mathcal{T}$ is approximately independent of excitation energy, hence the first generation matrix $P(E, E_\gamma)$ can be factorised as follows:

$$P(E, E_\gamma) \sim \mathcal{T}(E_\gamma)\rho(E - E_\gamma)\,, \tag{1}$$

where $\rho(E - E_\gamma)$ is the level density at the excitation energy after the first γ-ray is emitted. This allows us to extract the level density and the γ-ray transmission coefficient simultaneously.

3.Level Densities and γ-ray Strength Functions

The level density obtained by applying equation (1) to the first-generation matrix determines only the functional form of $\mathcal{T}$ and ρ. If one solution of $\mathcal{T}$ and ρ is known, it is possible to construct infinitely many fits to the $P(E, E_\gamma)$ matrix by the following equations:

$$\tilde{\rho}(E - E_\gamma) = A\exp[\alpha(E - E_\gamma)]\rho(E - E_\gamma) \tag{2}$$

$$\tilde{\mathcal{T}}(E_\gamma) = B\exp(\alpha E_\gamma)\mathcal{T}(E_\gamma)\,. \tag{3}$$

In order to normalize the two functions, one needs information to fix the A, α, and B parameters. For the level density, two fixed points are used to determine A and α. The procedure is shown in Figure 5 for ^{145}Nd. The level density is normalized to the known discrete levels [9] at low excitation energy. At high excitation energy, we make use of the neutron resonance spacing D_0 at the neutron separation energy S_n. Further descriptions of the Oslo method are given in references [2, 3].

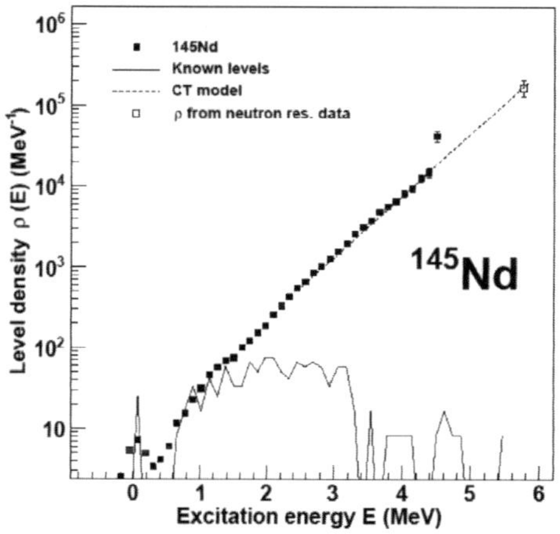

Figure 5: Normalization procedure for experimental level density. The level density of ^{145}Nd is normalized to known discrete levels at low energies and to (S_n) at the binding energy, which is calculated from neutron resonance spacing data.

The normalization procedure is performed by using the Constant Temperature (CT) formula at above $E = 2\Delta$ energy, where Δ is the pairing gap parameter. Here, the CT formula is as follows:

$$\rho_{CT}(E_x) = \frac{1}{T_{CT}} \exp\frac{E_x - E_0}{T_{CT}}, \tag{4}$$

where the shift in excitation energy E_0 is given by:

$$E_0 = S_n - T_{CT} \ln[\rho(S_n) T_{CT}]. \tag{5}$$

To normalize the γ-ray strength function, we also need the parameter B. We can determine this parameter by using the known average total radiation width $\langle \Gamma_\gamma \rangle$ at the neutron separation energy. By assuming that dipole transitions dominate the decay and equal parity holds for all excitation energies and spins, it can be found [2,10]:

$$\langle \Gamma_\gamma(S_n) \rangle = \frac{1}{2\pi\rho(S_n)} \sum_{I_f} \int_0^{S_n} dE_\gamma B\tau(E_\gamma)\rho(S_n - E_\gamma, I_f)\,. \tag{6}$$

Finally, as seen in figure 6, we obtain the level densities (left panel) and the γ-ray strength functions (right panel) for ^{145,149,151}Nd nuclei.

The parameters used for the Oslo method are displayed in Table 1 for the ^{145,149,151}Nd nuclei. The resonance spacing and total gamma width parameters are taken from RIPL-3 database [11]. The level density and back-shift parameters are taken from the systematic study of Egidy and Bucurescu [12] . The level density at S_n is calculated by using the following formula:

$$\rho(S_n) = \frac{2\sigma^2}{D_0} \frac{1}{(I+1)\ \exp[-\frac{(I+1)^2}{2\sigma^2}] + I exp[-{I^2}/{2\sigma^2}]}\ . \tag{7}$$

Here, I is the spin of the target nucleus. For the spin distribution $g(E_x\,, I)$ Gilbert and Cameron's work is used [13], and for the spin cut-off parameter σ, Egidy and Bucurescu's rigid body moment of inertia approach [12] is used in the following formulas:

$$g(E_x\,, I) = \frac{2I+1}{2\sigma^2(E_x)} \exp[-\left(I + \tfrac{1}{2}\right)^2 / 2\sigma^2] \tag{8}$$

$$\sigma^2(E_x) = 0.0146 A^{2/3} \frac{1+\sqrt{4aU(E_x)}}{2a}\ , \tag{9}$$

where, A is the mass number of the nucleus, a is the level density parameter, $U(E_x) = E_x - E_1$ is the intrinsic excitation energy and E_1 is the back-shift parameter.

Table 1: Parameters used to calculate level density of $^{145, 149, 151}$Nd.

Nucleus	Neutron binding energy	Level density parameter	Back-shift parameter	Spin cut-off parameter.	Total Gamma width	Resonance spacing parameter	Level density at S_n
	S_n (MeV)	a (MeV^{-1})	E_1 (MeV)	σ	Γ (meV)	D_0 (eV)	ρ (S_n)(MeV^{-1})
^{145}Nd	5755	15.942	-0.052	6.09	47	450	1.673E+05
^{149}Nd	5.039	17.867	-0.446	5.97	40	155	4.662E+05
^{151}Nd	5.335	17.820	-0.596	6.153	67	165	4.278E+05

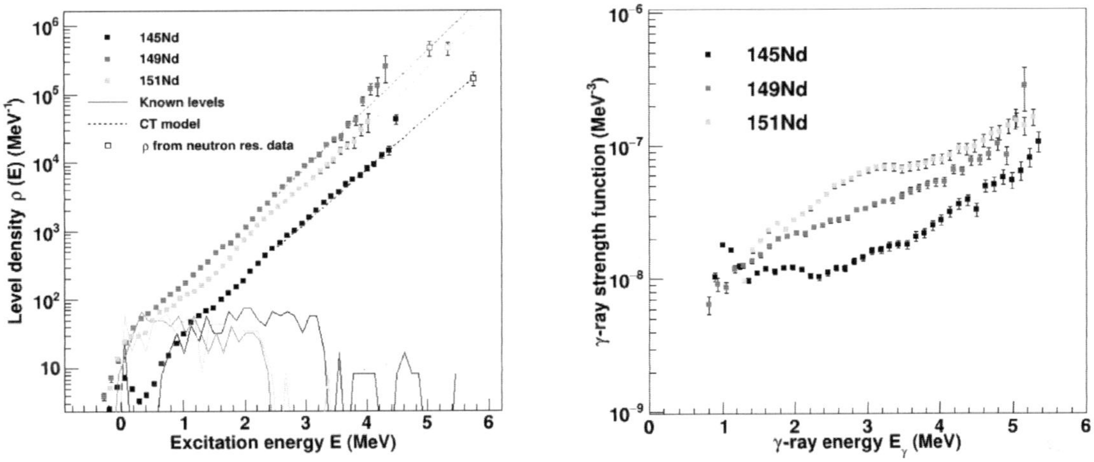

Figure 6: Level densities of ^{145,149,151}Nd nuclei (left panel) and γ-ray strength functions of the ^{145,149,151}Nd nuclei (right panel).

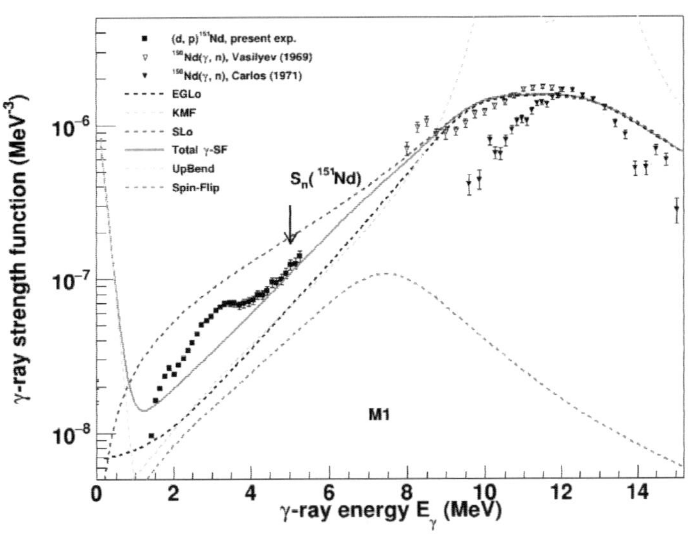

Figure 7: The γ-ray strength function for ^{151}Nd compared to models with systematic parameters.

Journal of Physics: Conference Series **766** (2016) 012027 doi:10.1088/1742-6596/766/1/012027

As seen in figure 6 (left), level density for ^{145}Nd is lower than those for the other two nuclei. This is because, ^{145}Nd is closer to the N=82 shell closure than the other two nuclei. Also, when we plot the γ-ray strength functions (figure 6, right), we see that the scissors resonance around 2.5 – 3.5 MeV increases as we move from ^{145}Nd to ^{151}Nd, which is due to the increasing deformation. For ^{145}Nd, it seems that the missing scissors strength reveals a low-energy enhancement.

In figure 7, γ-ray strength function for ^{151}Nd is shown together with the γ-ray strength functions obtained by Vasiliyev [14] and Carlos [15]. As seen, although there is a region with no data between ~6-8 MeV, our results are compatible with the measurements of Vasiliev. The γ-ray strength function used as a "background" to our data is described by the sum of three strength functions (purple) (see figure 7).

4.Conclusions

Nuclear level densities and γ-ray strength functions were obtained for ^{145,149,151}Nd isotopes by using the so-called Oslo method. Preliminary results show that scissors mode resonance strength increases with deformation.

For our future work, we plan to study the ^{144}Nd$(d,d'\gamma)^{144}$Nd, ^{148}Nd$(d,d'\gamma)^{148}$Nd, and ^{150}Nd$(d,d'\gamma)^{150}$Nd channels and to conduct ^{142,146}Nd$(d,p\gamma)^{143,147}$Nd experiments. Thus we will be able to study the level densities and γ-ray strength functions for a wide range of neodymium isotopes, i.e. $^{142-151}$Nd isotopes, systematically in detail.

Acknowledgements

We would like to thank J. Müller, A. Semchenkov and J. Wikne at the Oslo Cyclotron Laboratory for providing the stable and high-quality deuterium beam. This work was supported by the Scientific and Technological Research Council of Turkey (TUBITAK) under the project number 115F196 and Oslo Cyclotron Laboratory, Norway.

References:

[1] S. Goriely, et al., *Phys. Rev. Lett.* 102, 152503 (2009).
[2] A. Schiller et al., *Nucl. Instrum. Methods in Phys. Res. A* 447, 494 (2000).
[3] A.C. Larsen et al., *Phys. Rev. C* 83, 034315 (2011).
[4] M. Guttormsen et al., *Nucl. Instrum. Methods Phys. Res. A* 648, 168 (2011).
[5] M. Guttormsen et al., *Phys. Scr. T* 32, 54 (1990).
[6] M. Guttormsen et al., *Nucl. Instrum. Methods in Phys. Res. A* 374, 371 (1996).
[7] M. Guttormsen et al., *Nucl. Instrum. Methods in Phys. Res. A* 255, 518 (1987).
[8] D.M. Brink, Ph.D. thesis, Oxford University, 1955.
[9] Data extracted from the ENDSF database using NNDC's Chart of Nuclides, http://www.nndc.bnl.gov/chart/.
[10] J. Kopecky and M. Uhl. *Physical Review C* 41, 1941, May. 1990.
[11] RIPL-3 Handbook for calculation of nuclear reaction, (2009); available at http://www-nds.iaea.org/RIPL-3/
[12] T. von Egidy and Bucurescu, *Phys. Rev. C*, 72(4), 044311 (2005), 73, 049901 (2006).
[13] A. Gilbert and G.W. Cameron, *Canadian Journal of Physics*, 43(8), 1446-1496, (1965).
[14] P. Carlos et al., *Nuclear Physics A*, 172, 437- 448 (1971)
[15] O.V. Vasilijev et al., *Physics Letters*, 30B, 2 (1969)

Journal of Physics: Conference Series **766** (2016) 012028 doi:10.1088/1742-6596/766/1/012028

Numerical solution of the complex modified Korteweg-de Vries equation by DQM

Ali Başhan[1], Yusuf Uçar[2], N. Murat Yağmurlu[2], Alaattin Esen[2]

[1]Department of Mathematics, Faculty of Science and Art, Bulent Ecevit University, Zonguldak, 67100, Turkey
[2]Department of Mathematics, Faculty of Science and Art, Inonu University, Malatya, 44200, Turkey

E-mail: alibashan@gmail.com

Abstract. In this paper, a method based on the differential quadrature method with quintic B-spline has been applied to simulate the solitary wave solution of the complex modified Korteweg-de Vries equation (CMKdV). Three test problems, namely single solitary wave, interaction of two solitary waves and interaction of three solitary waves have been investigated. The efficiency and accuracy of the method have been measured by calculating maximum error norm L_∞ for single solitary waves having analytical solutions. Also, the three lowest conserved quantities and obtained numerical results have been compared with some of the published numerical results.

1. Introduction

In nature, various problems are modeled by partial differential equations. Being one of the well-known natural phenomena it is also a model for nonlinear evolution of plasma waves [1], propagation of transverse waves in a molecular chain model [2], and in a generalized elastic solid [3, 4]. Because of its importance, many researchers have dealt with the Complex Modified Korteweg-de Vries (CMKdV) equation given in the following form

$$\frac{\partial W(x,t)}{\partial t}+\alpha\frac{\partial\left(|W(x,t)|^2 W(x,t)\right)}{\partial x}+\frac{\partial^3 W(x,t)}{\partial x^3}=0,\quad -\infty<x<\infty,\quad t>0, \tag{1}$$

where W is a complex valued function of the the spatial coordinate x and the time t, α is a constant parameter.To avoid complex computation, we need to transformation of CMKdV equation (1) into a nonlinear coupled system by decomposing $W(x,t)$ into its real and imaginary parts

$$W(x,t)=U(x,t)+iV(x,t)\,,\quad i=\sqrt{-1}$$

and obtain the real valued-modified Korteweg-de Vries equation system,

$$U_t+\alpha\left[3U^2U_x+V^2U_x+2UVV_x\right]+U_{3x} = 0, \tag{2}$$

$$V_t+\alpha\left[3V^2V_x+U^2V_x+2UVU_x\right]+V_{3x} = 0, \tag{3}$$

where $U(x,t)$ and $V(x,t)$ are real functions.

2. Quintic B-spline DQM

Bellman *et al.*[8] first introduced DQM in 1972 where partial derivative of a function with respect to a coordinate direction is expressed as a linear weighted sum of all the functional values at all mesh points along that direction[9]. Let's take the grid distribution $a = x_1 < x_2 < \cdots < x_N = b$ of a finite interval $[a, b]$ into consideration. Provided that any given function $U(x)$ is smooth enough over the domain, its derivatives with respect to x at a grid points x_i can be approximated by a linear summation of all the functional values in the domain, namely,

$$U_x^{(r)}(x_i) = \frac{d^{(r)}U}{dx^{(r)}}|_{x_i} = \sum_{j=1}^{N} w_{ij}^{(r)} U(x_j), \quad i = 1, 2, ..., N, \quad r = 1, 2, ..., N-1 \tag{4}$$

where r denotes the order of derivative, $w_{ij}^{(r)}$ represent the weighting coefficients of the $r-th$ order derivative approximation, and N denotes the number of grid points in the solution domain. Here, the index j represents the fact that $w_{ij}^{(r)}$ is the corresponding weighting coefficient of the functional value $U(x_j)$. The quintic B-splines used as a base functions which are defined as given in [10].

2.1. Weighting coefficients of the first order derivative

From Eq.(4) with value of $r = 1$, and using quintic B-splines as test functions we have obtained the following equations

$$Q_k'(x_i) = \sum_{j=k-2}^{k+2} w_{i,j}^{(1)} Q_k(x_j), \tag{5}$$

For example, for the first grid point x_1 (5), we get the following equation

$$Q_k'(x_1) = \sum_{j=k-2}^{k+2} w_{1,j}^{(1)} Q_k(x_j), \tag{6}$$

By substituting the values of quintic basis functions into Eq.(6) and using four additional equations obtained from the derivative of Eq.(6) at four different B-spline Q_k $(k = -1, 0, N+1, N+2)$ and eliminating four unknown terms from the system of equations, we obtain the following system of equations

$$\begin{bmatrix} 37 & 82 & 21 & & & & & & & \\ 8 & 33 & 18 & 1 & & & & & & \\ 1 & 26 & 66 & 26 & 1 & & & & & \\ & 1 & 26 & 66 & 26 & 1 & & & & \\ & & \ddots & \ddots & \ddots & \ddots & \ddots & & & \\ & & & & 1 & 26 & 66 & 26 & 1 & \\ & & & & & 1 & 26 & 66 & 26 & 1 \\ & & & & & & 1 & 18 & 33 & 8 \\ & & & & & & & 21 & 82 & 37 \end{bmatrix} \begin{bmatrix} w_{1,-1}^{(1)} \\ w_{1,0}^{(1)} \\ w_{1,1}^{(1)} \\ w_{1,2}^{(1)} \\ w_{1,3}^{(1)} \\ w_{1,4}^{(1)} \\ \vdots \\ w_{1,N+1}^{(1)} \\ w_{1,N+2}^{(1)} \end{bmatrix} = \begin{bmatrix} -\frac{109}{2h} \\ -\frac{29}{h} \\ 0 \\ \frac{50}{h} \\ \frac{5}{h} \\ 0 \\ \vdots \\ 0 \\ 0 \end{bmatrix} \tag{7}$$

Similarly, using the value of quintic basis functions at x_i, $(2 \leq i \leq N)$ grid points, respectively, the equation systems is obtained which are used to determine the weighting coefficients.

So, weighting coefficients $w_{i,j}^{(1)}$ which are related to the x_i, $(i = 1, 2, ..., N)$ are found quite easily by solving the obtained equation systems with Thomas algorithm. Determining of the weighting coefficients of the third order derivatives have same process.

Journal of Physics: Conference Series **766** (2016) 012028 doi:10.1088/1742-6596/766/1/012028

3. Numerical discretization

We discritize the equations (2) − (3) separately by using forward finite difference and Crank-Nicolson.

$$2U^{n+1}+\Delta t\left[U_{3x}^{n+1}+\alpha\left(3\left(U^2U_x\right)^{n+1}+\left(V^2U_x\right)^{n+1}+2\left(UVV_x\right)^{n+1}\right)\right]$$

$$=2U^n+\Delta t\left[-U_{3x}^n+\alpha\left(3\left(U^2U_x\right)^n+\left(V^2U_x\right)^n+2\left(UVV_x\right)^n\right)\right] \tag{8}$$

Then, Rubin and Graves linearization technique[11] is used at the left side of the Eq. (8) to linearize the nonlinear terms so we obtained

$$2U^{n+1}+\Delta t[U_{3x}^{n+1}+3\alpha\left(\left(U^2\right)^nU_x^{n+1}+2U^nU_x^nU^{n+1}\right)+\alpha\left(\left(V^2\right)^nU_x^{n+1}+2V^nU_x^nV^{n+1}\right)$$

$$+2\alpha\left(U^{n+1}V^nV_x^n+U^nV^{n+1}V_x^n+U^nV^nV_x^{n+1}\right)]$$

$$=2U^n+\Delta t\left[-U_{3x}^n+3\alpha\left(U^2\right)^nU_x^n+\alpha\left(V^2\right)^nU_x^n+2\alpha U^nV^nV_x^n\right] \tag{9}$$

Let us define some terms to use in Eq. (9) as,

$$A_i^n=\sum_{j=1}^N w_{ij}^{(1)}U_j^n=U_{x_i}^n,\quad B_i^n=\sum_{j=1}^N w_{ij}^{(3)}U_j^n=U_{3x_i}^n, C_i^n=\sum_{j=1}^N w_{ij}^{(1)}V_j^n=V_{x_i}^n,\quad D_i^n=\sum_{j=1}^N w_{ij}^{(3)}V_j^n=V_{3x_i}^n \tag{10}$$

where A_i^n and B_i^n are the first and third-order derivative approximations of U functions at the n-th time level on x_i points, respectively. And C_i^n and D_i^n are the first and third order derivative approximations of V function at the $n-th$ time level on x_i points, respectively. By the substitution of definition (10) at Eq. (9) we obtained

$$2U_i^{n+1}+\Delta t[\sum_{j=1}^N w_{ij}^{(3)}U_j^{n+1}+3\alpha\left((U_i^n)^2\sum_{j=1}^N w_{ij}^{(1)}U_j^{n+1}+2U_i^nA_i^nU_i^{n+1}\right) \tag{11}$$

$$+\alpha\left((V_i^n)^2\sum_{j=1}^N w_{ij}^{(1)}U_j^{n+1}+2V_i^nC_i^nU_i^{n+1}\right)+2\alpha\left(V_i^nA_i^nV_i^{n+1}+U_i^nC_i^nV_i^{n+1}+U_i^nV_i^n\sum_{j=1}^N w_{ij}^{(1)}V_j^{n+1}\right)]=f_i^n$$

where

$$f_i^n=2U_i^n+\Delta t\left[-B_i^n+\alpha\left(3\left(U_i^n\right)^2A_i^n+\left(V_i^n\right)^2A_i^n+2U_i^nV_i^nC_i^n\right)\right], \tag{12}$$

we reorganised Eq. (11) for each grid points as,

$$\left[2+\Delta t\left(w_{ii}^{(3)}+\alpha\left(3\left(U_i^n\right)^2w_{ii}^{(1)}+6U_i^nA_i^n+\left(V_i^n\right)^2w_{ii}^{(1)}+2V_i^nC_i^n\right)\right)\right]U_i^{n+1}$$

$$+\left[\sum_{j=1,i\neq j}^N\Delta t\left(w_{ij}^{(3)}+\alpha\left(3\left(U_i^n\right)^2w_{ij}^{(1)}+\left(V_i^n\right)^2w_{ij}^{(1)}\right)\right)U_j^{n+1}\right]$$

$$+\left[2\alpha\Delta t\left(V_i^nA_i^n+U_i^nC_i^n+U_i^nV_i^nw_{ii}^{(1)}\right)\right]V_i^{n+1}+\left[\sum_{j=1,i\neq j}^N\left(2\alpha\Delta tU_i^nV_i^nw_{ij}^{(1)}\right)V_j^{n+1}\right]=f_i^n \tag{13}$$

By the same process, the Eq. (3) is discritized, linearized and organised. Then, boundary conditions have been applied to system of equations and the first and last equations are eliminated from each systems and solved by Gauss elimination method easily.

4. Numerical examples

The accuracy of the numerical method is checked by using the error norm L_∞ and three lowest invariants:

$$L_\infty \simeq \max_j \left|U_j^{exact} - (U_N)_j\right|, \; I_1 = \int_{-\infty}^{\infty} w dx,$$
$$I_2 \simeq \sum_{j=1}^{N} h_j |w_j^n|^2, \; I_3 \simeq \sum_{j=1}^{N} h_j \left[\frac{\alpha}{2}|w_j^n|^4 - (w_x^n)_j^2\right]$$

4.1. Single soliton

The analytically solution of complex mKdV equation is given in[5] as:

$$W(x,t) = \sqrt{\frac{2c}{\alpha}} \text{sech}\left[\sqrt{c}\,(x - x_0 - ct)\right] \exp(i\theta) \tag{14}$$

where soliton standing at x_0 position initially and moving to the right hand with constant c velocity and satisfies the boundary conditions $W \to 0$ as $x \to \pm\infty$. We first take $\alpha = 2$, $\theta = 0, c = 1, x_0 = 0$ in $[-20, 40]$ and at $t = 0$, we obtain the initial condition.

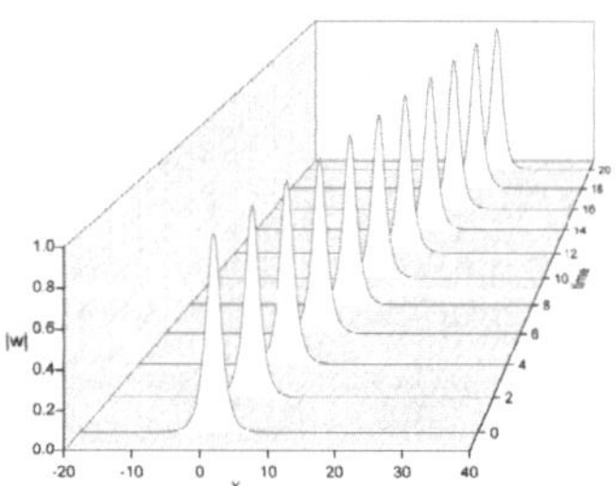

Figure 1. Single soliton

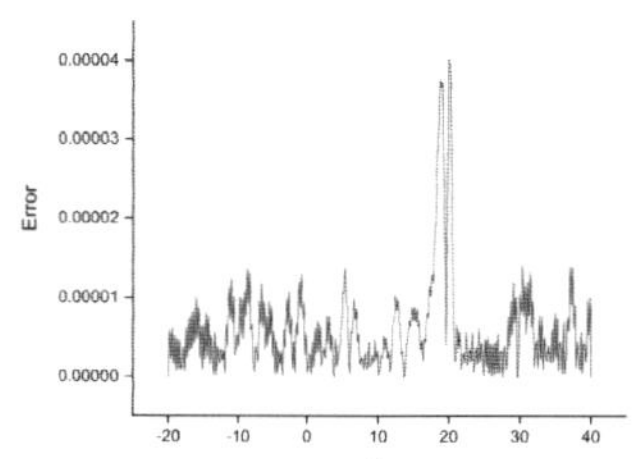

Figure 2. Absolute error

Table 1. L_∞ error norms and invariants for $\theta = 0$ and $\Delta t = 0.01$.

	Present, $N = 336$				P-G FEM [6] $N = 600$			
t	L_∞	I_1	I_2	I_3	L_∞	I_1	I_2	I_3
0	0.000000	3.141590	2.000000	0.666667	0.000000	3.141590	2.000000	0.669765
5	0.000065	3.141578	1.999999	0.666666	0.000057	3.141592	2.000000	0.669764
10	0.000070	3.141606	1.999999	0.666666	0.000108	3.141592	1.999999	0.669764
15	0.000068	3.141588	1.999999	0.666666	0.000163	3.141592	1.999999	0.669763
20	0.000066	3.141572	2.000000	0.666667	0.000218	3.141592	1.999999	0.669763

As it is seen from the Figure 1, with the time run up to the $t = 20$ the amplitude and velocity of wave do not change as a result of properties of solitons. As it is seen straightforward from Table 1 and 2 the present error norms L_∞ are smaller than earlier works [6, 7]. It is seen from Figure 2 that the maximum absolute error at time $t = 20$ is found 4.02×10^{-5} at $x = 19.94$.

4.2. Interaction of two solitary waves

The sum of two solitary waves which initial condition is given in [7] as:

$$W(x,0) = \sqrt{\frac{2c_1}{\alpha}} \text{sech}\left[\sqrt{c_1}\,(x - x_1)\right] \exp(i\theta_1) + \sqrt{\frac{2c_2}{\alpha}} \text{sech}\left[\sqrt{c_2}\,(x - x_2)\right] \exp(i\theta_2)$$

Table 2. L_∞ error norms and invariants for $\theta = \pi/4$ and $\Delta t = 0.01$.

Time	Present, $N = 336$			Coll. FEM [7] $N = 600$		
	L_∞	I_2	I_3	L_∞	I_2	I_3
0.0	0.000000	4.000000	1.333334	0.000000	4.000000	1.339529
5.0	0.000049	4.000000	1.333334	0.000078	3.999999	1.339529
10.0	0.000050	3.999999	1.333333	0.000154	3.999998	1.339528
15.0	0.000051	3.999998	1.333332	0.000231	3.999998	1.339528
20.0	0.000039	3.999994	1.333329	0.000308	3.999997	1.339526

where $x_1 = 25$ and $x_2 = 50$ are initial positions of two solitary waves, respectively in $[0, 100]$.

We investigated the interaction of two ortogonally polarized solitary waves which are interact with $y-$polarized ($\theta_1 = 0$) and $z-$polarized ($\theta_2 = \pi/2$) then the interaction of two $y-$polarized solitary waves which are interact with $y-$polarized ($\theta_1 = \theta_2 = 0$). We used fix value of $\alpha = 2, c_1 = 2, c_2 = 0.5$, and both simulations time run up to $t = 30$. As it seen clearly from Figure 3 and Figure 4 that at simulation of two ortogonally polarized solitary waves after the interaction a tail appeared behind the shorter wave and in opposition to two ortogonally polarized solitary waves there is not any tail appeared after the interaction of two $y-$polarized solitary waves. The obtained invariants given in Table 3 is acceptable good.

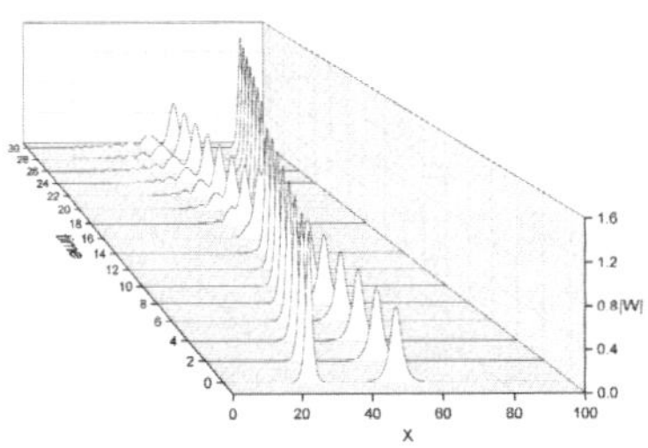

Figure 3. Two ortogonally polarized solitary waves

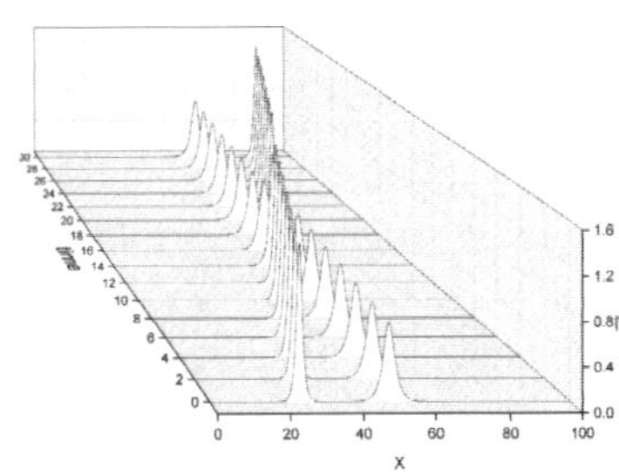

Figure 4. Two y-polarized solitary waves

Table 3. Invariants of two orthogonally and two $y-$polarized solitons for $\Delta t = 0.01$.

t	two ortogonally polarized solitary waves				two y-polarized solitary waves	
	$I_1(\mathrm{Real})$	$I_1(\mathrm{Imag})$	I_2	I_3	I_2	I_3
0	3.141593	3.141591	4.242640	2.121332	4.242640	2.121333
5	3.141630	3.141593	4.242590	2.121228	4.242587	2.121222
10	3.141667	3.141593	4.242537	2.121124	4.242529	2.121112
15	3.141534	3.141590	4.242371	2.120882	4.242330	2.120804
20	3.141566	3.141598	4.242489	2.121043	4.242473	2.121001
25	3.141812	3.142181	4.242424	2.120901	4.242431	2.120914
30	3.138532	3.135895	4.242476	2.120740	4.242377	2.120808

4.3. Interaction of three solitons

Our third test problem is interaction of three solitons which is given in [7] as follows

$$W(x,0) = \sqrt{\frac{2c_1}{\alpha}}\operatorname{sech}\left[\sqrt{c_1}\,(x - x_1)\right]\exp(i\theta_1) + \sqrt{\frac{2c_2}{\alpha}}\operatorname{sech}\left[\sqrt{c_2}\,(x - x_2)\right]\exp(i\theta_2)$$

$$+\sqrt{\frac{2c_3}{\alpha}}\operatorname{sech}\left[\sqrt{c_3}\left(x-x_3\right)\right]\exp\left(i\theta_3\right)$$

where $x_1 = 10$, $x_2 = 30$ and $x_3 = 50$ are initial positions of three single solitons, respectively. We investigated the interaction of three solitons which are interact with $y-$polarized ($\theta_1 = \theta_2 = \theta_3 = 0$) by using $\alpha = 2, c_1 = 1, c_2 = 0.5$ and $c_3 = 0.3$ time up to $t = 80$.

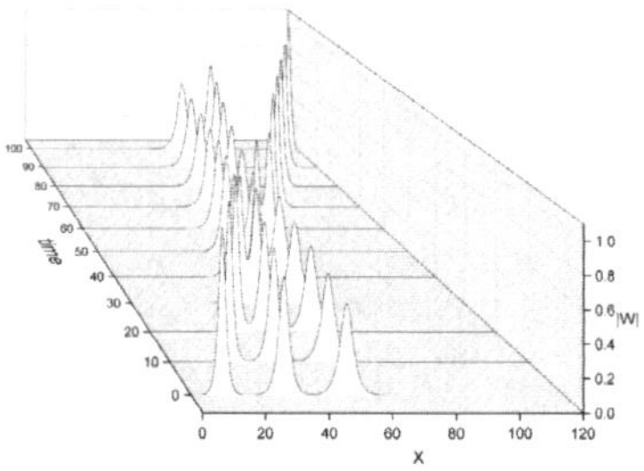

Figure 5. Three y-polarized solitons

Table 4. Invariants of three $y-$polarized solitons for $\Delta t = 0.01$.

	QBDQM $N = 501$		Coll.FEM[7] $N = 1200$	
t	I_2	I_3	I_2	I_3
0	4.510004	1.012099	4.510006	1.015897
20	4.510002	1.012098	4.510005	1.015783
40	4.509997	1.012103	4.510006	1.015031
60	4.510002	1.012097	4.510008	1.015660
80	4.510003	1.012096	4.510005	1.015752

5. Conclusion

In this work, we have implemented DQM based on quintic B-splines for numerical solution of complex mKdV equation. One of the main characteristics of the present method is to be able to obtain good results by using less number of grid points. As can be observed by the comparison between the obtained results of present method and earlier works, QBDQM results are acceptable good. The obtained results show that QBDQM can be used to produce reasonable accurate numerical solutions of the complex mKdV equation. So, QBDQM is a reliable one for getting the numerical solutions of some physically important nonlinear problems.

6. References

[1] Karney C F F, Sen A. and Chu F Y F 1979 *Phys. Fluids* **22** 940
[2] Gorbacheva O B and Ostrovsky L A , 1983 *Physica D.* **8** 223
[3] Erbay S and Suhubi E S, 1989 *Int. J. Engng.Sci.* **27** 915
[4] Erbay H A,1998 *Physica Scripta* **58** 9
[5] Muslu G M and Erbay H A, 2003 *Comput. Math. Appl.* **45** 503
[6] Ismail M S 2008 *Appl. Math. Comput.* **202** 520
[7] Ismail M S 2009 *Commun. Nonlinear Sci. Numer. Simul.* **14** 749
[8] Bellman R, Kashef B G and Casti J 1972 *J.of Comp. Phys.* **10** 40
[9] Shu C 2000 *Differential Quadrature and its application in engineering* (Springer-Verlag London Ltd.)
[10] Prenter P M 1975 *Splines and Variational Methods* (New York:John Wiley)
[11] Rubin S G and Graves R A 1975 *National aeronautics and space administration, Tech. Rep.*(Washington)

Journal of Physics: Conference Series **766** (2016) 012029 doi:10.1088/1742-6596/766/1/012029

A note on Berwald eikonal equation

Cumali Ekici, Muradiye Çimdiker

Eskisehir Osmangazi University, Mathematics-Computer Department, Eskisehir 26480, TURKEY
Kirklareli University, Mathematics Department, Kirklareli 39100, TURKEY

E-mail: cekici@ogu.edu.tr; muradiye.1001@hotmail.com

Abstract. In this study, firstly, we generalize Berwald map by introducing the concept of a Riemannian map. After that we find Berwald eikonal equation through using the Berwald map. The eikonal equation of geometrical optic that examining light reflects, refracts at smooth, plane interfaces is obtained for Berwald condition.

1. Introduction

Finsler manifolds can be thought as generalization of Riemannian manifolds; tangent spaces carry Minkowski norms instead of inner products and geometric objects on tangent vectors depend not only on the base but also on the fibre component. Finsler manifold have an intrinsic geometrical significance and also they have been used to model a variety of problems from dynamics, optics and relativity, cf. eg. [4] and [8].

One of the relevant figures working on that period on Finsler geometry was L. Berwald, who introduced a connection and a class of spaces sharing his name [7]. Especially, positive definite Berwald manifolds constitute the conceptually simplest and the best understood class of Finsler manifolds.

The eikonal equation is frequently solved using characteristic. This involves solving the ray equations in a combined coordinate and ray parameter phase space and then integrating the traveltime along these rays [10]. Physically rays are the trajectories along which high-frequency energy flows. A number of efficient methods for solving the ray equations have been developed [3]. Using the method of characteristics, traveltimes were computed along rays and approximate solutions of the eikonal equation could also obtained by R. L. Nowack [9].

∇f is used many areas of science such as mathematical physics and geometry. For example, the Riemannian condition $\|\nabla f\|^2 = 1$ is precisely the eikonal equation of geometrical optics. In the geometrical optical interpretation, the level sets of f are interpreted as wave fronts. The characteristics of the eikonal equation are then the solutions of the gradient flow equation for f,

$$x' = \operatorname{grad} f(x) \tag{1}$$

which are geodesics of M orthogonal to the level sets of f, and which are parametrized by arc length. These geodesics can be interpreted as light rays orthogonal to the wave fronts [1].

The concept of a Riemannian map was introduced by Fischer and it is shown that these maps are solutions of the eikonal equation [1]. The notion of semi-Riemannian map was stated, using the map, the solution of the eikonal equation was obtained by Garcio et al [5].

In this paper, we generalize these maps to Berwald manifolds by introducing the concept of a Riemannian map. We find Berwald eikonal equation though using the Berwald map.

2. Preliminaries

Let M be a connected, $n-$dimensional, C^∞ manifold and $TM=\cup_{x\in M} T_xM$ be the tangent bundle of M, where T_xM is the tangent space at $x \in M$. We denote a typical point in TM by $(x,\ y)$. Set $TM_0 = TM \setminus \{0\}$, where $\{0\}$ stands for $\{(x,0) : x \in X,\ 0 \in T_xM\}$. A Finsler metric on M is a function $F : V \to [0,\ \infty)$ with the following properties:

i) F is C^∞ on TM_0.

ii) At each point $x \in M$, the restriction $F_x : F\mid_{T_xM}$ is a Minkowski norm on T_xM. So the pair $(M,\ F)$ is called a Finsler manifold [11].

A Finsler metric F on a manifold M is called Berwald metric if in a standard local coordinate system $(x^i,\ y^i)$ in TM_0, the Christoffel symbols $\Gamma^i_{jk} = \Gamma^i_{jk}(x)$ are functions of $x \in M$ only, in which case, $G^i = \frac{1}{2}\Gamma^i_{jk}(x)y^jy^k$ are quadratics in $y = y^i\frac{\partial}{\partial x^i}\mid_x$. Riemannian metrics and Minkowski metrics are trivial Berwald metrics.

The Legendre transformation $l : TM \to T^*M$ is defined as

$$l(Y) = \begin{cases} g_Y(Y,\cdot), & Y \neq 0 \\ 0, & Y = 0, \end{cases}$$

where g is the symmetric bilinear form.

Let $f = M \to \mathbb{R}$ be a smooth function on M. The gradient of f is defined as $\nabla f = l^{-1}(df)$, where $l^{-1} : T^*M \to TM$ is the inverse Legendre transformation. Thus we have

$$df(X) = g_{\nabla f}(\nabla f,\ X),\ X \in TM. \tag{2}$$

[2].

Let M and M' be a Finsler manifolds with metric functions $F(x,\ y)$ and $F'(x',\ y')$. A differentiable map $f : M \to M'$ is local isometry if each of the tangent maps $f_* : T_xM \to T_{f(x)}M'$ is a vector space isomorphism which preserves the metric. A local isometry is an isometry if it is one to one and onto [6].

Let $(M,\ F)$ be a Berwald space. Given any parallel vector field V along a curve σ in M. We can express the Finslerian norm of V as

$$F(V) = \sqrt{g_V(V,\ V)} \tag{3}$$

where $g_V = g_{ij(\sigma,V)}dx^idx^j$ [4].

Let $f : (M,\ g_M) \to (N,\ g_N)$ be a smooth map between smooth finite dimensional Riemannian manifolds $(M,\ g_M)$ and $(N,\ g_N)$. Let

$$f_* : V_x \oplus H_x \to range f_* \oplus range(f_*)^\perp$$

denote the tangent map at $x \in M$. Here $V_x = \ker f_* \subseteq T_xM$ denotes the vertical subspace of T_xM and $H_x = \ker(f_*)^\perp \subseteq T_xM$ denotes the horizontal subspace of T_xM. If the horizontal restriction $(f_*)^h : H_x \to range f_*$ is a linear isometry between the inner product spaces $(H_x,\ g_M(x)\mid_{H_x})$ and $(range f_*,\ g_N(y)\mid_{range f_*})$, $y = f(x)$. The map is a Riemannian map if f is Riemannian at each $x \in M$ [1].

Theorem 2.1. Let $f : (M,\ g_M) \to (N,\ g_N)$ be a smooth map between smooth finite dimensional Riemannian manifolds $(M,\ g_M)$ and $(N,\ g_N)$, such that M is a connected manifold.

Then $rankf$ is constant on M and the norm squared $\|f_*\|^2$ of f_* satisfies

$$\|f_*\|^2 = rankf.$$

If in the above theorem $(N, g_N) = (\mathbb{R},\ 1) = \mathbb{R}$ is the real line with its Euclidean metric, and if $f : (M,\ g_M) \to \mathbb{R}$ is a real valued Riemannian map, then $rankf$ is constant on M equal to either zero or one. If $rankf = 0$, then f is a constant map, and if $rankf = 1$, then f satisfies the eikonal equation

$$\|f_*\|^2 = 1$$

of geometrical optics [1].

3. Berwald map and Berwald eikonal equation

Let M_1 and M_2 be differentiable, connected manifolds of dimensionals n_1 and n_2, respectively. Let $f : (M_1,\ F_1) \to (M_2,\ F_2)$ be a map, where F_1 and F_2 are Berwald metrics. Let $f_* : T_xTM_1 \to T_yTM_2$ denotes the tangent map at $x \in M_1,\ y = f(x) \in M_2$. Then T_xTM_1 and T_yTM_2 split orthogonally as

$$T_xTM_1 = V_xTM_1 \oplus H_xTM_1$$

and

$$T_yTM_2 = rangef_* \oplus range(f_*)^{\perp}.$$

Here

$$V_xTM_1 = \ker f_* \subseteq T_xTM_1$$

denotes the vertical subspace of T_xTM_1 and

$$H_xTM_1 = \ker(f_*)^{\perp} \subseteq T_xTM_1$$

denotes the horizontal subspace of T_xTM_1. Thus, we have

$$f_* : V_xTM_1 \oplus H_xTM_1 \to rangef_* \oplus range(f_*)^{\perp}.$$

Using these expression, we can write the following definitions:

Definition 3.1. A smooth map $f : (M_1,\ F_1) \to (M_2,\ F_2)$ between smooth finite dimensional $(M_1,\ F_1)$ and $(M_2,\ F_2)$ is Berwald at $x \in M_1$ if the horizontal restriction

$$(f_*)^h : H_xTM_1 \to rangef_*$$

is a linear isometry between $(H_xTM_1,\ F_1(x)\ |_{H_x})$ and $(rangef_*,\ F_2(y)\ |_{rangef_*})$, $y = f(x)$. The map is a Berwald map if f is Berwald at each $x \in M_1$.

Definition 3.2. Let $f : (M_1,\ F_1) \to (M_2,\ F_2)$ be a map. Then the square norm of f_* is the map $\|f_*\|^2 : M_1 \to \mathbb{R}^+$ defined by $\|f_*\|^2 (p_1) = \|f_{*p_1}\|^2$.

Note that, since $\|f_*\|^2 = \sum_{i=1}^{n_1} F_1^2(X_i) F_2^2(f_*(X_i))$, where $\{X_1, ..., X_{n_1}\}$ is an orthonormal local frame of TTM_1.

Journal of Physics: Conference Series **766** (2016) 012029 doi:10.1088/1742-6596/766/1/012029

Let us compute the square norm of a map $f:(M,\ F)\to(\mathbb{R}^+,\ dt\otimes dt)$. For this, let $\{X_1,...,X_n\}$ be an orthonormal local frame. We have

$$\|f_*\|^2=\sum_{i=1}^{n}F^2(X_i)(dt\otimes dt)(f_*X_i,\ f_*X_i)$$

From here, using the expression (2), then we find

$$\begin{aligned}\|f_*\|^2 &= \sum_{i=1}^{n}F^2(X_i)(g_{\nabla f}(\nabla f,\ X_i))^2(dt\otimes dt)\left(\frac{d}{dt}\circ f,\ \frac{d}{dt}\circ f\right)\\ &= g_{\nabla f}(\nabla f,\ \nabla f)\end{aligned}$$

From the equation (3), we have $\|f_*\|^2=F^2(\nabla f)$. The generalized eikonal equations for $f:(M,\ F)\to(\mathbb{R}^+,\ dt\otimes dt)$ become $\|f_*\|=F(\nabla f)=1$. This extension enables a fast calculation of geodesic paths.

Remark 3.1. According to the Theorem 2.1, for Berwald, this case is a geometrical optical condition which it describes light propagation in terms of rays.

Lemma 3.1. Let $f:(M,\ F)\to(\mathbb{R}^+,\ dt\otimes dt)$ be a map. Then

$$\nabla(F^2(\nabla f))=2\nabla_{\nabla f}\nabla f,$$

where ∇f and $\nabla(F^2(\nabla f))$ denote the gradients of f and $F^2(\nabla f)$ on Berwald manifolds.

Proof. If $X\in\Gamma TM$, then we have

$$g(\nabla(F^2(\nabla f)),\ X) = 2g_{\nabla f}(\nabla_X\nabla f,\ \nabla f)$$

Since $h_f(X)=\nabla_X\nabla f$, we find

$$g(\nabla(F^2(\nabla f)),\ X) = 2g_{\nabla f}(\nabla_{\nabla f}\nabla f,\ X).$$

Hence, this completes the proof.

Proposition 3.1. Let $f:(M,\ F)\to(\mathbb{R}^+,\ dt\otimes dt)$ be a map. If f satisfies a Berwald eikonal equation $F^2(\nabla f)=1$, then ∇f is a geodesic vector field on $(M,\ F)$.

Proof. For Berwald eikonal equation, we have $F^2(\nabla f)=1$. From Lemma 3.1., we find

$$\nabla(F^2(\nabla f))=2\nabla_{\nabla f}\nabla f=0. \tag{4}$$

Therefore, ∇f is a geodesic vector field on $(M,\ F)$. Namely, integral curves of the gradient flow of f are geodesics of the Berwald manifolds.

Proposition 3.2. Let $f:(M,\ F)\to(\mathbb{R}^+,\ dt\otimes dt)$ be a map. Then

i) The rank of f is constant on M.

ii) $F^2(\nabla f)$ is constant on M.

Proof. The rank of f is constant on M if and only if for every $p\in M$ either $\nabla f(p)=0$ or $\nabla f(p)\neq 0$. Here, we will only notice that $\nabla f(p)\neq 0$ for every $p\in M$.

i) Let $p,\ q\in M$ and $\gamma:[a,\ b]\to M$ be a curve with $\gamma(a)=p,\ \gamma(b)=q$. Since ∇f is a parallel vector field on $(M,\ F)$, $(\nabla f)\circ\gamma$ is parallel vector field along γ. By the uniqueness of parallel vector fields along a curve with respect to initial condition, $\nabla f(p)\neq 0\Leftrightarrow\nabla f(q)\neq 0$. Thus, we have $\nabla f(p)\neq 0$ for every $p\in M$.

ii) According to the expression (4), $F^2(\nabla f)$ is constant on M.

Journal of Physics: Conference Series **766** (2016) 012029 doi:10.1088/1742-6596/766/1/012029

4. Conclusion

In this work, we introduce Berwald map and study to explain the eikonal equation of geometrical optic for Berwald condition. In future work, we will consider affine solution of Berwald eikonal equation. First of all, the physical implications of affine solution of Berwald eikonal equation should be investigated. Last, an obvious next step would be to show how to use this theorem where generalizations of Berwald metric are employed in order to construct the eikonal equation in pseudo-Finsler space.

5. References

[1] A. E. Fischer, 1992, Riemannian Map Between Riemannian Manifolds, *Contemporary Math*, **182**, 342.
[2] B. Y. Wu, 2013, Comparison Theorems in Finsler geometry with Weighted Curvature Bounds and Related Results, *J. Korean Math. Soc.*, **52**(3), 603-624.
[3] B. Julian and D. Gubbins, 1977, Three Dimensional Seismic Ray Tracing, Geophys. J. Int., **43**, 95-113.
[4] D. Bao, S. S. Chern and Z. Shen, 2000, An Introduction to Riemann- Finsler Geometry, Springer-Verlag, Newyork.
[5] E. Garcio and D. N. Küpeli, 1999, Semi- Riemannian Maps and Their Applications, Kluwer Academic Publishers, London.
[6] J. K. Beem, 1971, Motions in Two Dimensional Indefinite Finsler Spaces, Indiana University Mathematics Journal, **21**(6).
[7] L. Berwald, 1925, Untersuchung der Krümmmung Allgemeiner Metrischen Raume auf Grund des in Ihnen Herrschenden Parallelismus, *Math. Zeit*, **25,** 40-73.
[8] P. L. Antonelli, R. S. Ingarden and M. Matsumoto, 1993, The Theory of Sprays and Finsler Spaces with Applications in Physics and Biology, Kluwer Academic Publishers.
[9] R. L. Nowack, 1992, Wavefronts and Solutions of the Eikonal Equation, Geophys. J. Int., **110**, 55-62.
[10] V. Cerveny, I. A. Molotkov and I. Psencik, 1977, Ray Method in Seismology, Praha, Univ. Karlova.
[11] Z. Shen, 2004, Landsberg Curvature, $S-$Curvature and Riemann Curvature, In: A Sampler of Finsler Geometry, MSRI Publications, **50**, Cambridge Univ. Press.

Journal of Physics: Conference Series **766** (2016) 012030 doi:10.1088/1742-6596/766/1/012030

Numerical Solutions of the General Rosenau–RLW Equation Using Meshless Kernel Based Method of Lines

Murat Arı, Yılmaz Dereli

Anadolu University, Mathematics Department, 26470
Eskişehir, Turkey

E-mail: murat_ari@anadolu.edu.tr, ydereli@anadolu.edu.tr

Abstract. In this paper, the General Rosenau–RLW equation is solved by using meshless kernel based method of lines. To test the accuracy of the method some numerical experiments are presented. Obtained numerical results are compared with some earlier works. It is seen that the method is very efficient and reliable due to obtained numerical results are very satisfactory.

1. Introduction

The General Rosenau RLW equation has following form

$$u_t - u_{xxt} + u_{xxxxt} + u_x + (u^p)_x = 0 \tag{1}$$

where $p \geq 2$ is an integer.

If $p = 2$, then Eq. (1) is called the usual Rosenau–RLW equation, [1, 2]. Moreover, if $p = 3$, then Eq. (1) is called the modified Rosenau–RLW equation and if $p \geq 4$ then it is called as the General Rosenau–RLW equation.

It is known that the solitary wave solution os Eq.(1) is

$$\begin{aligned} u(x,t) &= e^{ln\{(p+3)(3p+1)(p+1)/[2(p^2+3)(p^2+4p+7)]\}/(p+1)} \\ &\times \operatorname{sech}^{4/(p+1)}[\tfrac{(p-1)}{\sqrt{4p^2+8p+20}}(x-ct)] \end{aligned} \tag{2}$$

where $p \geq 2$ is an integer and

$$c = (p^4 + 4p^3 + 14p^2 + 20p + 25)/(p^4 + 4p^3 + 10p^2 + 12p + 21).$$

In the literature there a lot of studies about the Rosenau RLW equation see also references [3] - [10]. In this study as different from other publications evaluated numerical solutions of the General Rosenau–RLW equation by using the meshless kernel based method of lines (MKBMOL) will be presented. In this paper, we consider the following initial–boundary value problem of the general Rosenau–RLW equation with an initial condition:

$$u(x,0) = u_0(x) = 0, x_l \leq x \leq x_r \tag{3}$$

Journal of Physics: Conference Series **766** (2016) 012030 doi:10.1088/1742-6596/766/1/012030

and boundary conditions

$$u(x_l, t) = u(x_r, t) = 0, \tag{4}$$

The initial–boundary value problem possesses the following conservative properties:

$$Q(t) = \int_{x_l}^{x_r} u(x,t)dx = \int_{x_l}^{x_r} u_0(x,0)dx = Q(0),$$

and

$$E(t) = \|u\|_{L_2}^2 + \|u_x\|_{L_2}^2 + \|u_{xx}\|_{L_2}^2 = E(0).$$

When $-x_l \gg 0$ and $x_r \gg 0$, the initial boundary value problem (1)-(3) is consistent, so the boundary condition (3) is reasonable.

2. Application of the MKBMOL

In this section, the MKBMOL will be used to obtain the numerical solution of the Eq. (1) with initial and boundary conditions (3) and (4). In the method, time discretization isn't necessary and radial basis functions are used as kernel function. The method of lines approximates the solution u by a linear combination [11]

$$u(x,t) = \sum_{j=1}^{N} \alpha_j(t) v_j(x), \tag{5}$$

where $\alpha_j(t)$ is unknown and $v_j(x)$ is obtained by kernel functions such as Multiquadric [12], Gaussian or Wendland's [13] functions.
Multiquadric and Gaussian radial basis functions are defined as follows:

Multiquadric(MQ) $\quad \phi(r) = \sqrt{(\varepsilon r)^2 + 1}$

Gaussian(GA) $\quad \phi(r) = \exp(-r^2/\varepsilon^2)$

where $r = |x - x_j|$ is the Euclidean distance between collocation points x and x_j. The general form of compactly supported radial basis functions are defined as follows:

$$\phi_{l,k}(r) = (1-r)_+^n p(r) \tag{6}$$

where p is a prescribed polynomial for $k \geq 1$ with following conditions:

$$(1-r)_+^n = \begin{cases} (1-r)^n, & \text{if } 0 \leq r < 1 \\ 0, & \text{if } r \geq 1 \end{cases} \tag{7}$$

In our algorithms used following Wendland's functions are as follows:

$$\begin{aligned} \phi_{6,4}(r) &= (1-r)_+^{10}(5 + 50r + 210r^2 + 450r^3 + 429r^4), \\ \phi_{7,5}(r) &= (1-r)_+^{12}(9 + 108r + 566r^2 + 1644r^3 + 2697r^4 + 2048r^5). \end{aligned} \tag{8}$$

Derivatives in (1) with respect to time and space variables can be expressed as follows

$$u_t(x,t) = \sum_{j=1}^{N} \alpha_j'(t) v_j(x), \tag{9}$$

$$u_x(x,t) = \sum_{j=1}^{N} \alpha_j(t) v_j'(x), \tag{10}$$

$$u_{xxt}(x,t) = \sum_{j=1}^{N} \alpha_j'(t) v_j''(x), \tag{11}$$

$$u_{xxxxt}(x,t) = \sum_{j=1}^{N} \alpha_j'(t) v_j^{iv}(x), \tag{12}$$

Substituting the Eqs.(9-12) in to (1) we get

$$\begin{gathered}\sum_{j=1}^{N} \alpha_j'(t) v_j(x) + \sum_{j=1}^{N} \alpha_j'(t) v_j^{iv}(x) - \sum_{j=1}^{N} \alpha_j'(t) v_j''(x) + \sum_{j=1}^{N} \alpha_j(t) v_j'(x) \\ +[\sum_{j=1}^{N} \alpha_j(t) v_j'(x)]^p = 0\end{gathered} \tag{13}$$

The Eq.(13) can be written as follows

$$\sum_{j=1}^{N} (v_j(x) + v_j^{iv}(x) - v_j''(x))\alpha_j'(t) = -\sum_{j=1}^{N} \alpha_j(t) v_j(x) - [\sum_{j=1}^{N} \alpha_j(t) v_j'(x)]^p \tag{14}$$

The Eq.(13) is written as follows by using MATLAB notations

$$(V + V^{iv} - V'') * \alpha'(t) = -(V * \alpha(t)) - (V' * \alpha(t)). \wedge p \tag{15}$$

Also the Eq. (15) is rewritten as follows:

$$\alpha'(t) = -(V + V^{iv} - V'')^{-1}[(V * \alpha(t)) + (V' * \alpha(t)). \wedge p] \tag{16}$$

since matrices V and others are invertible. $V + V^{iv} - V''$ and vectors $\alpha(t)$ and $\alpha'(t)$ are defined as follows

$$\begin{aligned} &V := v_j(x_k) \\ &V' := v_j'(x_k) \\ &V'' := v_j''(x_k) \\ &V^{iv} = v_j^{iv}(x_k) \\ &\alpha(t) = [\alpha_1(t), \alpha_2(t), ...\alpha_n(t)]^T \\ &\alpha'(t) = [\alpha_1'(t), \alpha_2'(t), ...\alpha_n'(t)]^T \end{aligned} \tag{17}$$

for $1 \leq k \leq N$ and $1 \leq j \leq N$. Where the symbol "$*$" defines the pointwise product, the symbol "$. \wedge p$" defines the component-by-component multiplication of vectors. Eq.(16) is an ordinary differential equation with respect to $\alpha(t)$ and it is solved by using any ordinary differential equation solver in MATLAB.

3. Numerical examples

In this section, numerical experiments on a test problem to confirm and illustrate the accuracy of our proposed method will be presented. The accuracy of the method is measured by the root mean square error L_2 and maximum error L_∞. In the tables comparisons with numerical solutions references [9]-[10] are presented. Where L_2 and L_∞ are calculated as follows:

$$L_2 = \sqrt{h \sum_{j=1}^{N} \left| u_j^{exact} - u_j^{num.} \right|^2}, \quad L_\infty = \max_{1 \le j \le N} \left| u_j^{exact} - u_j^{num.} \right|.$$

The results in term of errors for different values of p at time $T = 40$ for $\Delta t = 0.1$ at the interval $(x_l, x_r) = (-60, 120)$ with $h = 0.5$ are reported in Tables 1-3. Obtained results with the MQ, G and Wendland's functions are compared with some previous results. The method by using MQ, G and Wendland's functions provides higher accuracy then compared with other numerical results in [9] and [10]. Especially method with the W(7,5) function provided least error and use of the MQ provided the highest errors between the used kernel functions interms of the error norms.

The invariants remain almost constant when compared with analytical values of invariants. Calculated numerical results are very satisfactorily. The motion of single solitary waves are plotted in Figures 1-3. The program is run up to time T=40 over the solution domain. Initially the peak of the solitary wave was positioned at $x = 0$ and at the end of running time location of peak position of the wave moves to the right-side with the preserved amplitude and shape while time increasing. In Figures 1-3, the motion of single solitary wave motion can be seen.

Table 1. Comparison of invariants and error norms with p=4 and h=0.5

Method	h	Δt	T	L_2	L_∞	Q	E
Analytical						6.265806	2.8676946
Wendland 7-5	0.5	0.1	40	0.00093510	0.00035069	6.266377	2.868226
Wendland 6-4	0.5	0.1	40	0.00002355	0.00001031	6.265844	2.867735
Gaussian	0.5	0.1	40	0.00066954	0.00029706	6.265806	2.867684
Multiquadric	0.5	0.1	40	0.00110452	0.00042250	6.265992	2.867617
[9]	0.5	0.1	40	0.00447881	0.00171122		
[10]	0.5	0.1	40	0.07451730	0.02787120		

Table 2. Comparison of invariants and error norms with p=8 and h=0.5

Method	h	Δt	T	L_2	L_∞	Q	E
Analytical						9.742086	4.735164
Wendland 7-5	0.5	0.1	40	0.00038078	0.00013784	9.742126	4.735346
Wendland 6-4	0.5	0.1	40	0.00007522	0.00002949	9.742181	4.735225
Gaussian	0.5	0.1	40	0.00170387	0.00062856	9.742146	4.735302
Multiquadric	0.5	0.1	40	0.00127623	0.00047892	9.742227	4.735082
[9]	0.5	0.1	40	0.00431841	0.00161891		
[10]	0.5	0.1	40	0.08037300	0.02953370		

Table 3. Comparison of invariants and error norms with p=16 and h=0.5

Method	h	Δt	T	L_2	L_∞	Q	E
Analytical						17.148841	8.375324
Wendland 7-5	0.5	0.1	40	0.00233340	0.00044109	17.168699	8.375376
Wendland 6-4	0.5	0.1	40	0.00231987	0.00044493	17.169258	8.375400
Gaussian	0.5	0.1	40	0.00302310	0.00053860	17.172776	8.375393
Multiquadric	0.5	0.1	40	0.00762184	0.00227095	17.116828	8.375272
[9]	0.5	0.1	40	0.00357250	0.00118750		
[10]	0.5	0.1	40	0.06130440	0.02254710		

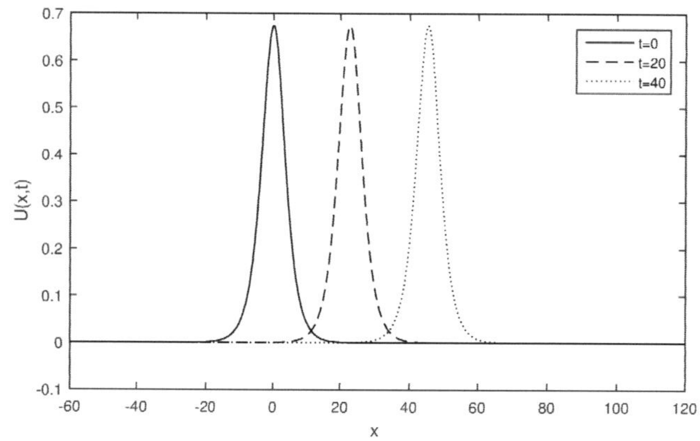

Figure 1. Motion of the single solitary wave for $p = 4$.

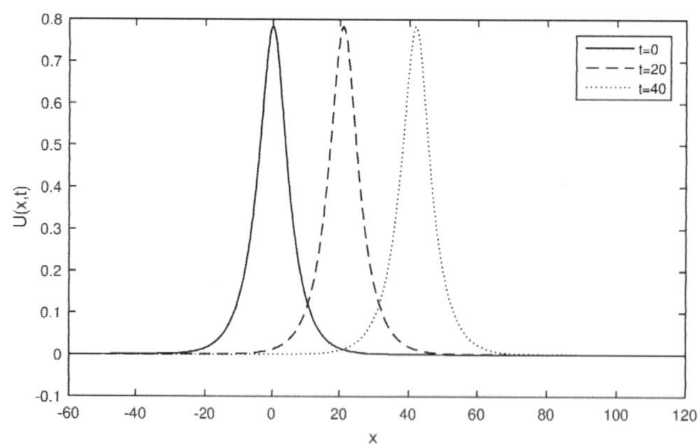

Figure 2. Motion of the single solitary wave for $p = 8$.

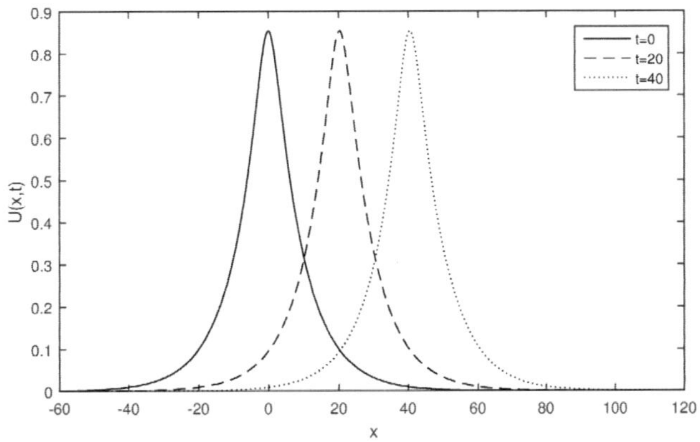

Figure 3. Motion of the single solitary wave for $p = 16$.

Conclusions

In this study the General Rosenau–RLW equation was solved by using a meshless technique which was MKBMOL. The main advantage of this technique there was no linearization to the nonlinear part of given equation. This equation was converted to a ordinary differential equation system and solved by using MATLAB codes. Here we present the numerical results of the Adams-Bashford multi-step method but any one can be present some numerical results by using other one step or multi step methods since MATLAB uses a ode solver package. From numerical results it is seen that the technique gives very reliable and sensitive result with all used kernel functions. Therefore it is said that this method is very reliable and applicable to this type equations. Also for different form of this type evaluation equations and systems this method can be applied at the future works.

References

[1] Rosenau P 1986 A quasi-continuous description of a nonlinear transmission line *Physica Scripta* **34** pp 827-29
[2] Rosenau P 1988 Dynamics of dense discrete systems *Progress of Theoretical Physics* **79** pp 1028-42
[3] Park M A 1990 On the Rosenau equation *Mathematica Aplicada e Computacional* **9** pp 145-52
[4] Park M A 1992 Pointwise decay estimate of solutions of the generalized Rosenau equation *Journal of the Korean Mathematical Society* **29** pp 26180
[5] Zuo J M Zhang Y M Zhang T D and Chang F 2010 A new conservative difference scheme for the general Rosenau-RLW equation *Boundary Value Problems* **2010** p 13
[6] Pan X Zhang L 2012 On the convergence of a conservative numerical scheme for the usual Rosenau-RLW equation *Applied Mathematical Modelling* **36** pp 3371-78
[7] Pan X Zheng K Zhang L 2013 Finite difference discretization of the Rosenau-RLW equation *Applicable Analysis* **92** pp 257889
[8] Atouani N Omrani K 2013 Galerkin finite element method for the Rosenau-RLW equation *Computers and Mathermatics with Applications* **66** pp 289-303
[9] Wongsaijai B Poochinapan K and Disyadej T 2014 A compact finite difference method for solving the general Rosenau-RLW equation *IAENG International Journal of Applied Mathematics* p 8
[10] X. Pan and L. Zhang 2012 Numerical simulation for general Rosenau-RLW equation: an average linearized conservative scheme *Mathematical Problems in Engineering* **2012** p 15
[11] Schaback, R. The Meshless Kernel-Based Method of Lines for Solving Nonlinear EvolutionEquations. Preprint, Gt-tingen, 2008.
[12] Rolland H L 1971 Multiquadric equations of topography and other irregular surfaces *J. Geophys. Res.* **176** pp 1905-15
[13] Wendland H 1995 Piecewise polynomial, positive definite and compactly supported radial functions of minimal degree *Advances in Comput.Math.* **4** pp 389-96

Journal of Physics: Conference Series **766** (2016) 012031 doi:10.1088/1742-6596/766/1/012031

Wave Propagation by Way of Exponential B-Spline Galerkin Method

M Zorsahin Gorgulu[1], **I Dag**[2] **and D Irk**[1]

[1] Department of Mathematics-Computer and [2] Department of Computer Engineering, Eskisehir Osmangazi University, 26480, Eskisehir, Turkey.

E-mail: mzorsahin@ogu.edu.tr, idag@ogu.edu.tr, dirk@ogu.edu.tr

Abstract. In this paper, the exponential B-spline Galerkin method is set up for getting the numerical solution of the Burgers' equation. Two numerical examples related to shock wave propagation and travelling wave are studied to illustrate the accuracy and the efficiency of the method. Obtained results are compared with some early studies.

1. Introduction

Burgers' equation in which convection and diffusion play an important role arises in applications such as meteorology, turbulent flows, modelling of the shallow water. Burgers' equation is considered to be useful model for many physical problems. Thus it is often studied for testing of both real life problems and computational techniques. Not only does exact solutions of nonlinear convective problem develops discontinuities in finite time, and might display complex structure near discontinuities. Efficient and accurate methods are in need to be tackled the complex solutions of the Burgers' equation. Though analytical solutions of the Burgers' equation exist for simple initial condition, the numerical techniques are of interest to meet requirement of the wide range of solutions of the Burgers' equation. Some variants of the spline methods have set up to find the numerical solutions of the Burgers' equation such as Galerkin finite element method [3, 8, 12], least square method [2], collocation method [4–7, 11], method based on the cubic B-spline quasi interpolant [9, 10], etc.

Finite element methods are mainly used methods to have good functional approximate solutions of the differential equations. The accuracy of the finite element solutions are increased by the selection of suitable basis function for the approximate function over the finite intervals. The exponential B-splines are suggested to interpolate data and function exhibiting sharp variations [1], since polynomial B-splines based interpolation cause unwanted osculation for interpolation. Some solutions of the Burgers' equation show sharpness. Thus we will construct the finite element method together with the exponential B-splines to have solutions of the Burgers' equation. A few exponential B-spline numerical methods have suggested for some partial differential equations such as [13, 14].

In this study, we will consider the Burgers' equation

$$u_t + uu_x - \nu u_{xx} = 0 \tag{1}$$

where subscripts x and t are space and time parameters, respectively and ν is the viscosity coefficient. Boundary conditions of the Eq. (1) are chosen from

$$u(a,t) = \beta_1, \ u(b,t) = \beta_2, \tag{2}$$

and initial condition is

$$u(x,0) = f(x), \ x \in [a,b]. \tag{3}$$

$f(x)$ and β_1, β_2 constants are described in the computational section.

2. Exponential B-splines Galerkin Finite Element Solution

Divide spatial interval $[a,b]$ in N subintervals of length $h = \frac{b-a}{N}$ and $x_i = x_0 + ih$ at the knots x_i, $i = 0, .., N$ and time interval $[0,T]$ in M interval of length Δt.

Let $\phi_i(x)$ be the exponential B-splines defined at the knots x_i, $i = 0, \ldots, N$, together with fictitious knots x_i, $i = -3, -2, -1, N+1, N+2, N+3$ outside the interval $[a,b]$. The $\phi_i(x)$, $i = -1, \ldots, N+1$ can be defined as

$$\phi_i(x) = \begin{cases} b_2\left[(x_{i-2} - x) - \frac{1}{p}(\sinh(p(x_{i-2} - x)))\right] & \text{if } x \in [x_{i-2}, x_{i-1}]; \\ a_1 + b_1(x_i - x) + c_1 e^{p(x_i - x)} + d_1 e^{-p(x_i - x)} & \text{if } x \in [x_{i-1}, x_i]; \\ a_1 + b_1(x - x_i) + c_1 e^{p(x - x_i)} + d_1 e^{-p(x - x_i)} & \text{if } x \in [x_i, x_{i+1}]; \\ b_2\left[(x - x_{i+2}) - \frac{1}{p}(\sinh(p(x - x_{i+2})))\right] & \text{if } x \in [x_{i+1}, x_{i+2}]; \\ 0 & \text{otherwise} \end{cases} \tag{4}$$

where

$$p = \max_{0 \le i \le N} p_i, \ s = \sinh(ph), \ c = \cosh(ph), \ a_1 = \frac{phc}{phc - s}, \ b_1 = \frac{p}{2}\left[\frac{c(c-1) + s^2}{(phc - s)(1-c)}\right],$$
$$b_2 = \frac{p}{2(phc - s)}, \ c_1 = \frac{1}{4}\left[\frac{e^{-ph}(1-c) + s(e^{-ph} - 1)}{(phc - s)(1-c)}\right], \ d_1 = \frac{1}{4}\left[\frac{e^{ph}(c-1) + s(e^{ph} - 1)}{(phc - s)(1-c)}\right].$$

The $\phi_i(x)$, $i = -1, \ldots, N+1$ forms a basis for functions defined on the interval $[a,b]$. The Galerkin method consists of seeking approximate solution in the following form:

$$u(x,t) \approx U(x,t) = \sum_{i=-1}^{N+1} \phi_i(x)\delta_i(t) \tag{5}$$

where $\delta_i(t)$ are time dependent unknowns to be determined from the boundary conditions and Galerkin approach to the Eq. (1). The approximate solution and the first two derivatives at the knots can be found from the (4-5) as

$$\begin{aligned} U_i &= U(x_i,t) = \alpha_1\delta_{i-1} + \delta_i + \alpha_1\delta_{i+1}, \\ U_i' &= U'(x_i,t) = \alpha_2\delta_{i-1} - \alpha_2\delta_{i+1}, \\ U_i'' &= U''(x_i,t) = \alpha_3\delta_{i-1} - 2\alpha_3\delta_i + \alpha_3\delta_{i+1} \end{aligned} \tag{6}$$

where $\alpha_1 = \frac{s-ph}{2(phc-s)}, \alpha_2 = \frac{p(1-c)}{2(phc-s)}, \alpha_3 = \frac{p^2 s}{2(phc-s)}$.

Over the sample interval $[x_m, x_{m+1}]$, applying Galerkin approach to Eq. (1) with the weight function $\phi_j(x)$ yields

$$\int_{x_m}^{x_{m+1}} \phi_j(x)(u_t + uu_x - \nu u_{xx})\,dx. \tag{7}$$

Substitution of the (6) into the integral equation lead to

$$\sum_{i=m-1}^{m+2}\left(\int_{x_m}^{x_{m+1}}\phi_j\phi_i dx\right)\overset{\bullet}{\delta}_i+\left(\int_{x_m}^{x_{m+1}}\phi_j\left(\sum_{k=m-1}^{m+2}\delta_k\phi_k\right)\phi_i' dx\right)\delta_i-\nu\left(\int_{x_m}^{x_{m+1}}\phi_j\phi_i'' dx\right)\delta_i, \quad (8)$$

where i, j and k take only the values $m-1, m, m+1, m+2$ for $m = 0, 1, \ldots, N-1$ and the notation $\bullet$ denotes time derivative.

If we denote A^e_{ji}, $B^e_{jki}(\delta^e)$ and C^e_{ji} by

$$A^e_{ji}=\int_{x_m}^{x_{m+1}}\phi_j\phi_i dx,\quad B^e_{jki}(\delta)=\int_{x_m}^{x_{m+1}}\phi_j\left(\sum_{k=m-1}^{m+2}\delta_k\phi_k\right)\phi_i' dx,\quad C^e_{ji}=\int_{x_m}^{x_{m+1}}\phi_j\phi_i'' dx \quad (9)$$

where $\mathbf{A}^e$ and $\mathbf{C}^e$ are the element matrices of which dimensions are 4×4 and $\mathbf{B}^e(\delta^e)$ is the element matrix with the dimension $4\times 4\times 4$, (8) can be written in the matrix form as

$$\mathbf{A}^e\overset{\bullet}{\delta^e}+\left(\mathbf{B}^e(\delta^e)-\nu\mathbf{C}^e\right)\delta^e, \quad (10)$$

where $\delta^e = (\delta_{m-1}, \ldots, \delta_{m+2})^T$.

Gathering the systems (10) over all elements, we obtain global system

$$\mathbf{A}\overset{\bullet}{\delta}+\left(\mathbf{B}(\delta)-\nu\mathbf{C}\right)\delta=0 \quad (11)$$

where $\mathbf{A}, \mathbf{B}(\delta), \mathbf{C}$ are derived from the corresponding element matrices $\mathbf{A}^e, \mathbf{B}^e(\delta^e), \mathbf{C}^e$, respectively and $\delta = (\delta_{-1}, \ldots, \delta_{N+1})^T$ contains all elements parameters.

The unknown parameters δ are interpolated between two time levels n and $n+1$ with the Crank-Nicolson method

$$\delta=\frac{\delta^{n+1}+\delta^n}{2},\quad \overset{\bullet}{\delta}=\frac{\delta^{n+1}-\delta^n}{\Delta t}.$$

Then we obtain iterative formula for the time parameters δ^n

$$\left[\mathbf{A}+\frac{\Delta t}{2}\left(\mathbf{B}(\delta^{n+1})-\nu\mathbf{C}\right)\right]\delta^{n+1}=\left[\mathbf{A}-\frac{\Delta t}{2}\left(\mathbf{B}(\delta^n)-\nu\mathbf{C}\right)\right]\delta^n. \quad (12)$$

The set of equations consist of $(N+3)$ equations with $(N+3)$ unknown parameters. Boundary conditions must be adapted into the system. Because of the this requirement, initially the first and last equations are eliminated from the (12) and parameters δ^{n+1}_{-1} and δ^{n+1}_{N+1} are substituted in the remaining system (12) by using following equations

$$u(a,t)=\alpha_1\delta^{n+1}_{-1}+\delta^{n+1}_0+\alpha_1\delta^{n+1}_1=\beta_1,\quad u(b,t)=\alpha_1\delta^{n+1}_{N-1}+\delta^{n+1}_N+\alpha_1\delta^{n+1}_{N+1}=\beta_2$$

which are obtained from the boundary conditions. Thus we obtain a septa-diagonal matrix with the dimension $(N+1)\times(N+1)$. Since the system (12) is an implicit system together with the nonlinear term $\mathbf{B}(\delta^{n+1})$, we have used the following inner iteration at each time step $(n+1)\Delta t$ to work up solutions:

$$(\delta^*)^{n+1}=\delta^n+\frac{(\delta^n-\delta^{n-1})}{2}. \quad (13)$$

We use the above iteration three times to find the new approximation $(\delta^*)^{n+1}$ for the parameters δ^{n+1} to recover solutions at time step $(n+1)\Delta t$.

Journal of Physics: Conference Series **766** (2016) 012031 doi:10.1088/1742-6596/766/1/012031

To start evolution of the iterative system for the unknown δ^n, the vector of initial parameters δ^0 must be determined by using the following initial and boundary conditions:

$$u'(x_0,0)=\frac{p(1-c)}{2(phc-s)}\delta_{-1}+\frac{p(c-1)}{2(phc-s)}\delta_1,\ u'(x_N,0)=\frac{p(1-c)}{2(phc-s)}\delta_{N-1}+\frac{p(c-1)}{2(phc-s)}\delta_{N+1}$$
$$u(x_m,0)=\frac{s-ph}{2(phc-s)}\delta_{m-1}+\delta_m+\frac{s-ph}{2(phc-s)}\delta_{m+1},\ m=0,...,N. \tag{14}$$

The solution of matrix equation (14) with the dimensions $(N+1)\times(N+1)$ is obtained by the way of Thomas algorithm. Once δ^0 is determined, we can start the iteration of the system to find the parameters δ^n at time $t^n=n\Delta t$. Approximate solutions at the knots are found from the (6).

3. Test Problems

The robustness of the algorithm is shown by studying two test problems. Error is measured by the maximum error norm;

$$L_\infty=\left\|u^{\text{exact}}-u^{\text{numeric}}\right\|_\infty=\max_{0\le j\le N}\left|u_j^{\text{exact}}-u_j^{\text{numeric}}\right|. \tag{15}$$

The free parameter p of the exponential B-spline is found by scanning the predetermined interval with very small increment.

(a) A shock propagation solution of the Burgers' equation is

$$u(x,t)=\frac{x/t}{1+\sqrt{t/t_0}\exp(x^2/(4\nu t))},\qquad t\ge 1, \tag{16}$$

where $t_0=\exp(1/(8\nu))$. The sharpness of the solutions increase with selection of the smaller ν.

Substitution of the $t=1$ in (16) gives the initial condition. The boundary conditions $u(0,t)=0$ and $u(1,t)=0$ are used. Computations are performed with parameters $\nu=0.0005$, 0.005, 0.01, $h=0.02$, 0.005 and $\Delta t=0.01$ over the solution domain [0, 1]. As time increases, shock evaluation is observed and some graphical solutions are drawn in Figs. 1 and 2 for various viscosity values and space steps. For $\nu=0.01$, algorithm produces smoother shock during run time. With decreasing values of ν, as seen in the Fig. 2 the steepening occurs. For the smaller viscosity constant $\nu=0.0005$, the sharper shock is observed and steepness of numerical solution is kept almost unchanged during the program run. The results obtained by present scheme can be compared with previous ones through the computation of error norm L_∞ at various times in the Table 1.

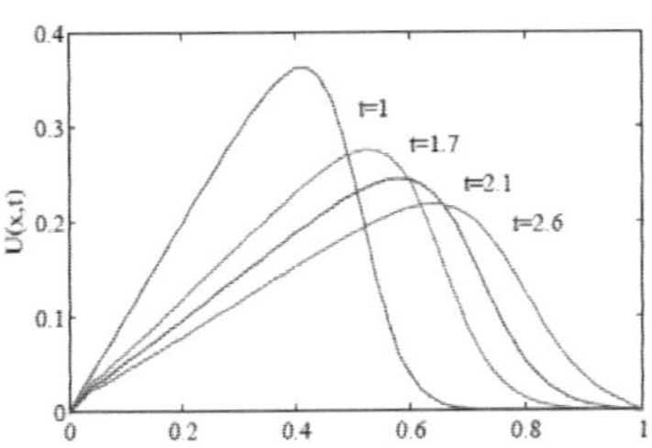

Figure 1. Solutions for $\nu=0.01$, $h=0.02$, $p=0.005111$.

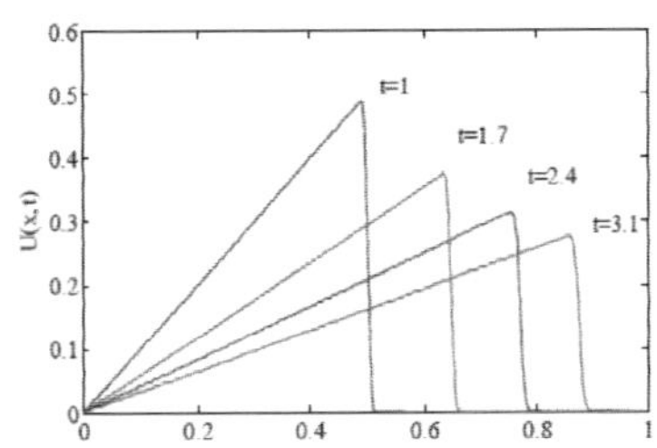

Figure 2. Solutions for $\nu=0.0005$, $h=0.005$, $p=0.005941$.

Table 1. Comparison of numerical results at different times.

	$L_\infty \times 10^3$	$L_\infty \times 10^3$	$L_\infty \times 10^3$
$h = 0.005$, $\nu = 0.005$	$t = 1.7$	$t = 2.4$	$t = 3.1$
Present ($p = 0.005941$)	3.15776	2.33757	4.79061
Ref. [3] (QBGM)	1.20755	0.80187	4.79061
Ref. [6] (QBCM1)	0.06192	0.05882	4.43469
Ref. [7] (QBCA1)	1.21175	0.80771	4.79061
Ref. [8]	0.04284	0.06464	4.79061
$h = 0.02$, $\nu = 0.01$	$t = 1.7$	$t = 2.1$	$t = 2.6$
Present ($p = 0.005111$)	8.08651	7.53518	8.06798
Ref. [5]	3.13476	2.66986	8.06798
Ref. [6] (QBCM1)	0.40431	0.86363	6.69425
Ref. [7] (QBCA1)	0.47456	1.14759	8.06798
Ref. [8]	0.09592	1.14760	8.06799

The absolute error distributions between the analytical and numerical solutions are drawn in Figs. 3 and 4 for various viscosity values and space steps.

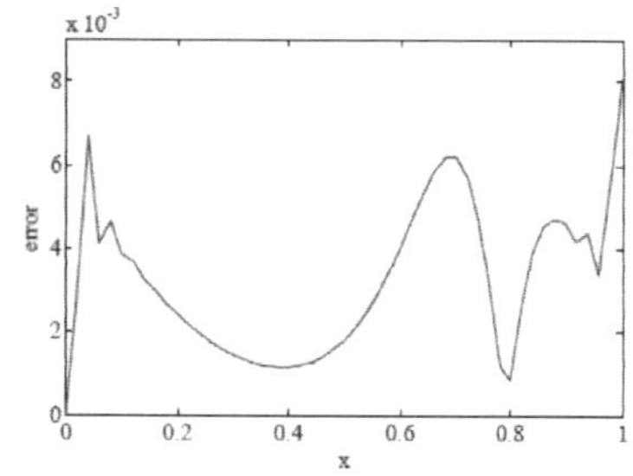

Figure 3. Absolute error for $\nu = 0.01$, $h = 0.02$, $p = 0.005111$.

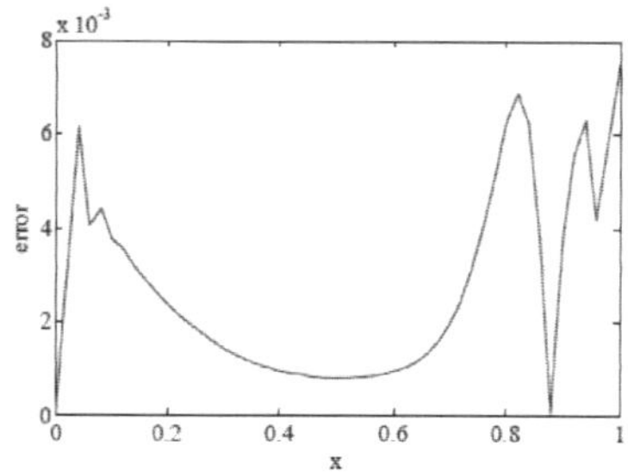

Figure 4. Absolute error for $\nu = 0.005$, $h = 0.02$, $p = 0.000739$.

(b) A well-known analytical solution of Burgers' equation is

$$u(x,t) = \frac{\alpha + \mu + (\mu - \alpha)\exp\eta}{1 + \exp\eta}, \quad 0 \le x \le 1,\ t \ge 0, \tag{17}$$

where $\eta = \dfrac{\alpha\,(x - \mu t - \gamma)}{\nu}$. α, μ and γ are constants. Parameters $\alpha = 0.4$, $\mu = 0.6$ and $\gamma = 0.125$ are used to coincide with the some previous studies. This solution involves a travelling wave and move to the right with speed μ. Initial condition is obtained from (17) when $t = 0$. The boundary conditions are $u(0,t) = 1$, $u(1,t) = 0.2$ for $t > 0$.

The calculation is performed with time step $\Delta t = 0.01$, space step $h = 1/36$ and viscosity coefficient $\nu = 0.01$. The program is run up to time $t = 0.5$. We have found $L_\infty = 6.73543978 \times 10^{-4}$ for the exponential B-spline Galerkin method at time $t = 0.5$ documented in Table 2 with results of the quadratic B-spline Galerkin method [3], the quartic B-spline collocation method [6], the quintic B-spline collocation method [7] and the quartic B-spline Galerkin method [8].

The numerical solution obtained by the present scheme gives better results than the others. The profiles of initial wave and solution at some times are depicted in Fig. 5. Error variations of the scheme are given in Fig. 6 at time $t = 0.5$.

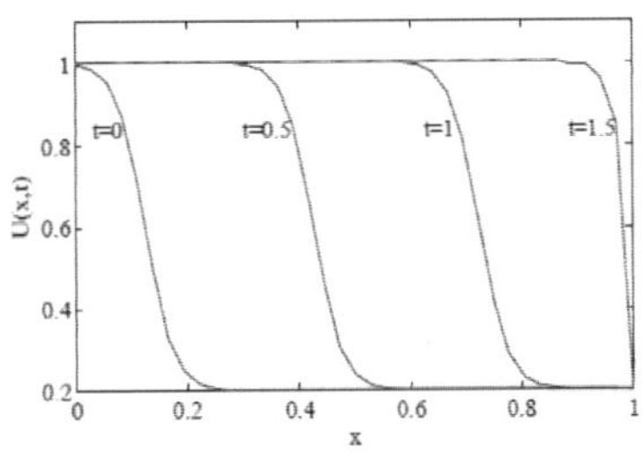

Figure 5. Solutions for $\nu = 0.01$.

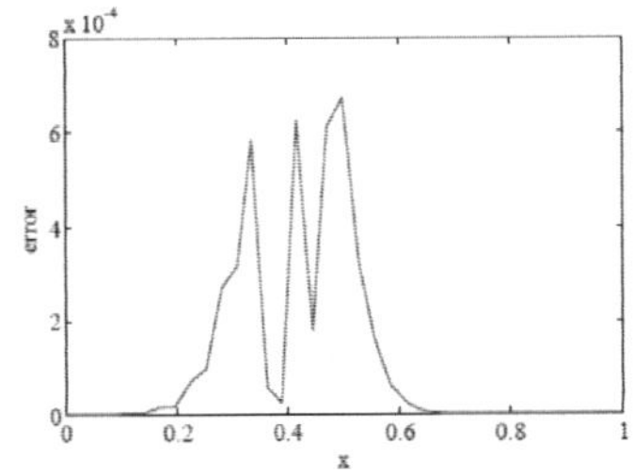

Figure 6. Absolute error for $\nu = 0.01$.

Table 2. Comparison of results at $t = 0.5$ for $h = 1/36$, $\nu = 0.01$.

	Present ($p = 0.002323$)	Ref. [3] (QBGM)	Ref. [6] (QBCM1)	Ref. [7] (QBCA1)	Ref. [8] (QBGM)
$L_\infty \times 10^3$	0.67354	6.35489	3.03817	5.78454	1.44

4. Conclusion

In this paper, we investigate the utility of the exponential B-spline in the Galerkin algorithm for solving the Burgers' equation. The efficiency of the method is tested for a shock propagation solution and a travelling solution of the Burgers' equation. For the first test problem, solutions found with the present methods are in good agreement with the results obtained by previous studies. In the second test problem, present method leads to accurate results than all of the others. In conclude, the numerical algorithm in which the exponential B-spline functions are used, performs well compared with other existing numerical methods for the solution of Burgers' equation.

Acknowledgments

The authors are grateful to The Scientific and Technological Research Council of Turkey for financial support given for the project 113F394. This paper is presented at the conference of ICQSA 2016.

References

[1] McCartin B J 1991 *J. Approx. Theory.* **66** 1–23
[2] Kutluay S, Esen A and Dag I 2004 *J. Comput. Appl. Math.* **167** 21–33
[3] Dag I, Saka B and Boz A 2005 *Appl. Math. Comput.* **166** 506–22
[4] Dag I, Irk D and Sahin A 2005 *Math. Probl. Eng.* **2005** 521–38
[5] Ramadan M A, El-Danaf T S and Abd Alaal F E I 2005 *Chaos. Soliton. Fract.* **26** 795–804
[6] Saka B and Dag I 2007 *Chaos. Soliton. Fract.* **32** 1125–37
[7] Saka B and Dag I 2008 *J. Franklin. I.* **345** 328–48
[8] Saka B and Dag I 2009 *Appl. Math. Comput.* **215** 746–58
[9] Zhu C and Wang R 2009 *Appl. Math. Comput.* **208** 260–72
[10] Jiang Z and Wang R 2010 *J. Inform. Comput. Sci.* **7** 1013–21
[11] Mittal R C and Jain R K 2012 *Appl. Math. Comput.* **218** 7839–55
[12] Chapani H V, Pradhan V H and Mehta M N 2012 *Int. J. Appl. Math. Mech.* **8** 18–32
[13] Mohammadi R 2013 *Appl. Math.* **4** 933–44
[14] Dag I and Ersoy O 2015 *AIP Conf. P.* **1648** 370008

Journal of Physics: Conference Series **766** (2016) 012032 doi:10.1088/1742-6596/766/1/012032

Cubic Trigonometric B-spline Galerkin Methods for the Regularized Long Wave Equation

Dursun Irk and Pınar Keskin

Eskisehir Osmangazi University, Department of Mathematics-Computer, 26480, Eskisehir, Turkey.

E-mail: `dirk@ogu.edu.tr;pinarkeskin.math@gmail.com`

Abstract. A numerical solution of the Regularized Long Wave (RLW) equation is obtained using Galerkin finite element method, based on Crank Nicolson method for the time integration and cubic trigonometric B-spline functions for the space integration. After two different linearization techniques are applied, the proposed algorithms are tested on the problems of propagation of a solitary wave and interaction of two solitary waves.

1. Introduction

The RLW equation

$$u_t + u_x + \varepsilon u u_x - \mu u_{xxt} = 0, \tag{1}$$

was formulated by Peregrine as an alternative to KdV equation for studying soliton phenomenon with boundary conditions $u \to 0$ as $x \to \pm\infty$ [1]. In the equation, ε and μ are positive constants and u, x, t denote the amplitude, spatial and time coordinate, respectively. In this study the RLW equation will be considered with boundary condition over the space region as

$$u(a,t) = u(b,t) = u_x(a,t) = u_x(b,t) = 0, \quad t \in (0,T] \tag{2}$$

and the initial condition

$$u(x,0) = f(x) \tag{3}$$

will be defined in the numerical experiments section.

Since the RLW equation has been solved analytically only for restricted set of boundary and initial conditions [2], the numerical solution of the RLW equation has been the subject of many papers over the last few years [3]-[7]. In this paper, the purpose of this study is to obtain numerical solution of the RLW equation based on the application of Crank Nicolson methods for the time discretization and cubic trigonometric B-spline Galerkin method for the space discretization using different two linearization techniques.

2. Cubic Trigonometric B-spline Galerkin Method

For computational work, the space-time plane is discretized by grid with the time step Δt and space step h. The exact solution of the unknown function at the grid points is denoted by

$$u(x_m, t_n) = u_m^n, \; m = 0, 1, \ldots, N; \qquad n = 0, 1, 2, \ldots$$

where $x_m = a + mh$, $t_n = n\Delta t$ and the notation U_m^n is used to represent the numerical value of u_m^n. For the space discretization, the space domain $[a, b]$ is discretized into partitions of N finite elements of equal length h by the knots

Using the Crank-Nicolson method for the time discretization of the RLW equation, we have

$$u^{n+1} + \frac{\Delta t}{2}(u_x)^{n+1} + \frac{\Delta t}{2}\varepsilon u^{n+1}(u_x)^{n+1} - \mu(u_{xx})^{n+1} = u^n - \mu(u_{xx})^n - \frac{\Delta t}{2}(u_x)^n - \frac{\Delta t}{2}\varepsilon u^n(u_x)^n. \tag{4}$$

Using the recurrence relation given in [8, 9], the cubic trigonometric B-spline functions are defined at these knots as

$$T_m(x) = \frac{1}{\theta}\begin{cases} g^3(x_{m-2}) & ,[x_{m-2}, x_{m-1}) \\ -g^2(x_{m-2})g(x_m) - g(x_{m-2})g(x_{m+1})g(x_{m-1}) - g(x_{m+2})g^2(x_{m-1}) & ,[x_{m-1}, x_m) \\ g(x_{m-2})g^2(x_{m+1}) + g(x_{m+2})g(x_{m-1})g(x_{m+1}) + g^2(x_{m+2})g(x_m) & ,[x_m, x_{m+1}) \\ -g^3(x_{m+2}) & ,[x_{m+1}, x_{m+2}) \\ 0 & ,\text{otherwise} \end{cases} \tag{5}$$

where

$$\theta = \sin(\frac{h}{2})\sin(h)\sin(\frac{3h}{2}), \quad g(x_m) = \sin\left(\frac{x - x_m}{2}\right).$$

Since from (5) each cubic trigonometric B-splines covers 4 intervals, each element $[x_m, x_{m+1}]$ is covered by four splines. Therefore over the element $[x_m, x_{m+1}]$, an approximation to the exact solution $u(x, t)$ in terms of cubic trigonometric B-splines can be written as

$$U(x,t) = \sum_{j=m-1}^{m+2} T_j(x)\delta_j(t) = T_{m-1}\delta_{m-1} + T_m\delta_m + T_{m+1}\delta_{m+1} + T_{m+2}\delta_{m+2}. \tag{6}$$

Using (5) and the trial solution (6) the values of $U_m = U(x_m, t)$ and their first two space derivatives at knots can be written in terms of the parameters δ_m as

$$U_m = \sin^2\left(\frac{h}{2}\right)\csc(h)\csc\left(\frac{3h}{2}\right)(\delta_{m-1} + \delta_{m+1}) + \left(\frac{2}{1 + 2\cos(h)}\right)\delta_m, \tag{7}$$

$$U'_m = \left(\frac{3}{4}\csc(\frac{3h}{2})\right)(-\delta_{m-1} + \delta_{m+1}), \tag{8}$$

$$U''_m = \left(\frac{3\left(3\cos^2\left(\frac{h}{2}\right) - 1\right)}{4\left(\sin(h)\sin\left(\frac{3h}{2}\right)\right)}\right)(\delta_{m-1} + \delta_{m+1}) - \left(\frac{3\cot^2\left(\frac{h}{2}\right)}{(2 + 4\cos(h))}\right)\delta_m. \tag{9}$$

2.1. Linearization Technique 1 (CN1)

Applying Galerkin method to Eq. (4) with weight function $W(x)$ and identifying the weight function $W(x)$ with the cubic trigonometric B-spline T_m and using the expression (6), a fully discrete approximation for the RLW equation is obtained over the element $[x_m, x_{m+1}]$ as

$$\begin{aligned}&\sum_{j=m-1}^{m+2}\left\{\left(\int_{x_m}^{x_{m+1}} T_iT_j dx\right)\delta_j^{n+1} - \mu\left(\int_{x_m}^{x_{m+1}} T_iT''_j dx\right)\delta_j^{n+1} + \frac{\Delta t}{2}\left(\int_{x_m}^{x_{m+1}} T_iT'_j dx\right)\delta_j^{n+1}\right.\\ &\left.+\frac{\Delta t}{2}\varepsilon\sum_{k=m-1}^{m+2}\left(\int_{x_m}^{x_{m+1}} T_iT_k\left(\delta_k^{n+1}\right)T'_j dx\right)\delta_j^{n+1}\right\} - \sum_{j=m-1}^{m+2}\left\{\left(\int_{x_m}^{x_{m+1}} T_iT_j dx\right)\delta_j^n\right.\\ &\left.-\mu\left(\int_{x_m}^{x_{m+1}} T_iT''_j dx\right)\delta_j^n - \frac{\Delta t}{2}\left(\int_{x_m}^{x_{m+1}} T_iT'_j dx\right)\delta_j^n - \frac{\Delta t}{2}\varepsilon\sum_{k=m-1}^{m+2}\left(\int_{x_m}^{x_{m+1}} T_iT_k\left(\delta_k^n\right)T'_j dx\right)\delta_j^n\right\}\end{aligned} \tag{10}$$

where i, j and k take only the values $m-1$, m, $m+1$, $m+2$ for this typical element $[x_m, x_{m+1}]$.
(10) can be written in the matrices form as

$$\left[\mathbf{A}^e - \mu\mathbf{D}^e + \frac{\Delta t}{2}\mathbf{B}^e + \frac{\Delta t}{2}\varepsilon\mathbf{C}^e\left((\boldsymbol{\delta}^e)^{n+1}\right)\right](\boldsymbol{\delta}^e)^{n+1} - \left[\mathbf{A}^e - \mu\mathbf{D}^e - \frac{\Delta t}{2}\mathbf{B}^e - \frac{\Delta t}{2}\varepsilon\mathbf{C}^e\left((\boldsymbol{\delta}^e)^{n}\right)\right](\boldsymbol{\delta}^e)^{n} \tag{11}$$

where the element matrices and element parameters are

$$\begin{aligned} A_{ij}^e &= \int_{x_m}^{x_{m+1}} T_i T_j dx, \quad B_{ij}^e = \int_{x_m}^{x_{m+1}} T_i T_j' dx, \quad C_{ij}^e\left((\boldsymbol{\delta}^e)^{n+1}\right) = \int_{x_m}^{x_{m+1}} T_i T_k\left(\boldsymbol{\delta}_k^{n+1}\right) T_j' dx, \\ D_{ij}^e &= \int_{x_m}^{x_{m+1}} T_i T_j'' dx, \quad (\boldsymbol{\delta}^e)^{n+1} = (\delta_{m-1}^{n+1}, \delta_m^{n+1}, \delta_{m+1}^{n+1}, \delta_{m+2}^{n+1})^T. \end{aligned}$$

The 4×4 element matrices $\boldsymbol{A}^e$, $\boldsymbol{B}^e$ and $\boldsymbol{D}^e$ are independent of the parameters $\boldsymbol{\delta}^e$ and $4 \times 4 \times 4$ element matrices $\boldsymbol{C}^e$ depends on the parameters $\boldsymbol{\delta}^e$.

Combining contributions from all elements lead to the nonlinear matrix equation

$$\left[\mathbf{A} - \mu\mathbf{D} + \frac{\Delta t}{2}\mathbf{B} + \frac{\Delta t}{2}\varepsilon\mathbf{C}\left(\boldsymbol{\delta}^{n+1}\right)\right]\boldsymbol{\delta}^{n+1} = \left[\mathbf{A} - \mu\mathbf{D} - \frac{\Delta t}{2}\mathbf{B} - \frac{\Delta t}{2}\varepsilon\mathbf{C}\left(\boldsymbol{\delta}^{n}\right)\right]\boldsymbol{\delta}^{n}. \tag{12}$$

After the first and last equations are deleted in the system (12), imposition of the boundary conditions $U(a, x) = U(b, x) = 0$ at the both ends of the region yields to eliminate δ_{-1}^{n+1} and δ_{N+1}^{n+1} from the above system. Therefore the solution of the seven band matrix equations with the dimensions $(N+1) \times (N+1)$ is obtained by way of Thomas algorithms. After initial vector $\mathbf{d}^0 = (\delta_{-1}^0, \ldots, \delta_{N-1}^0, \delta_N^0)$ are found with the help of the boundary and initial conditions, $\mathbf{d}^{n+1}, (n = 0, 1, \ldots)$ unknown vectors can be found repeatedly by solving the recurrence relation (12) using previous $\mathbf{d}^{n-1}$ unknown vector. Note that since the system (12) is an implicit system, we have used an inner iteration algorithm for all time steps to increase the accuracy of the system.

Step 1: Set $error = 1$ and $\delta_m^* = \delta_m^{n+1}$ in $\mathbf{C}\left(\boldsymbol{\delta}^{n+1}\right)$ and taking $\delta_m^* = \delta_m^n$ then find U_m^*,
Step 2: While $error \geq 10^{-10}$ do Steps 3–4,
Step 3: Find U_m^{n+1},
Step 4: Find $\max_m \left|U_m^{n+1} - U_m^*\right|$ and set $\delta_m^* = \delta_m^{n+1}$,
Step 5: Stop and go to next time step.

2.2. Linearization Technique 2(CN2)

The nonlinear term in (4) is approximated by the following formula based on Taylor series:

$$(UU_x)^{n+1} \approx U^{n+1}U_x^n + U^nU_x^{n+1} - U^nU_x^n.$$

Using this approximation and applying Galerkin method by using the expression (6) in (4), a fully discrete approximation is obtained over the element $[x_m, x_{m+1}]$ as

$$\begin{aligned} &\sum_{j=m-1}^{m+2}\left\{\left(\int_{x_m}^{x_{m+1}} T_iT_j dx\right)\delta_j^{n+1} - \mu\left(\int_{x_m}^{x_{m+1}} T_iT_j'' dx\right)\delta_j^{n+1} + \frac{\Delta t}{2}\left(\int_{x_m}^{x_{m+1}} T_iT_j' dx\right)\delta_j^{n+1} + \right. \\ &\left. + \frac{\Delta t}{2}\varepsilon\sum_{k=m-1}^{m+2}\left[\left(\int_{x_m}^{x_{m+1}} T_iT_k'\left(\delta_k^n\right)T_j dx\right)\delta_j^{n+1} + \left(\int_{x_m}^{x_{m+1}} T_iT_k\left(\delta_k^n\right)T_j' dx\right)\delta_j^{n+1}\right]\right\} - \\ &\sum_{j=m-1}^{m+2}\left\{\left(\int_{x_m}^{x_{m+1}} T_iT_j dx\right)\delta_j^{n} - \mu\left(\int_{x_m}^{x_{m+1}} T_iT_j'' dx\right)\delta_j^{n} - \frac{\Delta t}{2}\left(\int_{x_m}^{x_{m+1}} T_iT_j' dx\right)\delta_j^{n}\right\} \end{aligned} \tag{13}$$

where i, j and k take only the values $m-1$, m, $m+1$, $m+2$ for this typical element $[x_m, x_{m+1}]$. (13) can be written in the matrices form as

$$\left[\mathbf{A}^e-\mu\mathbf{D}^e+\frac{\Delta t}{2}\mathbf{B}^e+\frac{\Delta t}{2}\varepsilon\left(\mathbf{E}^e\left((\boldsymbol{\delta}^e)^n\right)+\mathbf{C}^e\left((\boldsymbol{\delta}^e)^n\right)\right)\right](\boldsymbol{\delta}^e)^{n+1}-\left[\mathbf{A}^e-\mu\mathbf{D}^e-\frac{\Delta t}{2}\mathbf{B}^e\right](\boldsymbol{\delta}^e)^n \quad (14)$$

where the element matrices and element parameters are

$$A_{ij}^e = \int_{x_m}^{x_{m+1}} T_iT_j dx, \quad B_{ij}^e = \int_{x_m}^{x_{m+1}} T_iT_j' dx, \quad C_{ij}^e(\boldsymbol{\delta}^n) = \int_{x_m}^{x_{m+1}} T_iT_k(\delta_k^n)T_j' dx,$$

$$D_{ij}^e = \int_{x_m}^{x_{m+1}} T_iT_j'' dx, \quad E_{ij}^e(\boldsymbol{\delta}^n) = \int_{x_m}^{x_{m+1}} T_iT_k'(\delta_k^n)T_j dx, \quad (\boldsymbol{\delta}^e)^n = (\delta_{m-1}^n, \delta_m^n, \delta_{m+1}^n, \delta_{m+2}^n)^T.$$

Assembling together contributions from all elements leads to the matrix equation

$$\left[\mathbf{A}-\mu\mathbf{D}+\frac{\Delta t}{2}\mathbf{B}+\frac{\Delta t}{2}\varepsilon\left(\mathbf{E}(\boldsymbol{\delta}^n)+\mathbf{C}(\boldsymbol{\delta}^n)\right)\right]\boldsymbol{\delta}^{n+1} = \left[\mathbf{A}-\mu\mathbf{D}-\frac{\Delta t}{2}\mathbf{B}\right]\boldsymbol{\delta}^n. \quad (15)$$

Eliminating δ_{-1}^{n+1} and δ_{N+1}^{n+1} from (15) just like in previous subsection, the solution of the seven band final matrix equations with the dimensions $(N+1)\times(N+1)$ is obtained by way of Thomas algorithms.

3. Numerical Experiments

Since an accurate numerical scheme must keep the conservation properties of the RLW equation, we will monitor the three invariants of numerical solution for the equation corresponding to conservation of mass, momentum and energy given by the following integrals [10]:

$$I_1 = \int_{-\infty}^{\infty} u dx \approx \int_a^b U dx, \quad I_2 = \int_{-\infty}^{\infty} (u^2+\mu(u_x)^2) dx, I_3 = \int_{-\infty}^{\infty} \left(u^3+3u^2\right) dx. \quad (16)$$

Integrals for the conservation invariants are computed approximately with the trapezoidal rule for the space interval at all time steps. For the first test problem, accuracy of the proposed algorithms is worked out by measuring error norm L_∞

$$L_\infty = \max_m |u_m - U_m|, \quad (17)$$

and the order of convergence is computed with fixed space step by the formula

$$\text{order} = \frac{\log\left|\dfrac{u-U_{\Delta t_m}}{u-U_{\Delta t_{m+1}}}\right|}{\log\left|\dfrac{\Delta t_m}{\Delta t_{m+1}}\right|}, \quad (18)$$

where u is the exact solution and $U_{\Delta t_m}$ is the numerical solution with time step Δt_m.

3.1. *Motion of Single Solitary Wave for the RLW Equation*

The solitary wave theoretical solution of the RLW equation is

$$u(x,t) = 3c\text{sech}^2(k[x - \tilde{x}_0 - vt]), \tag{19}$$

where $v = 1 + \varepsilon c$ is the wave velocity, $3c$ is amplitude of the solitary wave, $\tilde{x}_0$ is peak position of the initially centered wave and $k = \sqrt{\dfrac{\varepsilon c}{4\mu v}}$. This solution corresponds to a solitary wave of magnitude $3c$, initially centered on the position $\tilde{x}_0$ propagating towards the right across the interval $[a,b]$ over the up to the time T without change of shape at a steady velocity v.

After the program is run up to time $t = 20$ with fixed space and various time step, error norm L_∞, invariants, Cpu time and Order of convergence for the proposed algorithms are presented in Table 1.

Table 1: Error norms L_∞ and invariants for a single solitary wave at time $t = 20$ with $c = 0.1$, $-80 \le x \le 100$, $h = 0.01$ and various time step.

CN1	L_∞	I_1	I_2	I_3	Cpu	Order
Δt =0.1	8.79×10^{-5}	3.9799497	0.8104625	2.5790074	249.63	2.00
Δt =0.05	2.20×10^{-5}	3.9799497	0.8104624	2.5790074	359.13	2.00
Δt =0.02	3.52×10^{-6}	3.9799497	0.8104624	2.5790074	798.56	2.00
Δt =0.01	8.80×10^{-7}	3.9799497	0.8104624	2.5790074	1603.80	
CN2	L_∞	I_1	I_2	I_3	Cpu	Order
Δt =0.1	7.80×10^{-5}	3.9799497	0.8104625	2.5790074	31.40	2.00
Δt =0.05	1.95×10^{-5}	3.9799497	0.8104624	2.5790074	62.55	2.00
Δt =0.02	3.12×10^{-6}	3.9799497	0.8104624	2.5790074	156.68	2.00
Δt =0.01	7.81×10^{-7}	3.9799497	0.8104624	2.5790074	313.43	

Absolute error (analytical–numerical) distributions for all of the proposed methods are drawn at time $t = 20$ in Figure 1.

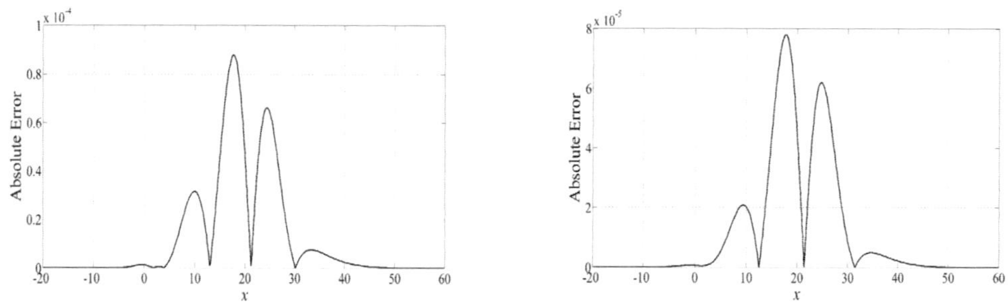

CN1 CN2

Figure 1: Absolute errors with $h = \Delta t = 0.1$

3.2. *Interaction of Two Solitary Waves for RLW Equation*

We consider interaction of two solitary waves using the following initial condition

$$u(x,0) = 3c_1\text{sech}^2(k_1[x - \tilde{x}_1]) + 3c_2\text{sech}^2(k_2[x - \tilde{x}_2]), \tag{20}$$

where $k_i = \sqrt{\dfrac{\varepsilon c_i}{4\mu(1+\varepsilon c_i)}}$, $i = 1, 2$. To ensure an interaction of two solitary waves, all of the computations are done for the parameters $\varepsilon = \mu = 1$, $c_1 = 0.2$, $c_2 = 0.1$, $\tilde{x}_1 = -177$ and $\tilde{x}_2 = -147$ over the region $-200 \le x \le 600$. These parameters provide two well separated solitary waves of magnitudes 0.6 and 0.3 and peak positions of them are located at $x = -177$ and -147.

The program is run until $t = 400$ with $h = 0.12$, $\Delta t = 0.1$ and numerical solutions of $U(x,t)$ at several times are drawn for visual views of the solutions. According to the figure, the nonlinear interaction takes place about time 200. Then, two solitary waves regain their original shapes after the interaction.

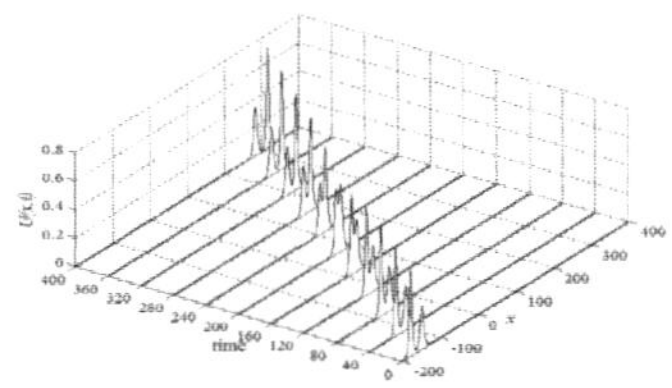

Figure 2: Interaction of two solitary waves.

Table 2 displays a comparison of the values of the invariants obtained by the proposed methods CN1, CN2 at some selected time t.

Table 2: Invariants for interaction of two solitary waves.

	CN1			CN2		
t =0	9.8582	3.2446	10.7783	9.8582	3.2446	10.7783
t =80	9.8582	3.2446	10.7783	9.8582	3.2446	10.7784
t =160	9.8581	3.2448	10.7785	9.8581	3.2448	10.7787
t =240	9.8580	3.2448	10.7785	9.8580	3.2449	10.7787
t =320	9.8579	3.2446	10.7781	9.8580	3.2446	10.7784
t =400	9.8577	3.2445	10.7779	9.8579	3.2446	10.7783

4. Conclusion

In this study, two numerical algorithms for the numerical solution of the RLW equation have been presented using Galerkin method based on cubic trigonometric B-splines as weight and trial functions for space discretization and Crank Nicolson method for time discretization. To compare all the proposed methods, CN2 gives accurate, reliable results and less Cpu time for the RLW equation. Also all of the methods have an advantage due to their small matrix operations. Therefore, the obtained results show that proposed algorithms, exhibit high accuracy and efficiency in both conservation of the invariants and error norm for the numerical solution of the RLW equation.

5. Acknowledgements

This work was supported by the Scientific Research Council of Eskişehir Osmangazi University under Project No. 2014-581.

6. References

[1] Peregrine D H 1966 *J. Fluid Mech.* **25** 321–330.
[2] Benjamin T B, Bona J L and Mahony J J 1972 *Philos. Trans. R. Soc. London A.* **272** 47–78.
[3] Gardner L R T and Gardner G A Gardner 1990 *J. Comput. Phys.* **91** 441–459.
[4] Irk D 2012 *Phys. Wave Phenom.* **20** 174–183.
[5] Esen A and Kutluay S 2006 *Appl. Math. Comput.* **174** 833–845.
[6] Dağ İ, Saka and Irk D 2006 *J. Comput. Appl. Math.* **190** 532-547
[7] Korkmaz A and Dağ İ 2013 *Arab. J. Sci. Eng.* **38** 1151-60.
[8] Lyche T and Winther R 1979 *J. Approx. Theory* **25** 266-279.
[9] Walz G 1997 *BIT* **37** 189-201.
[10] Olver P J 1979 *Math. Proc. Cambridge* **85** 143–159.

Journal of Physics: Conference Series **766** (2016) 012033 doi:10.1088/1742-6596/766/1/012033

Analytical solutions with the improved (G'/G)-expansion method for nonlinear evolution equations

Melike Kaplan, Ahmet Bekir, Arzu Akbulut

Eskisehir Osmangazi University, Art-Science Faculty, Mathematics-Computer Department, Eskisehir 26480, TURKEY

E-mail: mkaplan@ogu.edu.tr; abekir@ogu.edu.tr; ayakut1987@hotmail.com

Abstract. To seek the exact solutions of nonlinear partial differential equations (NPDEs), the improved $(G'/G)-$expansion method is proposed in the present work. With the aid of symbolic computation, this effective method is applied to construct exact solutions of the (1+1)-dimensional nonlinear dispersive modified Benjamin-Bona-Mahony equation and (3+1)-dimensional Kudryashov-Sinelshchikov equation. As a result, new types of exact solutions are obtained.

1. Introduction

Many important complex phenomena and dynamic processes in physics, mechanics, biology and chemistry can be described by NPDEs. Therefore, the investigation of the exact solutions for NPDEs have become more and more attractive in the study of soliton theory. Recently, a lot of direct methods have been proposed to construct exact solutions of these equations partly by virtue of the applicability of symbolic computation packages like Mathematica and Maple, which enables us to carry out the exact computation on computer [1-11].

The original $(G'/G)-$expansion method is a widely used method in soliton theory and mathematical physics. The keynote of this method is that the travelling wave solutions of NPDEs can be represented in terms of (G'/G) in which $G = G(\xi)$ satisfies the second order ordinary differential equation $G''(\xi) + \lambda G'(\xi) + \mu G(\xi) = 0$, where λ and μ are constants [5]. Comparably, in the improved $(G'/G)-$expansion method the travelling wave solutions of NPDEs can be represented in terms of (G'/G) in which $G = G(\xi)$ satisfies the second order ordinary differential equation $GG'' = DG^2 + EGG' + F(G')^2$, where D, E and F are real parameters [13].

1.1. *Algorithm of the improved $(G'/G)-$expansion method*

We consider a general partial differential equation, say in the independent variables x and t is given by

$$P(u, u_x, u_t, u_{xx}, u_{xt}, u_{tt}, ...) = 0, \tag{1}$$

where u is an unknown function, P is a polynomial in u and their various partial derivatives.

To find the exact solution of Eq.(1), we introduce the following travelling wave transformation:

$$u(x,t) = u(\xi), \xi = x - ct. \tag{2}$$

Here c denotes the wave velocity.

Employing Eq.(2), we can rewrite Eq.(1) in a nonlinear ordinary differential equation (ODE) as follows

$$Q(u, u', u'', u''', ...) = 0. \tag{3}$$

Here the prime denotes the derivation with respect to ξ. Eq.(3) is then integrated as many times as possible and setting the integration cosntant to zero. In the improved $(G'/G)-$expansion method, the solution $u(\xi)$ is considered in the finite series form

$$u(\xi) = \sum_{i=0}^{n} a_i \left(\frac{G'}{G}\right)^i. \tag{4}$$

Here the positive integer n denotes the balancing number, which is determined by considering the homogeneous balance principle. Namely, it can be calculated by balancing the highest order derivative term and nonlinear term appears in Eq.(3). Here $G(\xi)$ satisfies the second order auxiliary ODE in the form:

$$GG'' = DG^2 + EGG' + F(G')^2, \tag{5}$$

where D, E and F are real parameters. Also note that Eq.(5) reduces into following Riccati equation as:

$$\frac{d}{d\xi}\left(\frac{G'}{G}\right) = D + E\left(\frac{G'}{G}\right) + (F-1)\left(\frac{G'}{G}\right)^2. \tag{6}$$

From the general solutions of the Eq.(6), we have the following different cases:

Case 1: If $E \neq 0$ and $\Delta = E^2 + 4D - 4DF < 0$

$$\frac{G'(\xi)}{G(\xi)} = \frac{E}{2(1-F)} + \frac{E\sqrt{-\Delta}}{2(1-F)}\left(\frac{ic_1 \cos\left(\frac{\sqrt{-\Delta}}{2}\xi\right) - c_2 \sin\left(\frac{\sqrt{-\Delta}}{2}\xi\right)}{ic_1 \sin\left(\frac{\sqrt{-\Delta}}{2}\xi\right) + c_2 \cos\left(\frac{\sqrt{-\Delta}}{2}\xi\right)}\right) \tag{7}$$

Case 2: If $E \neq 0$ and $\Delta = E^2 + 4D - 4DF \geq 0$

$$\frac{G'(\xi)}{G(\xi)} = \frac{E}{2(1-F)} + \frac{E\sqrt{\Delta}}{2(1-F)}\left(\frac{c_1 \exp\left(\frac{\sqrt{\Delta}}{2}\xi\right) + c_2 \exp\left(\frac{-\sqrt{\Delta}}{2}\xi\right)}{c_1 \exp\left(\frac{\sqrt{\Delta}}{2}\xi\right) - c_2 \exp\left(\frac{-\sqrt{\Delta}}{2}\xi\right)}\right) \tag{8}$$

Case 3: If $E = 0$ and $\Delta = D(1-F) < 0$

$$\frac{G'(\xi)}{G(\xi)} = \frac{\sqrt{-\Delta}}{(1-F)}\left(\frac{ic_1 \cosh\left(\sqrt{-\Delta}\xi\right) - c_2 \sinh\left(\sqrt{-\Delta}\xi\right)}{-c_1 \sinh\left(\sqrt{-\Delta}\xi\right) - c_2 \cosh\left(\sqrt{-\Delta}\xi\right)}\right) \tag{9}$$

Case4: If $E = 0$ and $\Delta = D(1-F) \geq 0$

$$\frac{G'(\xi)}{G(\xi)} = \frac{\sqrt{\Delta}}{(1-F)}\left(\frac{c_1 \cos\left(\sqrt{\Delta}\xi\right) + c_2 \sin\left(\sqrt{\Delta}\xi\right)}{c_1 \sin\left(\sqrt{\Delta}\xi\right) - c_2 \cos\left(\sqrt{\Delta}\xi\right)}\right) \tag{10}$$

where $\xi = x - ct$ and D, E, F, c_1 and c_2 are arbitrary constants.

We substitute Eq.(4) into Eq.(3) along with Eq.(6) and collect the coefficients of $\left(\frac{G'(\xi)}{G(\xi)}\right)^i, (i = 0, 1, ...)$ then set each coefficient to zero to derive a set of algebraic equations for $a_i, (i = 0, 1, ..., n)$, D, E, F and c. We solve these set of algebraic equations with the aid of Maple packet program and substitute into Eq.(4) along with the general solutions of the Eq.(6) [13].

2. Implemantation of the improved $(G'/G)-$expansion method

In the current section, we apply our algoritm to the (1+1)-dimensional nonlinear dispersive modified Benjamin-Bona-Mahony equation and (3+1)-dimensional Kudryashov-Sinelshchikov equation.

2.1. (1+1)-dimensional nonlinear dispersive modified Benjamin-Bona-Mahony equation

The (1+1)-dimensional nonlinear dispersive modified Benjamin-Bona-Mahony equation was first derived to model an approximation for surface long waves in nonlinear dispersive media. This equation can also qualify the acoustic waves in inharmonic crystals, hydromagnetic waves in cold plasma and acoustic gravity waves in compressible fluids. It is given in the following form [12]:

$$u_t + u_x - \alpha u^2 u_x + u_{xxx} = 0, \tag{11}$$

where $u = u(\xi)$. Employing the travelling wave transformation (2), Eq.(11) reduced to an ODE and integrating the equation, we get

$$(1 - c)\, u - \frac{\alpha u^3}{3} + u'' = 0. \tag{12}$$

By balancing the highest order derivative terms and nonlinear terms in Eq.(12), we get the balancing number $m = 1$. According to the improved $(G'/G)-$expansion method, the exact solution takes the form:

$$u(\xi) = a_0 + a_1 \frac{G'(\xi)}{G(\xi)}. \tag{13}$$

Then we substitute Eq.(13) into the Eq.(12) and collect the coefficients of $\left(\frac{G'(\xi)}{G(\xi)}\right)^i, (i = 0, 1, 2, 3)$ then set each coefficient to zero to derive a set of algebraic equations. By solving this system with the aid of symbolic computation, we get the following results

$$a_0 = \pm\tfrac{E}{2}\sqrt{\tfrac{6}{\alpha}}, \quad a_1 = \pm\sqrt{\tfrac{6}{\alpha}}\,(F - 1), \quad c = 2DF - \tfrac{E^2}{2} + 1 - 2D, \quad D = D, \quad E = E, \quad F = F. \tag{14}$$

If we substitute these results into Eq.(13), we find the following cases for the exact solutions of (1+1) dimensional nonlinear dispersive modified Benjamin-Bona-Mahony equation.

Case 1: When $E \neq 0$ and $\Delta = E^2 + 4D - 4DF < 0$,

$$u_1(\xi) = \pm\frac{E}{2}\sqrt{\frac{6}{\alpha}} \pm \sqrt{\frac{6}{\alpha}}\,(F - 1)\left(\frac{E}{2(1-F)} + \frac{E\sqrt{-\Delta}}{2(1-F)}\left(\frac{ic_1\cos\left(\frac{\sqrt{-\Delta}}{2}\xi\right) - c_2\sin\left(\frac{\sqrt{-\Delta}}{2}\xi\right)}{ic_1\sin\left(\frac{\sqrt{-\Delta}}{2}\xi\right) + c_2\cos\left(\frac{\sqrt{-\Delta}}{2}\xi\right)}\right)\right) \tag{15}$$

Case 2: When $E \neq 0$ and $\Delta = E^2 + 4D - 4DF \geq 0$,

$$u_2(\xi) = \pm\frac{E}{2}\sqrt{\frac{6}{\alpha}} \pm \sqrt{\frac{6}{\alpha}}\,(C - 1)\left(\frac{E}{2(1-F)} + \frac{E\sqrt{\Delta}}{2(1-F)}\left(\frac{c_1\exp\left(\frac{\sqrt{\Delta}}{2}\xi\right) + c_2\exp\left(\frac{-\sqrt{\Delta}}{2}\xi\right)}{c_1\exp\left(\frac{\sqrt{\Delta}}{2}\xi\right) - c_2\exp\left(\frac{-\sqrt{\Delta}}{2}\xi\right)}\right)\right) \tag{16}$$

Case 3: When $E = 0$ and $\Delta = D(1 - F) < 0$,

$$u_3(\xi) = \pm\sqrt{\frac{6}{\alpha}}(F-1)\left(\frac{\sqrt{-\Delta}}{(1-F)}\left(\frac{ic_1\cosh\left(\sqrt{-\Delta}\xi\right) - c_2\sinh\left(\sqrt{-\Delta}\xi\right)}{-c_1\sinh\left(\sqrt{-\Delta}\xi\right) - c_2\cosh\left(\sqrt{-\Delta}\xi\right)}\right)\right) \tag{17}$$

Case 4: When $E = 0$ and $\Delta = D(1 - F) \geq 0$,

$$u_4(\xi) = \pm\sqrt{\frac{6}{\alpha}}(F-1)\left(\frac{\sqrt{\Delta}}{(1-F)}\left(\frac{c_1\cos\left(\sqrt{\Delta}\xi\right) + c_2\sin\left(\sqrt{\Delta}\xi\right)}{c_1\sin\left(\sqrt{\Delta}\xi\right) - c_2\cos\left(\sqrt{\Delta}\xi\right)}\right)\right) \tag{18}$$

Note that our solutions are different from the given ones in [12].

2.2. (3+1)-dimensional Kudryashov-Sinelshchikov equation

We handle with the (3+1)-dimensional Kudryashov-Sinelshchikov equation, which has been examined to model the physical characteristics of nonlinear waves in a bubbly liquid. It is given in the form

$$(u_t + uu_x + u_{xxx} - \gamma u_{xx})_x + \frac{1}{2}(u_{yy} + u_{zz}) = 0, \tag{19}$$

where $u = u(x, y, z, t)$ represents the density of the bubbly liquid, the scalar quantity γ depends on the kinematic viscosity of the bubbly liquid and the independent variables x, y and z are the scaled space coordinates, t is the scaled time coordinate [14].

By using the travelling wave transformation

$$\xi = x + y + z - ct, u(x, y, z, t) = u(\xi), \tag{20}$$

Eq.(19) can be reduced to a nonlinear ODE and integrating the equation twice with respect to ξ, and taking the integration constants as zero, we get

$$(1 - c)u + \frac{u^2}{2} + u'' - \gamma u' = 0. \tag{21}$$

Balancing the highest order derivative term with the nonlinear term appearing in Eq.(21), we find the balancing number as $n = 2$. By means of the improved $(G'/G)-$expansion method, the solutions takes the form as follows

$$u(\xi) = a_0 + a_1\frac{G'(\xi)}{G(\xi)} + a_2\left(\frac{G'(\xi)}{G(\xi)}\right)^2. \tag{22}$$

Then we substitute Eq.(22) into the Eq.(21) and collect the coefficients of $\left(\frac{G'(\xi)}{G(\xi)}\right)^i, (i = 0, 1, 2, 3, 4)$, then set each coefficient to zero to derive a set of algebraic equations. If we solve the set of algebraic equations above, we get different cases for the solutions of D, E, F, a_0, a_1, a_2 and c.

(I)

$$D = D, E = E, \quad c = \frac{6}{25}\gamma^2 + 1, \quad F = \frac{-\gamma^2 + 100D + 25E^2}{100D},$$
$$a_0 = -3E^2 + \frac{6}{5}\gamma E + \frac{9}{25}\gamma^2, \quad a_1 = -\frac{3(\gamma^3 - 25E^2\gamma - 5\gamma^2 E + 125E^3)}{125D}, \quad a_2 = -\frac{3(\gamma^4 - 50\gamma^2E^2 + 625E^4)}{2500D^2}. \tag{23}$$

Substituting these results into Eq.(22) we get the following cases for the exact solutions of (3+1)-dimensional Kudryashov-Sinelshchikov equation.

Case 1: When $E \neq 0$ and $\Delta = E^2 + 4D - 4DF < 0$,

$$\begin{aligned} u_1(\xi) &= -3E^2 + \tfrac{6}{5}\gamma E + \tfrac{9}{25}\gamma^2 \\ &- \tfrac{3(\gamma^3 - 25E^2\gamma - 5\gamma^2 E + 125E^3)}{125D}\left(\tfrac{E}{2(1-F)} + \tfrac{E\sqrt{-\Delta}}{2(1-F)}\left(\tfrac{ic_1\cos\left(\frac{\sqrt{-\Delta}}{2}\xi\right) - c_2\sin\left(\frac{\sqrt{-\Delta}}{2}\xi\right)}{ic_1\sin\left(\frac{\sqrt{-\Delta}}{2}\xi\right) + c_2\cos\left(\frac{\sqrt{-\Delta}}{2}\xi\right)}\right)\right) \\ &- \tfrac{3(\gamma^4 - 50\gamma^2E^2 + 625E^4)}{2500D^2}\left(\tfrac{E}{2(1-F)} + \tfrac{E\sqrt{-\Delta}}{2(1-F)}\left(\tfrac{ic_1\cos\left(\frac{\sqrt{-\Delta}}{2}\xi\right) - c_2\sin\left(\frac{\sqrt{-\Delta}}{2}\xi\right)}{ic_1\sin\left(\frac{\sqrt{-\Delta}}{2}\xi\right) + c_2\cos\left(\frac{\sqrt{-\Delta}}{2}\xi\right)}\right)\right)^2 \end{aligned} \tag{24}$$

Case 2: When $E \neq 0$ and $\Delta = E^2 + 4D - 4DF \geq 0$,

$$\begin{aligned} u_2(\xi) &= -3E^2 + \tfrac{6}{5}\gamma E + \tfrac{9}{25}\gamma^2 \\ &- \tfrac{3(\gamma^3 - 25E^2\gamma - 5\gamma^2 E + 125E^3)}{125D}\left(\tfrac{E}{2(1-F)} + \tfrac{E\sqrt{\Delta}}{2(1-F)}\left(\tfrac{c_1\exp\left(\frac{\sqrt{\Delta}}{2}\xi\right) + c_2\exp\left(\frac{-\sqrt{\Delta}}{2}\xi\right)}{c_1\exp\left(\frac{\sqrt{\Delta}}{2}\xi\right) - c_2\exp\left(\frac{-\sqrt{\Delta}}{2}\xi\right)}\right)\right) \\ &- \tfrac{3(\gamma^4 - 50\gamma^2E^2 + 625E^4)}{2500D^2}\left(\tfrac{E}{2(1-F)} + \tfrac{E\sqrt{\Delta}}{2(1-F)}\left(\tfrac{c_1\exp\left(\frac{\sqrt{\Delta}}{2}\xi\right) + c_2\exp\left(\frac{-\sqrt{\Delta}}{2}\xi\right)}{c_1\exp\left(\frac{\sqrt{\Delta}}{2}\xi\right) - c_2\exp\left(\frac{-\sqrt{\Delta}}{2}\xi\right)}\right)\right)^2 \end{aligned} \tag{25}$$

Case 3: When $E = 0$ and $\Delta = D(1-F) < 0$,

$$\begin{aligned} u_3(\xi) &= \tfrac{9}{25}\gamma^2 - \tfrac{3\gamma^3}{125D}\left(\tfrac{\sqrt{-\Delta}}{(1-F)}\left(\tfrac{ic_1\cosh(\sqrt{-\Delta}\xi) - c_2\sinh(\sqrt{-\Delta}\xi)}{-c_1\sinh(\sqrt{-\Delta}\xi) - c_2\cosh(\sqrt{-\Delta}\xi)}\right)\right) \\ &- \tfrac{3\gamma^4}{2500D^2}\left(\tfrac{\sqrt{-\Delta}}{(1-F)}\left(\tfrac{ic_1\cosh(\sqrt{-\Delta}\xi) - c_2\sinh(\sqrt{-\Delta}\xi)}{-c_1\sinh(\sqrt{-\Delta}\xi) - c_2\cosh(\sqrt{-\Delta}\xi)}\right)\right)^2 \end{aligned} \tag{26}$$

Case 4: When $E = 0$ and $\Delta = D(1-F) \geq 0$,

$$u_4(\xi) = \frac{9}{25}\gamma^2 - \frac{3\gamma^4}{2500D^2}\left(\frac{\sqrt{\Delta}}{(1-F)}\left(\frac{c_1\cos\left(\sqrt{\Delta}\xi\right) + c_2\sin\left(\sqrt{\Delta}\xi\right)}{c_1\sin\left(\sqrt{\Delta}\xi\right) - c_2\cos\left(\sqrt{\Delta}\xi\right)}\right)\right)^2 \tag{27}$$

(II)

$$\begin{aligned} &D = D, E = E, \qquad c = -\tfrac{6}{25}\gamma^2 + 1, \qquad F = \tfrac{-\gamma^2 + 100D + 25E^2}{100D}, \\ &a_0 = -3E^2 + \tfrac{6}{5}\gamma E - \tfrac{3}{25}\gamma^2, \quad a_1 = -\tfrac{3(\gamma^3 - 25E^2\gamma - 5\gamma^2E + 125E^3)}{125D}, \quad a_2 = -\tfrac{3(\gamma^4 - 50\gamma^2E^2 + 625E^4)}{2500D^2}. \end{aligned} \tag{28}$$

Substituting these results into Eq.(22) we get the following cases for the exact solutions of (3+1)-dimensional Kudryashov-Sinelshchikov equation.

Case 1: When $E \neq 0$ and $\Delta = E^2 + 4D - 4DF < 0$,

$$\begin{aligned} u_5(\xi) &= -3E^2 + \tfrac{6}{5}\gamma E - \tfrac{3}{25}\gamma^2 \\ &- \tfrac{3(\gamma^3 - 25E^2\gamma - 5\gamma^2 E + 125E^3)}{125D}\left(\tfrac{E}{2(1-F)} + \tfrac{E\sqrt{-\Delta}}{2(1-F)}\left(\tfrac{ic_1\cos\left(\frac{\sqrt{-\Delta}}{2}\xi\right) - c_2\sin\left(\frac{\sqrt{-\Delta}}{2}\xi\right)}{ic_1\sin\left(\frac{\sqrt{-\Delta}}{2}\xi\right) + c_2\cos\left(\frac{\sqrt{-\Delta}}{2}\xi\right)}\right)\right) \\ &- \tfrac{3(\gamma^4 - 50\gamma^2E^2 + 625E^4)}{2500D^2}\left(\tfrac{E}{2(1-F)} + \tfrac{E\sqrt{-\Delta}}{2(1-F)}\left(\tfrac{ic_1\cos\left(\frac{\sqrt{-\Delta}}{2}\xi\right) - c_2\sin\left(\frac{\sqrt{-\Delta}}{2}\xi\right)}{ic_1\sin\left(\frac{\sqrt{-\Delta}}{2}\xi\right) + c_2\cos\left(\frac{\sqrt{-\Delta}}{2}\xi\right)}\right)\right)^2 \end{aligned} \tag{29}$$

Case 2: When $E \neq 0$ and $\Delta = E^2 + 4D - 4DF \geq 0$,

$$u_6(\xi) = -3E^2 + \frac{6}{5}\gamma E - \frac{3}{25}\gamma^2 - \frac{3(\gamma^3 - 25E^2\gamma - 5\gamma^2 E + 125E^3)}{125D}\left(\frac{E}{2(1-F)} + \frac{E\sqrt{\Delta}}{2(1-F)}\left(\frac{c_1\exp\left(\frac{\sqrt{\Delta}}{2}\xi\right) + c_2\exp\left(\frac{-\sqrt{\Delta}}{2}\xi\right)}{c_1\exp\left(\frac{\sqrt{\Delta}}{2}\xi\right) - c_2\exp\left(\frac{-\sqrt{\Delta}}{2}\xi\right)}\right)\right) - \frac{3(\gamma^4 - 50\gamma^2E^2 + 625E^4)}{2500D^2}\left(\frac{E}{2(1-F)} + \frac{E\sqrt{\Delta}}{2(1-F)}\left(\frac{c_1\exp\left(\frac{\sqrt{\Delta}}{2}\xi\right) + c_2\exp\left(\frac{-\sqrt{\Delta}}{2}\xi\right)}{c_1\exp\left(\frac{\sqrt{\Delta}}{2}\xi\right) - c_2\exp\left(\frac{-\sqrt{\Delta}}{2}\xi\right)}\right)\right)^2 \quad (30)$$

Case 3: When $E = 0$ and $\Delta = D(1 - F) < 0$,

$$u_7(\xi) = -\frac{3}{25}\gamma^2 - \frac{3\gamma^3}{125A}\left(\frac{\sqrt{-\Delta}}{(1-F)}\left(\frac{ic_1\cosh(\sqrt{-\Delta}\xi) - c_2\sinh(\sqrt{-\Delta}\xi)}{-c_1\sinh(\sqrt{-\Delta}\xi) - c_2\cosh(\sqrt{-\Delta}\xi)}\right)\right) - \frac{3\gamma^4}{2500D^2}\left(\frac{\sqrt{-\Delta}}{(1-F)}\left(\frac{ic_1\cosh(\sqrt{-\Delta}\xi) - c_2\sinh(\sqrt{-\Delta}\xi)}{-c_1\sinh(\sqrt{-\Delta}\xi) - c_2\cosh(\sqrt{-\Delta}\xi)}\right)\right)^2 \quad (31)$$

Case 4: When $E = 0$ and $\Delta = A(1 - F) \geq 0$,

$$u_8(\xi) = -\frac{3}{25}\gamma^2 - \frac{3\gamma^3}{125A}\left(\frac{\sqrt{\Delta}}{(1-F)}\left(\frac{c_1\cos(\sqrt{\Delta}\xi) + c_2\sin(\sqrt{\Delta}\xi)}{c_1\sin(\sqrt{\Delta}\xi) - c_2\cos(\sqrt{\Delta}\xi)}\right)\right) - \frac{3\gamma^4}{2500D^2}\left(\frac{\sqrt{\Delta}}{(1-F)}\left(\frac{c_1\cos(\sqrt{\Delta}\xi) + c_2\sin(\sqrt{\Delta}\xi)}{c_1\sin(\sqrt{\Delta}\xi) - c_2\cos(\sqrt{\Delta}\xi)}\right)\right)^2 \quad (32)$$

Note that, our solutions are different from the given ones in [14].

3. Conclusion

In this paper, the improved (G'/G)−expansion method has been successfully applied to get analytical solutions two nonlinear evolution equation. In the original (G'/G)−expansion method, the auxiliary equation $G''(\xi) + \lambda G'(\xi) + \mu G(\xi) = 0$, has three different general solutions. But in the improved (G'/G)−expansion method, the auxiliary differential equations has $GG'' = DG^2 + EGG' + F(G')^2$ four different general solutions. By this way, the improved (G'/G)−expansion method can give more different solutions comparably the original (G'/G)−expansion method and it is suggested to get new and more general type analytical exact solutions. Accordingly, we use the improved (G'/G)−expansion method in this work. The adopted method also is a direct and powerful technique in obtaining the exact solutions of NPDEs. To our bestknowledge, the exact solutions obtained in this work will be significant to reveal the pertinent features of the physical phenomena.

4. References

[1] Ablowitz M J and Clarkson P A 1990, *Solitons, Nonlinear Evolution Equations and Inverse Scattering Transform* (Cambridge:Cambridge University Press)
[2] Wazwaz A M 2005 *Appl. Math. Comput.* **167** 1196
[3] Misirli E.and Gurefe Y 2011 *Mathematical and Computational Applications* **16** 258
[4] Tascan F and Bekir A 2009 *Appl. Math. Comput.* **207** 279
[5] Naher H, Abdullah F A and Akbar M A 2011 *Math. Prob. Eng.* **2011** 218216
[6] Kaplan M, Akbulut A and Bekir A 2015 *Z. Naturforsch. A* **70** 969
[7] Bekir A, Kaplan M and Guner O 2014 *AIP Conf. Proc.* **1611** 30
[8] Mirzazadeh M and Eslami M 2012 *Nonlinear Anal. Model. Control* **4** 481
[9] Zayed E M E 2010 *J Appl Math Informatics* **28** 383
[10] Pandir Y 2014 *Pramana-J Phys.* **82** 949
[11] Kaplan M, Bekir A and Ozer M N 2015 *Open Phys.* **13** 383
[12] Khan K,Akbar M A and Islam S M R 2014 *SpringerPlus* **3** 724
[13] Sahoo S and Saha Ray S 2016 *Physica A* **448** 265
[14] Yang H, Liu W, Yang B and He B 2015 *Commun Nonlinear Sci Numer Simulat* **27** 271

Journal of Physics: Conference Series **766** (2016) 012034 doi:10.1088/1742-6596/766/1/012034

Some Results About Concircular and Concurrent Vector Fields On Pseudo-Kaehler Manifolds

Sibel Sevinç, Gülşah Aydın Şekerci and A. Ceylan Çöken

Cumhuriyet University, Department of Mathematics, Sivas, TURKEY
Süleyman Demirel University, Department of Mathematics, Isparta, TURKEY
Akdeniz University, Department of Mathematics, Antalya, TURKEY

E-mail: ssevinc@cumhuriyet.edu.tr

E-mail: gulsahaydin@sdu.edu.tr

E-mail: ceylancoken@akdeniz.edu.tr

Abstract. Kaehler manifolds which are used in physics have a lot of application fields. In this study we only state concircular and concurrent vector field that are defined on these manifolds. A vector field on a pseudo-Riemannian manifold N is called concircular, if it satisfies $\nabla_X v = \mu X$ for any vector X tangent to N, where ∇ is the Levi-Civita connection of N. Furthermore, a concircular vector field v is called a concurrent vector field if the function μ is non-constant. So, we provide some results on submanifolds of pseudo-Kaehler manifolds with respect to a concircular vector field or a concurrent vector field. Morever, we investigate this problem for another manifolds and proof some theorems.

1. Introduction

The notion of concircular vector field on a Riemannian manifold was introduced by A. Fialkow [7] by

$$\nabla_X v = \mu X \tag{1}$$

where X is tangent to N, ∇ denotes the Levi-Civita connection of N and μ is a non-trivial function on N. A concircular vector field satisfying (1) is called non-trivial if the function μ is non-constant. Furthermore, a concircular vector field v is called a concurrent vector field if the function μ is equal to one [2]. For simplicity, we call as a concurrent vector field v if the function μ in (1) is a non-zero constant.

Concircular vector fields have an important role in the theory of projective and conformal transformations. Additionally, such vector fields have interesting applications in general relativity. Particularly, a Lorentzian manifold is a generalized Robertson-Walker spacetime if and only if it admits a timelike concircular vector field. For some further results related to concircular vector fields, you can see [7, 3, 8].

A pseudo-Riemannian metric g on a complex manifold (M, J) is called pseudo-Hermitian if the metric g and the complex structure J on M are given by

$$g(JX, JY) = g(X, Y), X, Y \in T_pM, p \in M \tag{2}$$

A pseudo-Hermitian manifold is called a pseudo-Kaehler manifold if its complex structure J is parallel with respect to its Levi-Civita connection ∇. So it is clear that $\nabla J = 0$.

Let N be an n-dimensional manifold with a 3-dimensional vector bundle Q consisting of three tensors J_1, J_2 and J_3 of type $(1,1)$ over N. Suppose, in any coordinate neighborhood U of N, there is a local basis $\{J_1, J_2, J_3\}$of Q such that

$$J_1^2 = J_2^2 = J_3^2 = -I$$

and

$$J_2J_3 = -J_3J_2 = J_1, J_3J_1 = -J_1J_3 = J_2, J_1J_2 = -J_2J_1 = J_3$$

Such a basis$\{J_1, J_2, J_3\}$is called a canonical local basis of the bundle Q in U. We say that the bundle Q has an almost quaternion structure in N and (N, Q) is called an almost quaternion manifold whose dimension is $n = 4m, (m \geq 1)$. In such a case, the given structure Q is called an integrable quaternionic structure. Suppose g is an indefinite metric on (N, Q) such that $g(\phi X, \phi Y) = g(X, Y), \forall X, Y \in TpN$, $p \in N$ and $\phi = J_1, J_2, J_3$, with$\{J_1, J_2, J_3\}$being a basis of Q at p. Then, (N, g, Q) is called an indefinite almost quaternion manifold.If the Levi-Civita connection ∇ of $\left(\tilde{M}, g, Q\right)$ satisfies

$$\begin{aligned} \nabla_X J_1 &= r(X) J_2 - q(X) J_3 \\ \nabla_X J_2 &= -r(X) J_1 - p(X) J_3 \\ \nabla_X J_3 &= q(X) J_1 - p(X) J_2 \end{aligned} \tag{3}$$

for any vector field X on $\tilde{M}$, then $\tilde{M}$ is called an indefinite quaternion Kaehler manifold, where p, q, r are certain local 1−forms and $\{J_1, J_2, J_3\}$ is a local canonical basis of Q. Let M be a lightlike submanifold of an indefinite quaternion Kaehler manifold N. We say that M is a QR-lightlike submanifold if the following conditions are fulfilled:

$$\begin{aligned} &J_a RadTM \cap RadTM = \{0\}, \\ &S(TM) = \{J_a RadTM \oplus \widetilde{D}\} \perp D_0, \\ &J_a(D_0) = D_0, J_a(L_1 \perp L_2) = \widetilde{D} \subset S(TM), \end{aligned}$$

where $L_1 = ltr(TM)$ and L_2 is a vector sub-bundle of $S(TM^{\perp})$. Now, we consider

$$D = \{RadTM \oplus_{orth} J_a RadTM\} \oplus_{orth} D_0,$$

where D_0 is non-degenerate distribution. It is easy to check that D_0 is invariant with respect to each J_a.Thus we have

$$TM = D \oplus \widetilde{D}$$

[6].

2. Some results for concurent vector fields

Assume that N is a indefinite quaternionic Kaehler manifold and M is a pseudo-Kaehler submanifold of N which admits a concurrent vector field v. We have the following:

Proposition 2.1. *Let N be a indefinite quaternionic Kaehler manifold and M be a submanifold with the indefinite metric of N. Then we have*
a) For any concurrent vector field v on M, $J_a v\ (a = 1, 2, 3)$ is never a concurrent vector field on M.
b) The complex distribution $D = Sp\{v, J_1v, J_2v, J_3v\}$ of N is always integrable.

Proof. a) Let v be a concurrent vector field on M. For $\bar{\nabla}$ Levi-Civita connection of N

$$\bar{\nabla}_X J_a v = J_a \bar{\nabla}_X v = cJ_a X + h(X, v) J_a W_i$$

which implies that $J_a v$ is not a concurrent vector field for $\bar{\nabla}$ (since $\bar{\nabla}_X Jv \neq cX$).

b) Due to the Gauss equation, for $W_i \in TM^{\perp}$

$$\begin{aligned} \bar{\nabla}_{J_a v} v &= \nabla_{J_a v} v + h(J_a v, v) W_i = cJ_a v + h(J_a v, v) W_i \\ \bar{\nabla}_v J_a v &= \nabla_v J_a v + h(v, J_a v) W_i \\ \bar{\nabla}_X J_a v &= J_a \nabla_X v = J_a cX = cJ_a X \end{aligned}$$

then we find

$$\begin{aligned} \bar{\nabla}_{J_a v} v - \bar{\nabla}_v J_a v &= cJ_a v + h(J_a v, v) W_i - cJ_a v - h(v, J_a v) W_i = 0 \\ &\Rightarrow [J_a v, v] = 0 \end{aligned} \tag{4}$$

which implies that for $\bar{\nabla}$ Levi-Civita connection of N, the distribution $\mathcal{D}$ is integrable. □

Corollary. *Let v is a concurrent vector field on an indefinite quaternionic Kaehler manifold N. Then v is a concurrent vector field on a submanifold of N if and only if $h(X, v) = 0$.*

Proof. Let N be an indefinite quaternionic Kaehler manifold and M be a submanifold of N. For $\bar{\nabla}$ Levi-Civita connection of N we can write $\bar{\nabla}_X v = cX$, then

$$\bar{\nabla}_X v = \nabla_X v + h(X, v) W_i = cX$$

and we find

$$h(X, v) = 0$$

Obviously, the opposite of the theorem is true. □

Now we define the Ricci tensor of N. Let $e_1, ..., e_{2n}$ be an orthonormal frame on M, then the Ricci tensor Ric of M is defined by

$$Ric(X, Y) = \sum_{i=1}^{2n} \varepsilon_i g(R(X, e_i) Y, e_i) + \sum_{j=2n}^{2m} \varepsilon_j g(R(X, e_j) Y, e_j) \tag{5}$$

where $g(e_i, e_j) = \varepsilon_i \delta_{ij}$. Let R denote the Riemannian curvature tensor of M which is given by

$$R(X, Y) Z = \nabla_X \nabla_Y Z - \nabla_Y \nabla_X Z - \nabla_{[X,Y]} Z \tag{6}$$

for vector fields X, Y, Z tangent to M. Furthermore, it is well-known that the Riemannian curvature tensor R satisfies

$$\begin{aligned} &i) R(X, Y) = -R(Y, X) \\ &ii) R(X, Y) JZ = J(R(X, Y) Z) \\ &iii) R(JX, JY) = R(X, Y) Z \\ &iv) g(R(X, Y) Z, W) = g(R(Z, W) X, Y) \end{aligned} \tag{7}$$

Since v is a concurrent vector field,

$$\begin{aligned} \bar{R}(X, v) v &= \bar{\nabla}_X \bar{\nabla}_v v - \bar{\nabla}_v \bar{\nabla}_X v - \bar{\nabla}_{[X,v]} v \\ &= \bar{\nabla}_X (cv + h(v, v) W_i) - \bar{\nabla}_v (cX + h(X, v) W_i) \\ &\quad -c[X, v] - h([X, v], v) W_i \\ &= X(h(v, v)) W_i + h(v, v) \nabla_X^{\perp} W_i - h(v, v) A_{W_i} X - v(h(X, v)) W_i \\ &\quad -h(X, v) \nabla_v^{\perp} W_i + h(X, v) A_{W_i} v - h([X, v], v) W_i \end{aligned} \tag{8}$$

Journal of Physics: Conference Series **766** (2016) 012034 doi:10.1088/1742-6596/766/1/012034

So, we can write

$$Ric(v,v)=\sum_{j=2n}^{2m}\varepsilon_j g(R(v,e_j)v,e_j)=-\sum_{j=2n}^{2m}\varepsilon_j g(R(e_j,v)v,e_j) \tag{9}$$

Multiplying with e_j we find

$$g(R(e_j,v)v,e_j)=g(-h(v,v)A_{W_i}X+h(X,v)A_{W_i}v,e_j) \tag{10}$$

From this,

$$Ric(v,v)=-\sum_{j=2n}^{2m}\varepsilon_j g(-h(v,v)A_{W_i}X+h(X,v)A_{W_i}v,e_j) \tag{11}$$

Consequently, we have the following:

Proposition 2.2. *Let N be a indefinite quaternionic Kaehler manifold and M be a submanifold with the indefinite metric of N which admits a concurrent vector field (e.i. M is Ricci-flat). If*

$$-h(v,v)\mathcal{A}_{W_i}X+h(X,v)\mathcal{A}_{W_i}v=0 \tag{12}$$

then N is Ricci-flat ($Ric=0$).

Theorem 2.3. *Let N be an indefinite quatenionic Kaehler manifold and M be a QR-lightlike submanifold of manifold N. Then, there are the followings.*
a) *For any concurent vector field v on M, $J_a v$ is never a concurent vector field on M.*
b) *The complex distribution $\mathcal{D}$ of N is always integrable.*

Proof. a) Let v be a concurent vector field on M. For $\bar{\nabla}$ Levi-Civita connection of N

$$\begin{aligned}\bar{\nabla}_X J_a v &= J_a\bar{\nabla}_X v=J_a\left(\nabla_X v+h^l(X,v)N_j+h^s(X,v)W_i\right)\\ &= J_a\left(cX+h^l(X,v)N_j+h^s(X,v)W_i\right)=cJ_aX+h^l(X,v)J_aN_j+h^s(X,v)J_aW_i\end{aligned}$$

where $N_j\in ltr(TM), W_i\in S(TM)^{\perp}$. Since $\bar{\nabla}_X J_a v\neq cX$, $J_a v$ is not a concurent vector field for $\bar{\nabla}$.

b)

$$\begin{aligned}\bar{\nabla}_{J_a v}v &= \nabla_{J_a v}v+h^l(X,v)N_j+h^s(X,v)W_i\\ \bar{\nabla}_v J_a v &= \nabla_v J_a v+h^l(X,v)N_j+h^s(X,v)W_i\end{aligned}$$

then we find

$$\bar{\nabla}_{J_a v}v-\bar{\nabla}_v J_a v=cJ_a v-\nabla_v J_a v=cJ_a v-J_a\nabla_v v=cJ_a v-J_a cv=0\Rightarrow[J_a v,v]=0$$

which implies that for $\bar{\nabla}$ Levi-Civita connection of N, the distribution $\mathcal{D}$ is integrable. □

Corollary. *Let v is a concurrent vector field on a indefinite quaternionic Kaehler manifold N. Then v is a concurrent vector field on a QR-lightlike submanifold of N if and only if $h^l(X,v)=0$ and $h^s(X,v)=0$*

Proof. Let N be a indefinite quaternionic Kaehler manifold and M be a submanifold of N. For $\bar{\nabla}$ Levi-Civita connection of N we can write $\bar{\nabla}_X v=cX$, then

$$\bar{\nabla}_X v=\nabla_X v+h^l(X,v)N_j+h^s(X,v)W_i=cX$$

and we find

$$h^l(X,v)=0 \text{ and } h^s(X,v)=0$$

Obviously, the opposite of the theorem is true. □

3. QR-lightlike hypersurfaces with concircular vector fields

Theorem 3.1. *Let M be a QR-lightlike hypersurface of an indefinite Kaehler manifold N. Then we have, $h(X,v)=0, X\in\Gamma(D_0)$ such that v is a concircular vector field.*

Proof. For v$\in\Gamma(TM)$ and $\xi\in\Gamma(RadTM)$,

$$\begin{aligned}\widetilde{g}\left(\tilde{\nabla}_X v, J_a\xi\right) &= \widetilde{g}\left(\nabla_X v + h(X,v)N, J_a\xi\right) \\ &= \widetilde{g}\left(\nabla_X v, J_a\xi\right) + \widetilde{g}\left(h(X,v)N, J_a\xi\right)\end{aligned} \tag{13}$$

Due to $N\in\Gamma(ltrTM)$ and $J_a\xi\in\Gamma(S(TM))$, $\widetilde{g}(N,J_a\xi)=0$. So,

$$\widetilde{g}\left(\tilde{\nabla}_X v, J_a\xi\right) = g(\mu X, J_a\xi) \tag{14}$$

By using this equation, we find

$$\widetilde{g}\left(\tilde{\nabla}_X v, J_a\xi\right) = 0 \tag{15}$$

From here, we get

$$\begin{aligned}\widetilde{g}\left(\tilde{\nabla}_X v, J_aJ_b\xi\right) &= \widetilde{g}\left(\nabla_X v + h(X,v)N, J_aJ_b\xi\right) \\ &= \widetilde{g}\left(\mu X, J_aJ_b\xi\right) + h(X,v)\,\widetilde{g}\left(N, J_aJ_b\xi\right)\end{aligned} \tag{16}$$

This equation must equal to zero. Then, let us find the needed conditions for this. For $a=b$,

$$\widetilde{g}\left(\tilde{\nabla}_X v, J_aJ_b\xi\right) = \widetilde{g}(\mu X, -\xi) + h(X,v)\,\widetilde{g}(N,-\xi) = h(X,v) \tag{17}$$

So, $h(X,v)=0$ is. For $a\neq b$,

$$\widetilde{g}\left(\tilde{\nabla}_X v, J_aJ_b\xi\right) = \widetilde{g}(\mu X, J_c\xi) + h(X,v)\,\widetilde{g}(N, J_c\xi) = 0 \tag{18}$$

where $J_c = J_aJ_b$. □

We say that M is a mixed geodesic QR-lightlike surface if $h(X,Z)=0$ for any $X\in\Gamma(D), Z\in\Gamma\left(\widetilde{D}\right)$.

Theorem 3.2. *Let M be a QR-lightlike hypersurface of an indefinite Kaehler manifold N.*
a) M is a mixed geodesic if $X(J_a)=0$ while Z is a concircular vector field.
b) M is a mixed geodesic if $\widetilde{g}([J_aN,X],\xi)+\widetilde{g}(A_NX,J_a\xi)=X(J_a)$ while X is a concircular vector field.

Proof. a) Let Z be a concircular vector field and $Z=J_aN$.

$$\begin{aligned}h(X,J_aN)N &= \tilde{\nabla}_X J_aN - \nabla_X J_aN \\ &= X(J_aN) + J_a\left(\tilde{\nabla}_X N\right) - \mu X \\ &= X(J_aN) + J_a\left(\nabla^{\perp}_X N\right) + J_a(A_NX) - \mu X\end{aligned}$$

So, we find that

$$\widetilde{g}\left(h(X,J_aN)N,\xi\right) = X(J_a)$$

b) Let X be a concircular vector field.

$$\begin{aligned} h(X, J_a N)\, N &= \tilde{\nabla}_{J_a N} X - \nabla_{J_a N} X \\ &= [J_a N, X] - \tilde{\nabla}_X J_a N - \nabla_{J_a N} X \end{aligned}$$

So, we find that

$$\begin{aligned} \widetilde{g}\left(h(X, J_a N)\, N, \xi\right) &= \widetilde{g}\left([J_a N, X], \xi\right) - \widetilde{g}\left(\tilde{\nabla}_X (J_a N), \xi\right) - \widetilde{g}\left(\nabla_{J_a N} X, \xi\right) \\ &= \widetilde{g}\left([J_a N, X], \xi\right) - \widetilde{g}\left(X(J_a)\, N, \xi\right) - \widetilde{g}\left(J_a\left(\tilde{\nabla}_X N\right), \xi\right) - \mu \widetilde{g}\left(J_a N, \xi\right) \\ &= \widetilde{g}\left([J_a N, X], \xi\right) - X(J_a) - \widetilde{g}\left(\tilde{\nabla}_X N, J_a \xi\right) - \mu \widetilde{g}\left(J_a N, \xi\right) \\ &= \widetilde{g}\left([J_a N, X], \xi\right) - X(J_a) - \widetilde{g}\left(A_N X, J_a \xi\right) \end{aligned}$$

□

4. References

[1] Barros M and Romero A 1982 *Math. Ann.* **261** 1
[2] Chen B Y 1973 *Geometry of Submanifolds* (New York: M. Dekker) p 298
[3] Chen B Y 2011 *Pseudo-Riemannian Geometry, δ−Invariants and Applications (Singapore: World Scientific Publishing Co. Pte. Ltd.) p 477*
[4] *Chen B Y 2015 Bull. Korean Math. Soc.* ***52*** *5*
[5] *Chen B Y 2016 Kragujevac J. Math.* ***40*** *1*
[6] *Duggal K L and Şahin B 2010 Differential Geometry of Lightlike Submanifolds(Basel-Boston-Berlin: Birkhauser Verlag AG) p 300*
[7] *Fialkow A 1939 Trans. Amer. Math. Soc.* ***45*** *3*
[8] *Kobayashi S and Nomizu K 1969 Foundations of Differantial Geometry Vol. 1 (New York: Interscience Publishers) p 329*
[9] *Küpeli N D 1996 Singular Semi-Riemannian Geometry (The Netherlands: Kluwer Academic Publishers) p 177*
[10] *O'Neill B 1983 Semi-Riemannian Geometry with Applications to Relativity (New York: Academic Press Inc.) p 482*
[11] *Romero A and Suh Y 2004 Extracta Mathematicae* ***19*** *3*
[12] *Yano K 1940 Proc. Imp. Acad.* ***16*** *6*

Exact solutions of some fractional differential equations by various expansion methods

Muammer Topsakal[a], **Ozkan Guner**[b], **Ahmet Bekir**[a], **Omer Unsal**[a]
[a]Eskisehir Osmangazi University, Department of Mathematics, Eskisehir 26480, TURKEY
[b]Cankiri Karatekin University, Department of International Trade, Cankiri 18100, TURKEY

E-mail: mtpskl@yahoo.com; ozkanguner@karatekin.edu.tr; abekir@ogu.edu.tr; ounsal@ogu.edu.tr

Abstract. In this paper, we construct the exact solutions of some nonlinear space-time fractional differential equations involving modified Riemann-Liouville derivative in mathematical physics and applied mathematics; namely the fractional modified Benjamin-Bona-Mahony (mBBM) and Kawahara equations by using $\left(\frac{G'}{G}\right)$ and $(G'/G, 1/G)$-expansion methods.

1. Introduction

Fractional calculus has been used to model physical and engineering sciences. Fractional differential equations (FDEs) are regarded as models of physical systems [1, 2]. We introduce the space-time fractional mBBM equation [3]

$$D_t^\alpha u + D_x^\alpha u - vu^2 D_x^\alpha u + D_x^{3\alpha} u = 0, \tag{1}$$

where v is a nonzero positive constant. We also consider the time fractional modified nonlinear Kawahara equation [4]

$$D_t^\alpha u + u^2 u_x + p u_{xx} + q u_{xxx} = 0, \tag{2}$$

where α is the parameter standing for the order of the fractional time derivative, and $0 < \alpha \leq 1$. The Jumarie's modified Riemann-Liouville derivative of order α is defined in [5]:

2. The $\left(\frac{G'}{G}\right)$-expansion method for FDEs

We consider the following general nonlinear fractional differential equations (FDEs) of the type

$$P(u, D_t^\alpha u, D_x^\beta u, D_t^\alpha D_t^\alpha u, D_t^\alpha D_x^\beta u, D_x^\beta D_x^\beta u, ...) = 0, \quad 0 < \alpha, \beta < 1 \tag{3}$$

where $u = u(x,t)$ is an unknown function. The traveling wave variable

$$u(x,t) = U(\xi), \xi = \frac{\tau x^\beta}{\Gamma(1+\beta)} + \frac{ct^\alpha}{\Gamma(1+\alpha)}, \tag{4}$$

where τ and c are nonzero arbitrary constants [6]. By using the chain rule

$$D_t^\alpha u = \sigma_t' \frac{dU}{d\xi} D_t^\alpha \xi, D_x^\alpha u = \sigma_x' \frac{dU}{d\xi} D_x^\alpha \xi \tag{5}$$

where $\sigma_t^{'}$ and $\sigma_x^{'}$ are called the sigma indexes see [7], without loss of generality we can take $\sigma_t^{'} = \sigma_x^{'} = l_0$, where l_0 is a constant. Eq. (3) can be reduced into an ODE;

$$Q(U, U^{'}, U'', U''', ...) = 0. \tag{6}$$

where the prime denotes the derivation with respect to ξ. Suppose the solution of equation (9) can be expressed by a polynomial in $\left(\frac{G'}{G}\right)$ as follows [8, 9]:

$$u(\xi) = \sum_{i=0}^{m} a_i \left(\frac{G'}{G}\right)^i, \quad a_m \neq 0, \tag{7}$$

where a_i $(i = 0, 1, ..., m)$ are constants, while $G(\xi)$ satisfies the following second order LODE

$$G''(\xi) + \lambda G'(\xi) + \mu G(\xi) = 0, \tag{8}$$

with λ and μ are being constants. The positive integer m can be determined by the homogeneous balance principle in equation (6). Substituting equation (7) into equation (6) and using equation (8) collecting all terms with the same order of $\left(\frac{G'}{G}\right)$ together. Then equating each coefficient of the resulting polynomial to zero, we obtain a set of algebraic equations for a_i $(i = 0, 1, ..., m)$, λ, μ, τ and c. Solving these equations, we can get a variety of exact solutions of equation (3).

3. Algorithm of $(G'/G, 1/G)$-expansion method

For the second order linear ordinary differential equation (LODE)

$$G''(\xi) + \lambda G(\xi) = \mu, \tag{9}$$

we choose

$$\phi = G'/G, \psi = 1/G \tag{10}$$

for simplicity here and after. (9) and (10) gives

$$\phi' = -\phi^2 + \mu\psi - \lambda, \psi' = -\phi\psi. \tag{11}$$

The details of the solution of LODE (9), concludes following three cases:

Case 1 If $\lambda < 0$, the general solution of the LODE (9) reads

$$G(\xi) = A_1 \sinh(\sqrt{-\lambda}\xi) + A_2 \cosh(\sqrt{-\lambda}\xi) + \frac{\mu}{\lambda} \tag{12}$$

and we have

$$\psi^2 = \frac{-\lambda}{\lambda^2\sigma + \mu^2}(\phi^2 - 2\mu\psi + \lambda), \sigma = A_1^2 - A_2^2. \tag{13}$$

Case 2 If $\lambda > 0$, the general solution of the LODE (9) gives

$$G(\xi) = A_1 \sin(\sqrt{\lambda}\xi) + A_2 \cos(\sqrt{\lambda}\xi) + \frac{\mu}{\lambda} \tag{14}$$

and corresponding relations are

$$\psi^2 = \frac{\lambda}{\lambda^2\sigma - \mu^2}(\phi^2 - 2\mu\psi + \lambda), \sigma = A_1^2 + A_2^2. \tag{15}$$

Case 3 If $\lambda = 0$, the general solution of the LODE (9) can be written as

$$G(\xi) = \frac{\mu}{2}\xi^2 + A_1\xi + A_2 \tag{16}$$

and we have

$$\psi^2 = \frac{1}{A_1^2 - 2\mu A_2}(\phi^2 - 2\mu\psi) \tag{17}$$

where A_1 and A_2 are two arbitrary constants. Suppose that the solution of ODE (6) can be expressed by a polynomial in terms of ϕ and ψ in the form of

$$u(\xi) = \sum_{i=0}^{N} a_i\phi^i + \sum_{i=1}^{N} b_i\phi^{i-1}\psi \tag{18}$$

where $G = G(\xi)$ is the solution of the second order LODE (9), a_i, $b_i(i = 1, ..., N)$, λ and μ are constants and the positive integer N can be determined by balancing principle in ODE (6). Employing (18) into Eq. (6), using (11) and (13) the left-hand side of (6) can be expressed a polynomial in terms of ϕ and ψ, in which the degree of ψ is not larger than 1. Coefficients of the polynomial give a system of algebraic equations in a_i, b_i $\lambda(\lambda < 0)$, μ, A_1 and A_2. Solving the algebraic equations and substituting the values of a_i, b_i, λ, μ, A_1 and A_2 into (18), we can obtain the travelling wave solutions expressed by the hyperbolic functions of Eq. (6). Substituting (18) into Eq. (6), using (11) and (15) (or (11) and (17)), we obtain the travelling wave solutions of Eq. (9) expressed by trigonometric and rational functions [10, 11].

4. The space-time fractional mBBM equation

Firstly, we consider the following transformations;

$$u(x,t) = U(\xi), \qquad \xi = \frac{kx^\alpha}{\Gamma(1+\alpha)} - \frac{ct^\alpha}{\Gamma(1+\alpha)}, \tag{19}$$

where k and c are nonzero constants. Eq. (1) can be reduced into an ODE;

$$(c+k)U - vk\frac{U^3}{3} + k^3U'' + \xi_0 = 0. \tag{20}$$

where ξ_0 is an integration constant. Using homogeneous balance principle, we have $m = 1$. Suppose that the solutions of (20) can be expressed by a polynomial in $\left(\frac{G'}{G}\right)$ as follows:

$$U(\xi) = a_0 + a_1\left(\tfrac{G'}{G}\right), a_1 \neq 0. \tag{21}$$

Substituting these expressions into Eq. (21), collecting the coefficients of $\left(\frac{G'}{G}\right)^i$ $(i = 0, ..., 3)$ and set it to zero we obtain a system. When this system of algebraic equations is solved, we can be written the following expansion.

$$U(\xi) = \pm\tfrac{3k\lambda}{\sqrt{6v}} \pm k\sqrt{\tfrac{6}{v}}\left(\tfrac{G'}{G}\right). \tag{22}$$

We have three types of travelling wave solutions of the space-time fractional mBBM equation as follows:

When $\lambda^2 - 4\mu > 0$,

$$U_{1,2}(\xi) = \pm\frac{k}{2}\sqrt{\frac{6\lambda^2 - 24\mu}{v}}\left(\frac{C_1\sinh\frac{1}{2}\sqrt{\lambda^2-4\mu}\xi + C_2\cosh\frac{1}{2}\sqrt{\lambda^2-4\mu}\xi}{C_1\cosh\frac{1}{2}\sqrt{\lambda^2-4\mu}\xi + C_2\sinh\frac{1}{2}\sqrt{\lambda^2-4\mu}\xi}\right), \tag{23}$$

When $\lambda^2 - 4\mu < 0$,

$$U_{3,4}(\xi) = \pm\frac{k}{2}\sqrt{\frac{24\mu - 6\lambda^2}{v}}\left(\frac{-C_1\sin\frac{1}{2}\sqrt{4\mu-\lambda^2}\xi + C_2\cos\frac{1}{2}\sqrt{4\mu-\lambda^2}\xi}{C_1\cos\frac{1}{2}\sqrt{4\mu-\lambda^2}\xi + C_2\sin\frac{1}{2}\sqrt{4\mu-\lambda^2}\xi}\right), \tag{24}$$

When $\lambda^2 - 4\mu = 0$,

$$U_{5,6}(\xi) = \pm k\sqrt{\frac{6}{v}}\frac{C_2}{C_1 + C_2\xi}, \tag{25}$$

where $\xi = \frac{kx^\alpha}{\Gamma(1+\alpha)} + \frac{(k^3\lambda^2-2k)t^\alpha}{2\Gamma(1+\alpha)}$. In particular, if $C_1 \neq 0, C_2 = 0, \lambda > 0, \mu = 0$, then $U_{1,2}(\xi)$ become

$$u_{1,2}(x,t) = \pm k\lambda\sqrt{\frac{3}{2v}}\tanh\left\{\frac{\lambda}{2}\left(\frac{kx^\alpha}{\Gamma(1+\alpha)} + \frac{(k^3\lambda^2 - 2k)t^\alpha}{2\Gamma(1+\alpha)}\right)\right\}, \tag{26}$$

which are the exact solutions of the space time fractional mBBM equation. According to $(G'/G, 1/G)$-expansion method, we look for the solution in the form of

$$u(\xi) = a_0 + a_1\phi + b_1\psi \tag{27}$$

where a_0, a_1,and b_1 are constants to be determined later. Substituting (27) into Eq. (20), using (11) and (13), (20) can be expressed as a polynomial in ϕ and ψ. Coefficients of equation yield a system of algebraic equations in $a_0, a_1, b_1, k, c, \sigma, \mu, \xi_0$ and λ. Solving the algebraic equations by Maple, we obtain hyperbolic, trigonometric and rational function solutions of Eq. (1) as follows:

$$u(\xi) = \sqrt{\frac{3}{2v}}\frac{k(A_1\cosh(\xi\sqrt{\lambda})\sqrt{-\lambda} + A_2\sinh(\xi\sqrt{-\lambda})\sqrt{-\lambda})}{A_1\sinh(\xi\sqrt{-\lambda}) + A_2\cosh(\xi\sqrt{-\lambda}) + \frac{\mu}{\lambda}} + \frac{\sqrt{-\frac{3\lambda^2\sigma+3\mu^2}{2v\lambda}}k}{A_1\sinh(\xi\sqrt{-\lambda}) + A_2\cosh(\xi\sqrt{-\lambda}) + \frac{\mu}{\lambda}} \tag{28}$$

where $\xi = \frac{kx^\alpha}{\Gamma(1+\alpha)} - \frac{k(2+k^2\lambda)t^\alpha}{2\Gamma(1+\alpha)}$ and $\sigma = A_1^2 - A_2^2$

$$u(\xi) = \sqrt{\frac{3}{2v}}\frac{k(A_1\cosh(\xi\sqrt{-\lambda})\sqrt{-\lambda} + A_2\sinh(\xi\sqrt{-\lambda})\sqrt{-\lambda})}{A_1\sinh(\xi\sqrt{-\lambda}) + A_2\cosh(\xi\sqrt{-\lambda}) + \frac{\mu}{\lambda}} + \frac{\sqrt{-\frac{3\lambda^2\sigma+3\mu^2}{2v\lambda}}k}{A_1\sinh(\xi\sqrt{-\lambda}) + A_2\cosh(\xi\sqrt{-\lambda}) + \frac{\mu}{\lambda}} \tag{29}$$

where $\xi = \frac{kx^\alpha}{\Gamma(1+\alpha)} - \frac{k(2+k^2\lambda)t^\alpha}{2\Gamma(1+\alpha)}$ and $\sigma = A_1^2 + A_2^2$

$$u(\xi) = \sqrt{\frac{3}{2v}}\frac{k(\mu\xi + A_1)}{(\frac{\mu\xi^2}{2} + A_1\xi + A_2)} + \frac{\sqrt{-\frac{-3A_1^2+6\mu A_2}{2v}}k}{(\frac{\mu\xi^2}{2} + A_1\xi + A_2)} \tag{30}$$

where $\xi = \frac{kx^\alpha}{\Gamma(1+\alpha)} - \frac{kt^\alpha}{\Gamma(1+\alpha)}$. It is easily can be checked that our solutions are new in the literature.

5. The time fractional modified nonlinear Kawahara equation

For our goal, we present the following transformation

$$u(x,t) = U(\xi), \xi = kx - \frac{ct^\alpha}{\Gamma(1+\alpha)}, \tag{31}$$

where c and k are non zero constants. Eq. (2) can be turned into an ODEs

$$-cU + k\frac{U^3}{3} + pk^2U' + qk^3U'' + \xi_0 = 0, \tag{32}$$

where ξ_0 is an integration constant. By considering the homogeneous balance $m = 1$. Suppose that the solutions of (32) can be expressed by a polynomial in $\left(\frac{G'}{G}\right)$ as follows:

$$U(\xi) = a_0 + a_1 \left(\frac{G'}{G}\right), a_1 \neq 0, \tag{33}$$

where a_0, a_1, λ and μ are constants that will be determined later. Substituting Eq. (33) into Eq. (32), collecting the coefficients of $\left(\frac{G'}{G}\right)^i$ $(i = 0, ..., 3)$ and set it to zero we obtain a system. When this system of algebraic equations is solved, we can be written the following expansion.

$$U(\xi) = \pm\tfrac{p-3qk\lambda}{\sqrt{-6q}} \pm k\sqrt{-6q}\left(\tfrac{G'}{G}\right). \tag{34}$$

By substituting general solutions of Eq. (8) into Eq. (34) we have three types of exact solutions of the time fractional modified nonlinear Kawahara equation as follows:

When $\lambda^2 - 4\mu > 0$,

$$U_{1,2}(\xi) = \pm\frac{p}{\sqrt{-6q}} \pm \frac{k\sqrt{24\mu q - 6q\lambda^2}}{2}\left(\frac{C_1 \cosh\frac{1}{2}\sqrt{\lambda^2 - 4\mu}\xi + C_2 \sinh frac12\sqrt{\lambda^2 - 4\mu}\xi}{C_1 \sinh\frac{1}{2}\sqrt{\lambda^2 - 4\mu}\xi + C_2 \cosh\frac{1}{2}\sqrt{\lambda^2 - 4\mu}\xi}\right), \tag{35}$$

where $\xi = kx - \frac{(12q^2k^3\mu - kp^2 - 3q^2k^3\lambda^2)t^\alpha}{6q\Gamma(1+\alpha)}$. In particular, if $C_1 = 0$, $C_2 \neq 0$, $\lambda > 0$, $\mu = 0$, then $U_{1,2}$ becomes

$$u_{1,2}(x,t) = \pm\frac{p}{\sqrt{-6q}} \pm \frac{k\lambda\sqrt{-6q}}{2}\tanh\left\{\frac{\lambda}{2}\left(kx + \frac{(kp^2 + 3q^2k^3\lambda^2)\,t^\alpha}{6q\Gamma(1+\alpha)}\right)\right\}. \tag{36}$$

If $C_1 \neq 0$, $C_2 = 0$, $\lambda > 0$, $\mu = 0$, then $U_{1,2}$ becomes

$$u_{3,4}(x,t) = \pm\frac{p}{\sqrt{-6q}} \pm \frac{k\lambda\sqrt{-6q}}{2}\coth\left\{\frac{\lambda}{2}\left(kx + \frac{(kp^2 + 3q^2k^3\lambda^2)\,t^\alpha}{6q\Gamma(1+\alpha)}\right)\right\}. \tag{37}$$

When $\lambda^2 - 4\mu < 0$,

$$U_{3,4}(\xi) = \pm\frac{p}{\sqrt{-6q}} \pm \frac{k\sqrt{6q\lambda^2 - 24\mu q}}{2}\left(\frac{-C_1 \sin\frac{1}{2}\sqrt{4\mu - \lambda^2}\xi + C_2 \cos\frac{1}{2}\sqrt{4\mu - \lambda^2}\xi}{C_1 \cos\frac{1}{2}\sqrt{4\mu - \lambda^2}\xi + C_2 \sin\frac{1}{2}\sqrt{4\mu - \lambda^2,\xi}}\right), \tag{38}$$

where $\xi = kx - \frac{(12q^2k^3\mu - kp^2 - 3q^2k^3\lambda^2)t^\alpha}{6q\Gamma(1+\alpha)}$.

When $\lambda^2 - 4\mu = 0$,

$$u_{5,6}(x,t) = \pm\frac{p}{\sqrt{-6q}} \pm \frac{k\sqrt{-6q}C_2}{C_1 + C_2\left(kx + \frac{kp^2t^\alpha}{6q\Gamma(1+\alpha)}\right)}, \tag{39}$$

which are the exact solutions of the time fractional modified nonlinear Kawahara equation. According to $(G'/G, 1/G)$-expansion method, we look for the solution in the form of

$$u(\xi) = a_0 + a_1\phi + b_1\psi \tag{40}$$

where a_0, a_1,and b_1 are constants to be determined later. We obtain hyperbolic, trigonometric, rational function solutions of Eq. (2) as follows:

$$u(\xi) = a_0 + \frac{\sqrt{-\frac{\lambda}{\lambda^2\sigma+\mu^2}}\,b_1(A_1\cosh(\xi\sqrt{-\lambda})\sqrt{-\lambda} + A_2\sinh(\xi\sqrt{-\lambda})\sqrt{-\lambda})}{A_1\sinh(\xi\sqrt{-\lambda}) + A_2\cosh(\xi\sqrt{-\lambda}) + \frac{\mu}{\lambda}} + \frac{b_1}{A_1\sinh(\xi\sqrt{-\lambda}) + A_2\cosh(\xi\sqrt{-\lambda}) + \frac{\mu}{\lambda}} \tag{41}$$

where $\xi = kx - \frac{k(3a_0^2\mu^2+3\lambda^2\sigma a_0^2+b_1^2\lambda^2)t^\alpha}{3(\lambda^2\sigma+\mu^2)\Gamma(1+\alpha)}$ and $\sigma = A_1^2 - A_2^2$

$$u(\xi) = a_0 + \frac{\sqrt{-\frac{\lambda}{-\lambda^2\sigma+\mu^2}}\,b_1(A_1\cos(\xi\sqrt{\lambda})\sqrt{\lambda} - A_2\sin(\xi\sqrt{\lambda})\sqrt{\lambda})}{A_1\sin(\xi\sqrt{\lambda}) + A_2\cos(\xi\sqrt{\lambda}) + \frac{\mu}{\lambda}} + \frac{b_1}{A_1\sin(\xi\sqrt{\lambda}) + A_2\cos(\xi\sqrt{\lambda}) + \frac{\mu}{\lambda}} \tag{42}$$

where $\xi = kx - \frac{k(3a_0^2\mu^2-3\lambda^2\sigma a_0^2+b_1^2\lambda^2)t^\alpha}{3(-\lambda^2\sigma+\mu^2)\Gamma(1+\alpha)}$ and $\sigma = A_1^2 + A_2^2$

$$u(\xi) = a_0 + \sqrt{\frac{3}{2v}}\frac{\sqrt{\frac{1}{A_1^2-2\mu A_2}}\,b_1(\mu\xi + A_1)}{(\frac{\mu\xi^2}{2} + A_1\xi + A_2)} + \frac{b_1}{(\frac{\mu\xi^2}{2} + A_1\xi + A_2)} \tag{43}$$

where $\xi = kx - \frac{a_0^2kt^\alpha}{\Gamma(1+\alpha)}$. It is easily can be checked that our solutions are new in the literature.

6. Conclusion

In this paper, we apply the (G'/G) and $(G'/G, 1/G)$-expansion methods to solve nonlinear fractional partial differential equations. These methods for nonlinear FDEs with fractional complex transform has its own advantages: direct, succinct, basic; and so it can also be applied to other FDEs where the homogeneous balancing principle is satisfied.

7. Acknowledgements

This work was supported by Eskisehir Osmangazi University Scientific Research Projects (Grant No: 2016-1043).

8. References

[1] I., Podlubny, Fractional Differential Equations, (Academic Press, California) 1999.

[2] A.A., Kilbas, H.M., Srivastava, J.J., Trujillo, Theory and Applications of Fractional Differential Equations, (Elsevier, Amsterdam) 2006.

[3] S.M., Ege, E., Mısırlı, The modified Kudryashov method for solving some fractional-order nonlinear equations, Advances in Difference Equations, 2014 (2014) 135.

[4] Atangana, A., Bildik, N., Noutchie, S.C.O., New Iteration Methods for Time-Fractional Modified Nonlinear Kawahara Equation, Abstract and Applied Analysis, 2014 (2014) 740248.

[5] G., Jumarie, Table of some basic fractional calculus formulae derived from a modified Riemann-Liouvillie derivative for nondifferentiable functions. Appl. Maths. Lett., 22 (2009) 378-385.

[6] J.H., He, S.K. Elagan, Z.B., Li, Geometrical explanation of the fractional complex transform and derivative chain rule for fractional calculus, Phys. Lett. A, 376 (2012) 257-259.

[7] T., Elghareb, S.K., Elagan, Y.S., Hamed, M., Sayed, An Exact Solutions for the Generalized Fractional Kolmogrove-Petrovskii Piskunov Equation by Using the Generalized (G'/G)-expansion Method, Int. Journal of Basic & Applied Sciences, 13 01 (2013) 19–22.

[8] M., Wang, X., Li, J., Zhang, The (G'/G)-expansion method and traveling wave solutions of nonlinear evolution equations in mathematical physics Phys. Lett. A, 372 (2008) 417-423.

[9] E.M.E., Zayed, K.A., Gepreel, The (G'/G)-expansion method for finding traveling wave solutions of nonlinear partial differential equations in mathematical physics J. Math. Phys., 50 (2009) 013502.

[10] Ling-Xiao, L., Er-Qiang, L., Ming-Liang, W., The $(G'/G, 1/G)$ -expansion method and its application to travelling wave solutions of the Zakharov equations, Appl. Math. J. Chinese Univ., 25 4 (2010) 454-462.

[11] S. Demiray, O. Unsal, A. Bekir, New Exact Solutions for Boussinesq Type Equations by Using (G/G, 1/G) and (1/ G)-Expansion Methods, Acta Physica Polonica A 125, 5 (2014) 1093-1098

Journal of Physics: Conference Series **766** (2016) 012036 doi:10.1088/1742-6596/766/1/012036

A Galerkin-like scheme to solve Riccati equations encountered in quantum physics

Şuayip Yüzbaşı[1] and Murat Karaçayır
[1]Akdeniz University, Department of Mathematics, Antalya 07058, Turkey
E-mail: syuzbasi@akdeniz.edu.tr

Abstract. In this study, a Galerkin-like approach is applied to numerically solve Riccati differential equations. In this method, inner product is applied to a set of monomials and an expression obtained from the equation in question. The resulting nonlinear system is then solved, yielding a polynomial as the approximate solution. Additionally, the technique of residual correction, whose aim is to increase the accuracy of the approximate solution, is discussed briefly. Lastly, the method and the residual correction technique are illustrated with two examples.

1. Introduction

Many problems of science and engineering utilize differential equations as an important modelling tool. In particular, the Riccati differential equation given by

$$u'(x) = p(x) + q(x)u(x) + r(x)u^2(x),\ u(0) = a \tag{1}$$

plays a significant role in many fields of applied science[1]. To name a few applications of Riccati equations, solitary wave solutions of a partial differential equation are closely related to a projective Riccati equation[2]. The Riccati differential equation also constitute an important tool in present day control theory[3]. Thirdly, applications of Riccati equations can be encountered in Kalman filtering systems such as orbitting satellites[4, 5]. Finally, the application which motivated the writing of this paper is the close relation of a one-dimensional static Schrödinger equation to a Riccati differential equation[6]. Namely, the fundamental solution of the Schrödinger equation given by

$$i\frac{\partial\psi}{\partial t} + \frac{1}{4}\frac{\partial^2\psi}{\partial x^2} + tx^2\psi = 0$$

can be obtained using the following substitution[7]:

$$\psi = A(t)e^{iS(x,y,t)} = \frac{1}{\sqrt{2\pi i\mu(t)}}e^{i\left(\alpha(t)x^2+\beta(t)xy+\gamma(t)y^2\right)}.$$

Here, the real-valued functions $\alpha(t), \beta(t)$ and $\gamma(t)$ satisfy the following system:

$$\begin{aligned} \alpha'(t) - t + \alpha(t)^2 &= 0, \\ \beta'(t) + \alpha(t)\beta(t) &= 0, \\ \gamma'(t) + \frac{1}{4}\beta(t)^2 &= 0. \end{aligned}$$

The first equation in this system is a Riccati equation with $q(t) = 0$.

Since Riccati differential equations are linked to many important problems of science and engineering, obtaining their solutions is a problem of great importance. As of today, there does not exist an analytical method that has the capability to yield the solution to any given Riccati equation. As a result, numerical techniques or approximate approaches has to be utilized for the purpose of obtaining its solutions. To name a few of such studies, Adomian Decomposition Method was used by El-Tawil et al.[8] and Tsai[9] to solve the Riccati equation approximately. Abbasbandy utilized He's Variational Iteration Method[10] and Homotopy Perturbation Method[11] to obtain its solutions. Gülsu and Sezer used Taylor matrix method[12] for the same purpose. Yüzbaşı used a collocation method[13] based on Bessel polynomials to obtain approximate solutions of Riccati-type differential equation systems. Finally, Yang et.al. used hybrid functions and Chebyshev polynomials to solve Riccati differential equations.

In this study, our aim is to find an approximate solution to the Riccati differential equation (1) by using a Galerkin-like scheme. The organization of the paper is as follows: The solution method is explained in Section 2. The subject of Section 3 is the technique of residual correction, whose aim is to derive better approximations from already obtained approximate solutions. Section 4 contains numerical examples. Finally, the conclusion of the paper is given in Section 5.

2. Method of Solution

In this section, the procedure we will use to solve Equation (1) is outlined. The same method was used in [14] to obtain approximate solutions of high-order Fredholm integro-differential equations with possibly singular kernel functions.

To begin with, we assume that the unique solution $u(x)$ of Equation (1) can be expressed as a power series of the form

$$u(x) = \sum_{k=0}^{\infty} a_k x^k.$$

We then truncate this power series after the $(N+1)$st term so that

$$u_N(x) = \sum_{k=0}^{N} a_k x^k = \mathbf{X}_N(x) \cdot \mathbf{A}$$

where

$$\mathbf{X}_N(x) = [\ 1 \quad x \quad x^2 \quad \ldots \quad x^N\ ],\ \mathbf{A} = [\ a_0 \quad a_1 \quad a_2 \quad \ldots \quad a_N\ ]^T.$$

Here, the purpose of the method is to determine the unknown coefficients a_i. The derivatives of $u_N(x)$ can be expressed as a product of matrices with the help of a specially defined matrix. Namely, if we define $\mathbf{B}$ to be the $(N+1)\times(N+1)$ matrix such that $\mathbf{B}_{i,i+1} = i$ for $i = 1, 2, \ldots, N$ and $\mathbf{B}_{i,j} = 0$ elsewhere, then the following equality holds:

$$u_N^{(i)}(x) = \mathbf{X}_N(x)\mathbf{B}^i\mathbf{A} = [\ 1 \quad x \quad x^2 \quad \ldots \quad x^N\ ] \begin{bmatrix} 0 & 1 & 0 & \ldots & 0 \\ 0 & 0 & 2 & \ldots & 0 \\ \vdots & \vdots & \vdots & \ddots & \vdots \\ 0 & 0 & 0 & \ldots & N \\ 0 & 0 & 0 & \ldots & 0 \end{bmatrix}^i \begin{bmatrix} a_0 \\ a_1 \\ a_2 \\ \vdots \\ a_N \end{bmatrix}$$

Next, substituting the matrix expressions for $u_N(x)$ and $u_N^{(i)}(x)$ into Equation (1) gives us the equation

$$\mathbf{G}(x)\mathbf{A} = p(x) \tag{2}$$

where

$$\mathbf{G}(x) = (\mathbf{X}_N(x)\mathbf{B} - q(x)\mathbf{X}_N(x) - r(x)\mathbf{X}_{2N}(x)\tilde{\mathbf{A}})\mathbf{A}$$

and where

$$\tilde{\mathbf{A}} = \begin{bmatrix} a_0 & 0 & 0 & \dots & 0 \\ a_1 & a_0 & 0 & \dots & 0 \\ a_2 & a_1 & a_0 & \dots & 0 \\ \vdots & \vdots & \vdots & \ddots & \vdots \\ a_N & a_{N-1} & a_{N-2} & \dots & a_0 \\ 0 & a_N & a_{N-1} & \dots & a_1 \\ 0 & 0 & a_N & \dots & a_2 \\ \vdots & \vdots & \vdots & \ddots & \vdots \\ 0 & 0 & 0 & \dots & a_N \end{bmatrix}.$$

Now, in order to convert Equation (2) into a system of nonlinear equations in the unknowns a_i, we take inner product of Equation (2) with the elements of the set $\mathbf{\Phi} = \{1, x, x^2, \dots, x^N\}$, where the inner product of two functions f, g from the Hilbert space $L^2[0, b]$ is defined by

$$< f, g >= \int_0^b f(x)g(x)\mathrm{d}x.$$

Each inner product will result in a nonlinear equation in the unknown coefficients a_i. Finally, we are left with a nonlinear system $\mathbf{WA} = \mathbf{G}$ where the $(N+1) \times (N+1)$ matrix $\mathbf{W}$ and the column matrix $\mathbf{G}$ of length $N+1$ are given by

$$\mathbf{W}_{i,j} =< x^{i-1}, \mathbf{G}(x)_{1,j} >, \ \mathbf{G}_{i,1} =< x^{i-1}, p(x) > .$$

More explicitly, $\mathbf{W}$ and $\mathbf{G}$ are given by

$$\mathbf{W} = \begin{bmatrix} < 1, \mathbf{G}(x)_{1,1} > & < 1, \mathbf{G}(x)_{1,2} > & \dots & < 1, \mathbf{G}(x)_{1,N+1} > \\ < x, \mathbf{G}(x)_{1,1} > & < x, \mathbf{G}(x)_{1,2} > & \dots & < x, \mathbf{G}(x)_{1,N+1} > \\ < x^2, \mathbf{G}(x)_{1,1} > & < x^2, \mathbf{G}(x)_{1,2} > & \dots & < x^2, \mathbf{G}(x)_{1,N+1} > \\ \vdots & \vdots & \vdots & \vdots \\ < x^N, \mathbf{G}(x)_{1,1} > & < x^N, \mathbf{G}(x)_{1,2} > & \dots & < x^N, \mathbf{G}(x)_{1,N+1} > \end{bmatrix},$$

$$\mathbf{G} = \begin{bmatrix} < 1, p(x) > & < x, p(x) > & < x^2, p(x) > & \dots & < x^N, p(x) > \end{bmatrix}^T.$$

Note that $\mathbf{W}$ contains terms which are linear in some a_i; therefore the system $\mathbf{WA} = \mathbf{G}$ is nonlinear. Solving it under the initial condition $u_N(0) = a$, which implies $a_0 = a$, we compute the matrix $\mathbf{A}$ of unknown coefficients and the approximate solution is then given by

$$u_N(x) = a_0 + a_1 x + \dots + a_N x^N.$$

3. Error Estimation and Residual Correction

In this section, the error estimation of our method is performed based on the residual function corrresponding to Equation (1). Then, starting with an approximate solution of Equation (1), the way of obtaining a better approximation from the original one is described.

Let us consider the residual function

$$R(x) = u'(x) - p(x) - q(x)u(x) - r(x)u^2(x) \tag{3}$$

of Equation (1). Substituting the approximate solution $u_N(x)$ in place of $u(x)$ we get

$$R_N(x) = u'_N(x) - p(x) - q(x)u_N(x) - r(x)u_N^2(x) \tag{4}$$

Subtracting Equation (3) from Equation (4) and rearranging yields

$$R_N(x) = -e'_N(x) + q(x)e_N(x) + 2r(x)u_N(x)e_N(x) - r(x)e_N^2(x), \tag{5}$$

which is similar to Equation (1) with $p(x)$ replaced by $-R_N(x)$, $q(x)$ replaced by $q(x) - 2r(x)u_N(x)$ and $e_N(x) = u_N(x) - u(x)$ is the error function. In addition, since $u(x)$ and $u_N(x)$ both satisfy the initial condition $u(x_0) = a$,

$$e_N(x_0) = 0$$

is the initial condition for Equation (5). Next, the method of Section 2 is applied to solve it for $e_N(x)$ with some choice of the parameter value M and the approximate solution $e_{N,M}(x)$ is obtained. Consequently, this new approximation is used to obtain a corrected solution

$$u_{N,M}(x) = u_N(x) + e_{N,M}(x)$$

of Equation (1). In what follows, $E_{N,M}(x)$ will denote the actual error of $u_{N,M}(x)$ given by $E_{N,M}(x) = u_{N,M}(x) - u(x)$.

4. Numerical Examples

In this section, the method of Section 2 is applied to several examples.

Example 1. Our first example is the following equation from [5]:

$$u'(x) = u(x) - 2u^2(x),\ u(0) = 1. \tag{6}$$

Here, the interval of interest is $0 \le x \le 0.5$. Equation (6) also has the property of being separable, thus its exact solution can be found to be $u(x) = \frac{e^x}{2e^x+1}$. We applied the method of Section 2 to this equation with several N values. For example, the approximate solution corresponding to $N = 3$ is obtained as follows:

$$u_3(x) = 1 - 0.9782168543x + 1.1840357357x^2 - 0.7141540751x^3.$$

Since the exact solution is at hand for this particular problem, we measure the accuracy of the approximate solutions looking at the absolute value of the actual error functions $E_N(x)$. Figure 1 depicts these absolute error functions for $N = 4, 5, 6$. As the figure shows, the absolute error decreases as we increase the parameter N. Another means of understanding the effectivity of a numerical method is comparisons with other methods, therefore such a comparison is given in Table 1. As the table shows, the present scheme gives rise to better results when compared to three other methods.

Example 2. Next, we consider the following equation:

$$u'(x) = u(x) - u^2(x),\ u(0) = 2. \tag{7}$$

This time, the problem is considered in the interval $0 \le x \le 1$. We applied the present scheme to Equation (7) with the choice of $N = 3, 4, 5, 6, 7$ and we considered the residual functions $R_N(x)$ for these N values. The results are depicted in Table 2. As apparent from the table, the acuracy of the approximate solutions improves with increasing N, just like the case in Example 1. We now attempt to improve the approximate solutions using the method of residual correction described in Section 3. For this purpose we choose $u_3(x)$ and $u_6(x)$ as the solutions to be improved. For $N = 3$, let us apply residual correction with $M = 5, 6, 8$ and for $N = 6$, let us apply it with $M = 7, 8, 9$. In order to understand how much improvement is satisfied by residual correction for these parameter values, one can examine Figure 2. As is clear from the figure, residual correction brings about great improvement in the accuracy of the appoximate solutions. This is the case both for $N = 3$ and $N = 6$. Furthermore, for fixed N, increasing M value yields better and better results.

Table 1. Comparison of the error function with three other methods in Example 1.

x	Euler method	TMM ($N=6$)	Tau method	Present ($N=4$)
0	0	0	0	0
0.1	0.013146	2.941E-6	5.270E-4	3.468E-5
0.2	0.0185470	1.487E-4	5.282E-4	4.794E-5
0.3	0.0204833	0.0019719	4.001E-4	6.858E-5
0.4	0.207262	0.0111989	3.272E-4	9.714E-6
0.5	0.0201354	0.0414865	2.749E-4	1.266E-8

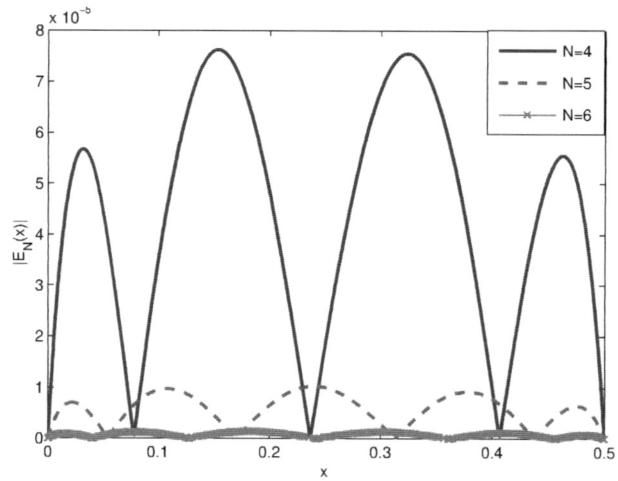

Figure 1. Graphics of the absolute errors for $u_N(x)$ for $N=4,5,6$ in Example 1.

Table 2. Absolute residuals for several values of N and at some points in Example 2.

x	$N=3$	$N=4$	$N=5$	$N=6$	$N=7$
0	0	0	0	0	0
0.2	0.0014	3.1731E − 4	4.5167E − 5	1.0794E − 5	2.9698E − 6
0.4	0.0071	2.4464E − 4	6.0264E − 5	1.8150E − 5	3.7031E − 6
0.6	0.0034	5.3748E − 5	6.9106E − 5	1.7900E − 5	2.8054E − 6
0.8	0.0054	2.3628E − 4	5.6255E − 5	1.4082E − 5	1.8500E − 6
1	1.3094E − 4	4.7008E − 8	5.7946E − 10	3.2620E − 10	6.0696E − 11

5. Conclusion
In this paper, we have presented a Galerkin-like approach for the approximate solutions of Riccati differential equations. It can be concluded that our approach gives fairly good results despite its simplicity. Another advantage of this approach is that the quality of the approximate solutions improves with increasing N values. Residual error correction technique to improve the accuracy of approximate solutions has also been discussed. Simulation results show that considerable

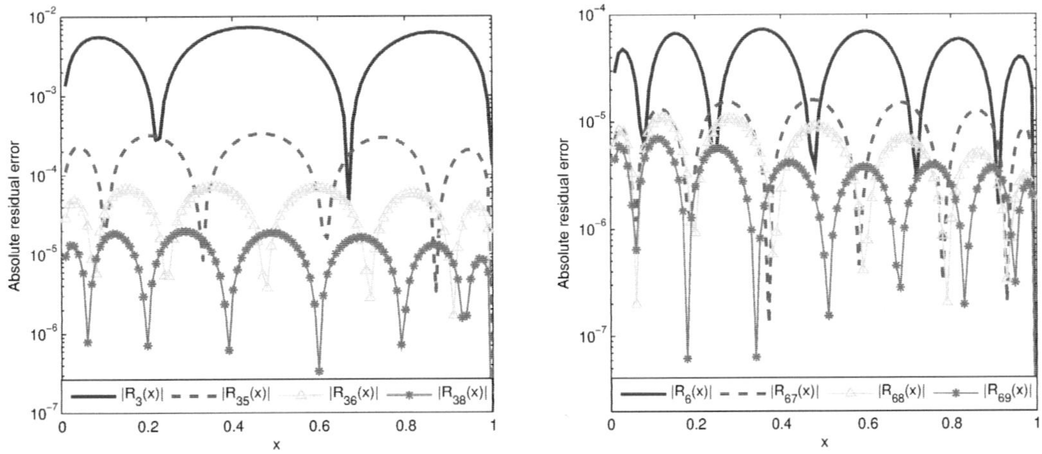

Figure 2. Graphics of the absolute residual errors of $u_{NM}(x)$ for several values of (M, N) in Example 2.

improvements in the approximate solutions can be achieved as a result of this technique.

Acknowledgments

The authors are supported by the *Scientific Research Project Administration of Akdeniz University.*

References

[1] Reid W T 1972 *Riccati Differential Equations* (New York, Academic Press)
[2] Carinena J F, Marmo G, Perelomov A M and Randa M F 1998 Related operators and exact solutions of Schrödinger equations *Int. J. Mod. Phys.* A **13** 4913-29
[3] Bittanti S, Colaneri P and Guardabassi G 1984 Periodic solutions of periodic Riccati equations *IEEE Trans. Autom. Control* **29** 665-7
[4] Anderson B D O and Moore J B 1979 *Optimal Filtering* (Englewood Cliffs: Prentice Hall)
[5] Yang C, Hou J and Qin B 2012 Numerical solution of Riccati differential equations by using hybrid functions and tau method *International Journal of Mathematical and Computational Sciences* **6** 216
[6] Lanfear N and Suslov S K 2009 The time-dependent Schrödinger equation, Riccati equation and Airy functions *Preprint* arXiv:0903.3608
[7] Cordero-Soto R, Lopez R M, Suazo E and Suslov S K 2008 Propagator of a charged particle with a spin in uniform magnetic and perpendicular electric fields *Lett. Math. Phys.* **84** 159-78
[8] El-Tawil M A, Bahnasawi A A and Abdel-Naby A 2004 Solving Riccati differential equation using Adomian's decomposition method *Appl. Math. Comput.* **157** 503-14
[9] Tsai P Y 2010 An approximate analytic solution of the nonlinear Riccati differential equation *J. Franklin Inst.* **347** 1850-62
[10] Abbasbandy S 2007 A new application of He's variational iteration method for quadratic Riccati differential equation by using Adomian's polynomials *J. Comput. Appl. Math.* **207** 59-63
[11] Abbasbandy S 2006 Homotopy perturbation method for quadratic Riccati differential equation and comparison with Adomians decomposition method *Appl. Math. Comput.* **172** 485-90
[12] Gülsu M and Sezer M 2006 On the solution of the Riccati equation by the Taylor matrix method *Appl. Math. Comput.* **176** 414-21
[13] Yüzbaşı Ş 2012 A collocation approach to solve the Riccati-type differential equation systems *Int. J. Comput. Math.* **89** 2180-97
[14] Türkyılmazoğlu M 2014 An effective approach for numerical solutions of high-order Fredholm integro-differential equations *Appl. Math. Comput.* **227** 384-98

AUTHOR INDEX